Lecture Notes in Control and Information Sciences 413

Editors: M. Thoma, F. Allgöwer, M. Morari

Evangelos Eleftheriou and
S.O. Reza Moheimani (Eds.)

Control Technologies for Emerging Micro and Nanoscale Systems

Springer

Editors

Dr. Evangelos Eleftheriou
IBM Research GmbH
Säumerstr. 4
8803 Rüschlikon
Switzerland
E-mail: ele@zurich.ibm.com

Professor S.O. Reza Moheimani
School of Electrical Engineering &
Computer Science
The University of Newcastle
Callaghan, NSW, 2308
Australia
E-mail: Reza.Moheimani@newcastle.edu.au

ISBN 978-3-642-22172-9 e-ISBN 978-3-642-22173-6

DOI 10.1007/978-3-642-22173-6

Lecture Notes in Control and Information Sciences ISSN 0170-8643

Library of Congress Control Number: 2011930523

Typeset & Cover Design: Scientific Publishing Services Pvt. Ltd., Chennai, India.

Printed on acid-free paper

9 8 7 6 5 4 3 2 1

springer.com

Preface

This book comprises a selection of the presentations made at the "Workshop on Dynamics and Control of Micro and Nanoscale Systems" held at IBM Research – Zurich, Switzerland, on the 10th and 11th of December 2009. The aim of the workshop was to bring together some of the leading researchers in the field of dynamics and control of micro- and nanoscale systems. It proved an excellent forum for discussing new ideas and approaches, and we thank all attendees for making the workshop such a success and the authors for contributing to this book.

In Chapter 1, Ando describes fundamental challenges that had to be overcome to achieve high-speed atomic force microscopes capable of recording dynamic processes such as those in live cells. In particular, he discusses the control and instrumentation complications that arise in high-speed AFM and proposes practical solutions. The chapter covers an impressive number of applications of the author's video-rate AFM in recording dynamic changes in biological samples.

In Chapter 2, Trumper, Hocken, Amin-Shahidi, Ljubicic, and Overcash report on the design, construction and characterization of a high-accuracy atomic force microscope for dimensional metrology investigations. The device is designed to work in conjunction with a sub-atomic measuring machine and to develop surface measurements with 0.1 nm resolution over a very large scan area. They use a self-sensing quartz-tuning-fork-based AFM probe, and describe specialized electronics that enable its operation in constant-amplitude self-resonance. They discuss control design issues, the FPGA implementation of the controller, and report experimental scans obtained from several test gratings.

Chapter 3 by Bazaei, Yong, Moheimani and Sebastian proposes a new control design framework for high-speed nanopositioning that avoids the adverse effects of sensor noise associated with high-bandwidth linear control loops. This is done by introducing nonlinear signal-shaping blocks into the feedback loop, making it possible to track high-speed triangular waveforms, as required in scanning probe microscopy, using a low-bandwidth feedback controller. The method is implemented on an AFM scanner, and high-bandwidth scans are obtained at extremely high positioning accuracy of 0.1 nm.

Chapter 4 by Lee, Mohan and Salapaka is concerned with design of two-degree-of-freedom controllers for nanopositioning systems. The idea is based on using a feedforward controller in conjunction with a feedback controller to improve the achievable tracking bandwidth. They report that this control design methodology enabled them to increase the bandwidth of an AFM scanner by 330% beyond what is achievable with a feedback controller alone.

Chapter 5 by Kuiper and Schitter looks at a number of possibilities to improve the imaging speed of an AFM. One method involves applying the concept of self-sensing, originally developed in the smart structures literature, to the AFM piezoelectric tube scanner. Once the piezoelectric strain signal has been measured, standard control design methods can be employed to improve the lateral positioning of the device. The authors discuss also other ideas, such as iterative learning control and dual actuation methods.

Chapter 6 by Chang and Andersson proposes a non-raster-scan method for AFM imaging of string-like objects such as biopolymers. Their algorithm uses the data measured by the AFM probe together with basic information available about the object to force the probe to remain in the region of interest. To illustrate the algorithm, they perform simulation experiments using data obtained from a standard raster-scan image of DNA.

Chapter 7 by Sahoo, Häberle, Sebastian, Pozidis and Eleftheriou discusses an approach to tapping-mode scanning probe microscopy employing low-stiffness electrostatic cantilevers. They explain how the oscillation amplitude is kept small to enable the probe to operate at high frequencies, which results in reduction of tip wear. They provide experimental results that show only minimal tip wear during a long-term scan operation.

In Chapter 8, Sebastian discusses a feedback-based modeling method to capture the dynamics of electro-thermal sensing in MEMS, where this a relatively new sensing technology. The sensors consist of silicon structures with integrated heating elements. Sebastian explains the remarkable linearity of these sensors and describes how the sensitivity and bandwidth of electro-thermal sensors can be improved by feeding back a part of the sensed current.

In Chapter 9, Xi, Yang, Lai, Song, Gao, Shi and Su discuss how an AFM can be used as a real-time nanomanipulator. They explain that the problem boils down to a multivatiable feedback control design problem and describe the implementation of the controller on a real-time control system. The modified AFM-based nanomanipulator is fast enough to allow a user to manipulate a surface in real time. They report a successful nanobiomanipulation experiment that involved dissection of the intercellular junctions of keratinocytes.

Chapter 10 by Fukuda and Nakajima presents an overview of nanorobotics-based manipulation methods and their use in nano-biotechnology. They discuss the issues that arise when a nanomanipulator is operated under various microscopes, particularly during investigations such as single-cell analysis and bottom-up fabrication of nanostructures. They describe the construction of an experimental apparatus for single-cell nanosurgery. The device works under an SEM and can be used for cell diagnosis, cutting, injection and insertion of micro/nano devices into a cell. The chapter includes a discussion of potential future research directions in nanorobotics for bio-nanotechnology

Chapter 11 by Fedder, Mukherjee and Pileggi is concerned with the concept of reconfigurability in CMOS MEMS. It exploits several examples, including reconfigurable RF MEMS capacitors, micro-instrumented scanning probe arrays for tip-based nanomanufacturing and self-healing RF microresonators to discuss some of the challenges — and opportunities — associated with MEMS research.

In Chapter 12, Knospe proposes a new method of actuation in MEMS, based on capillary forces. In this approach, the alteration of the capillary pressure inside an existing liquid bridge between a fixed and a movable electrode, via electrowetting, is used as a means of microscale actuation. The author provides simulation investigations that indicate that in this way a significantly larger stroke range can be achieved than with other MEMS actuation techniques.

In Chapter 13, Åström develops a methodology for designing sensors based on force feedback. The problem is formulated as an estimation problem in which both the feedback signal and the error signal are used to form an estimate of the measured variable. The method is applied to a MEMS accelerometer in which the mass deflection is measured by quantum tunneling.

In Chapter 14, Conway, Nie and Horowitz discuss two applications of the guaranteed cost control design methodology to dual-stage hard-disk drives. The methods are based on minimizing the upper bound on the worst-case H_2 norm of a discrete time-uncertain system with unstructured uncertainty. One of the solutions is based on solving linear matrix inequalities, the other on solving a number Riccati equations. Both methods are tested on a model of a dual-stage hard-disk drive, a two-input, one-output system.

Chapter 15 by Pantazi demonstrates control-systems enhancements that make it possible to increase the track densities in a tape drive. She describes a novel active-guiding concept that, in conjunction with a track-following controller, achieves a positioning accuracy of 128 nm.

We would like to express our gratitude to IBM Research – Zurich for supporting and hosting this workshop. Lisa Raissig is gratefully acknowledged for her help in organizing the workshop and for handling the logistics involved, and Charlotte Bolliger for her help in compiling this proceedings volume.

Rüschlikon, Switzerland, and Newcastle, Australia — Evangelos Eleftheriou
April 2011 — S.O. Reza Moheimani

Contents

Chapter 1
Techniques Developed for High-Speed AFM

Toshio Ando

Abstract. This article describes fundamental devices and control techniques that have materialized high-speed atomic force microscopy (AFM) capable of recording dynamic processes of individual biomolecules on video at an imaging rate of 10–25 frames/s, without disturbing their biological functions.

1.1 Introduction

There is no doubt that new research fields will be pioneered if we can attain a tool that allows directly observing nanometer-scale dynamic phenomena occurring in liquids or at the solid-liquid interfaces. Not only biomolecular phenomena of course, a vast array of reactions occur under liquid environments, including electrochemical reaction, corrosion, cleaning by detergent, non-biological catalytic reactions, and so on. Among available microscopy techniques, AFM has almost all capabilities that make such observations possible but lacks one capability, i.e., high temporal resolution.

The low imaging rate of AFM arises from its fundamental operation mechanism, i.e., mechanical scanning and mechanical sensing. The relative horizontal position between a cantilever tip and a sample is changed by raster scanning of either of a cantilever tip or the sample stage. A local portion of the sample surface is sensed by a cantilever tip interacting with the surface, and this interaction is transmitted to the cantilever, resulting in the deflection of the cantilever beam. When the cantilever is oscillated in the z-direction, the tip-sample interaction results in a change in the oscillation amplitude, phase, or resonant frequency. One of these mechanical signals is maintained constant during the xy-scanning by moving the scanner in the z-direction under feedback control. As a result, the movement of the scanner approximately traces the sample surface. Therefore, the sample height map over the

Toshio Ando
Department of Physics, Kanazawa University,
Kakuma-machi, Kanazawa 920-1192, Japan

E. Eleftheriou & S.O.R. Moheimani (Eds.): Cntrl. Tech. for Emerging Micro/Nanoscale Sys. LNCIS 413, pp. 1–16.
springerlink.com

xy-scan area is reconstructed from the signals driving the scanner in the z-direction at respective xy positions.

To increase the imaging rate of AFM, we first need to increase the speed of the mechanical scanning and sensing without generating unwanted vibrations. For biological applications, we further need to minimize the tip-sample interaction force, without deteriorating the imaging speed. In the past 20 years, many efforts have been carried out to materialize high-speed AFM. However, only a few groups have been participated in developing the entire system of high-speed AFM (1; 2; 3; 4). In this article, I focus on the technical development carried out in my group, and show a few results of high-speed AFM imaging of bacteriorhodopsin. For a historical view of the last 20 years in this field, see reviews (5; 6; 7).

1.2 Imaging Speed and Feedback Bandwidth

In bio-AFM imaging, it is mandatory to maintain the tip-sample distance (hence, tip-sample interaction force) constant under feedback control. When a sample is scanned over an area of $W \times W$ with scan velocity V_s in the x-direction and N scan lines in the y-direction, the image acquisition time T is given by $T = 2WN/V_s$. Supposing that the sample surface is characterized with a single spatial frequency $1/\lambda$, then the feedback scan is executed in the z-direction with frequency $f = V_s/\lambda$ to trace the sample surface. The feedback bandwidth f_B of the microscope should be equal to or higher than f, and thus, we obtain the relationship $T \geq 2WN/(\lambda f_B)$.

The feedback bandwidth is usually defined by a feedback frequency at which a closed-loop phase delay of $\pi/4$ occurs in tracing the sample surface. The phase delay in the closed loop (φ_{closed}) is approximately twice that in the open loop (φ_{open}), provided the feedback gain is maintained at ~ 1 (4). The phase delay is caused by the limited response speed of devices contained in the feedback loop and by the time delay due to a 'parachuting' effect mentioned below. By denoting the total time delay by $\Delta\tau_{total}$, φ_{open} is given by $\varphi_{open} = 2\pi f \Delta\tau_{total}$. The phase delay can be compensated by a factor of a $\alpha(2-3)$ using the (P + D) operation of the proportional-integral-derivative (PID) feedback controller. f_B is thus given by $f_B = \alpha/(16\Delta\tau_{total})$. Hence, the highest possible imaging rate (V_{max}) is given by $V_{max} = \alpha\lambda/(32WN\Delta\tau_{total})$, when the sample fragility is not taken into account. To materialize, for example, an imaging rate of 25 frames/s for $\alpha = 2.5$, $\lambda = 10\,\text{nm}$, $W = 200\,\text{nm}$, and $N = 100$, we have to achieve $f_B > 100\,\text{kHz}$ and thus $\Delta\tau_{total} < 1.6\,\mu\text{s}$. The time delays at the respective devices and at parachuting are described in the following sections.

1.3 Mechanical Devices in High-Speed AFM

1.3.1 Cantilevers

The response time of a cantilever τ_c is expressed by $\tau_c = Q_c/(\pi f_c)$, where Q_c and f_c are the quality factor and resonant frequency of a cantilever, respectively. To shorten the response time, we need a high f_c and a small Q_c. A high f_c together with a small spring constant k_c can be achieved by reducing the cantilever size (8). We developed small cantilevers, collaborating with Olympus (3; 4). The most recent small cantilevers made of Si_3N_4 are $\sim 6\,\mu m$ long, $2\,\mu m$ wide, and 90 nm thick (Fig. 1.1a), resulting in f_c in air 3.5 MHz, f_c in water 1.2 MHz, $k_c \sim 0.2\,N/m$, and Q_c in water 2–3 (9). Therefore, its response time in water is about $0.66\,\mu s$. Somewhat larger cantilevers (BL-AC10DS-A2, Olympus: f_c in air 1.5 MHz, f_c in water 600 kHz, $k_c \sim 0.1\,N/m$) are already commercially available (Atomic Force F&E GmbH).

In addition to the advantage of achieving a high imaging rate, small cantilevers have other advantages. The total thermal noise only depends on the spring constant and the temperature. Therefore, a cantilever with a higher resonant frequency has a lower noise density. In addition, shorter cantilevers have higher optical beam deflection (OBD) detection sensitivity, because the sensitivity follows $\Delta\varphi/\Delta z = 3/(2L)$, where Δz is the displacement and $\Delta\varphi$ is the change in the angle of a cantilever free-end. A high resonant frequency and a small spring constant result in a large ratio (f_c/k_c), which gives the cantilever high sensitivity to the gradient (k) of the force exerted between the tip and the sample, because the gradient of the force shifts the cantilever resonant frequency by approximately $-0.5kf_c/k_c$. Therefore, small cantilevers with large values of f_c/k_c are useful for phase-contrast imaging and frequency-modulation AFM (FM-AFM).

The tip apex radius of the small cantilevers developed by Olympus is $\sim 17\,nm$, which is not sufficiently small for the high-resolution imaging of biological samples. We usually attach a sharp tip on the original tip by electron-beam deposition (EBD) in phenol gas (4). A piece of phenol crystal (sublimate) is placed in a small container with small holes ($\sim 0.1\,mm$ diameter) in the lid. The container is placed in a scanning electron microscope (SEM) chamber under a low-vacuum condition, and cantilevers are placed immediately above the holes. A spot-mode electron beam is irradiated onto the cantilever tip, which produces a needle on the original tip at a growth rate of $\sim 1\,\mu m/min$. The newly formed tip has an apex radius of $\sim 25\,nm$ and is sharpened by plasma etching in argon or oxygen gas (Fig. 1.1b), which decreases the apex radius to $\sim 4\,nm$ in the best case. The mechanical durability of this sharp tip is not high but is still sufficient to be used to capture many images.

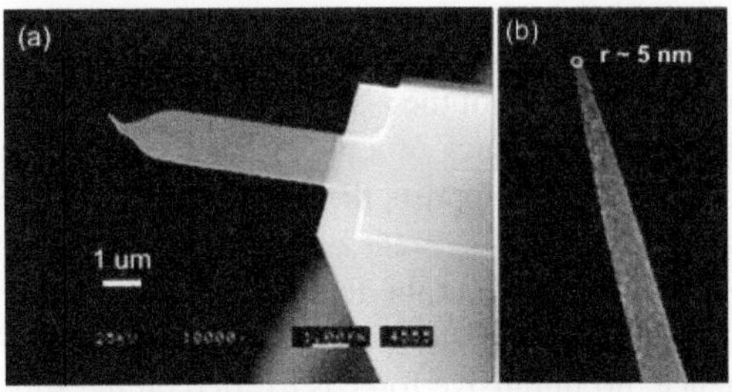

Fig. 1.1 SEM images of small cantilever **(a)** and EDB tip **(b)**

1.3.2 *Scanners*

Tube scanners that have been often used for conventional AFM are inadequate for high-speed AFM, as their long and thin structure lowers the resonant frequencies in the x-, y-, and z-directions. The structural resonant frequency can be enhanced by adopting a compact structure. However, a compact structure tends to produce interference (crosstalk) between the three scan axes. To minimize the crosstalk, we employ flexures (blade springs) that are sufficiently flexible to be displaced but sufficiently rigid in the directions perpendicular to the displacement axis (Fig. 1.2). In our scanner, the y-scanner moves the x-scanner, the x-scanner moves the z-scanner, and the z-scanner moves a sample stage (10). In the most recent x-scanner, a piezoactuator is inserted between two identical flexures so that the center of mass is unchanged while the piezoactuator is displaced (Fig. 1.2) (4). One of the flexures is connected with the z-scanner, while the other is connected with an identical dummy z-scanner for counterbalancing. The whole scanner framework is made by monolithically processing a single metal block to minimize the number of resonant elements. The lowest resonant frequency of the x-scanner is $\sim 60\,\text{kHz}$. The maximum displacements of the x- and y-scanners are $\sim 2\,\mu\text{m}$.

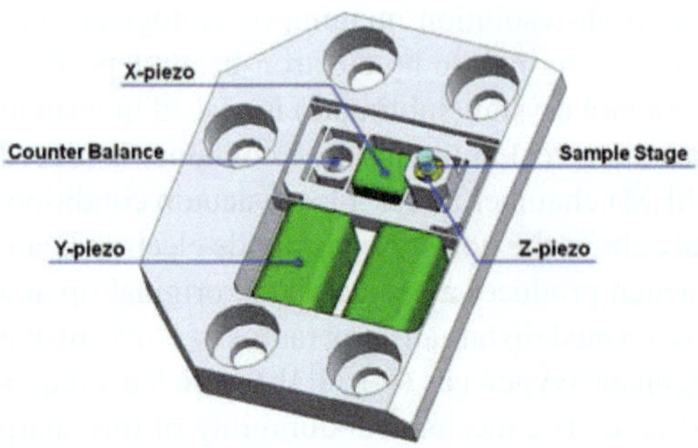

Fig. 1.2 Structure of high-speed scanner

The response time of the z-scanner (τ_z) is expressed by $\tau_z = Q_z/(\pi f_z)$, where Q_z and f_z are the quality factor and resonant frequency of the z-scanner, respectively. f_z is almost solely determined by the resonant frequency of a piezoactuator used and by how to hold the piezoactuator to completely suppress the structural resonance arising from the surrounding framework. In the most recent z-scanner, a piezoactuator with a resonant frequency of $\sim 450\,\text{kHz}$ is held at the four rims parallel to the displacement direction (Fig. 1.2) (11). In this way of holding, the center of mass of the piezoactuator is not displaced (hence, no impulsive force is exerted on the supporting mechanism), and importantly, the resonant frequency of the z-piezoactuator is nearly unchanged. The maximum displacement is $\sim 1\,\mu\text{m}$.

1.4 Active Damping Techniques

1.4.1 *Feedback Q-Control for z-Scanner*

In the diagram shown in Fig. 1.3a, the operator $H(s)$ converts the resonant system $G(s)$ to a target system $R(s)$ expressed as

$$R(s) = G(s)/[1 - G(s)H(s)] \,. \tag{1.1}$$

Here, let us consider the simplest case where $G(s)$ consists of a single resonant element with resonant frequency ω_1 and quality factor Q_1. The target system is expressed as a single resonator with resonant frequency ω_0 and quality factor Q_0 ($< Q_1$). For these systems, $H(s)$ is expressed as

$$H(s) = \frac{1}{G(s)} - \frac{1}{R(s)} = \left(\frac{s^2}{\omega_1^2} + \frac{s}{Q_1\omega_1} + 1\right) - \left(\frac{s^2}{\omega_0^2} + \frac{s}{Q_0\omega_0} + 1\right) . \tag{1.2}$$

To eliminate the second-order term from $H(s)$, ω_0 should equal ω_1, which results in

$$H(s) = -\left(\frac{1}{Q_0} - \frac{1}{Q_1}\right)\frac{s}{\omega_1} \,. \tag{1.3}$$

This $H(s)$ is identical to a derivative operator with gain $= -1$ at the frequency $\hat{\omega} = Q_0Q_1\omega_1/(Q_1 - Q_0)$. By adjusting the gain parameter of the derivative operator, we can arbitrarily change the target quality factor Q_0. This method is known as 'Q-control'. When this method is applied to the z-scanner, we must measure the displacement (i.e., the output of $G(s)$) or velocity of the z-scanner. However, it is difficult to do so. This problem is solved by using a mock z-scanner $M(s)$ (a second-order low-pass filter) characterized by a transfer function similar to that of the lowest resonant component of the real z-scanner (Fig. 1.3b) (12).

Since the real z-scanner usually has additional higher resonant components, this simple Q-control is not satisfactory. When elemental resonators are connected in series, we can use mock scanners, $M_1(s)$, $M_2(s)$, - - -, each of which is characterized by a transfer function representing each elemental resonator. Each mock scanner is controlled by a corresponding Q-controller. For example, when the scanner consists

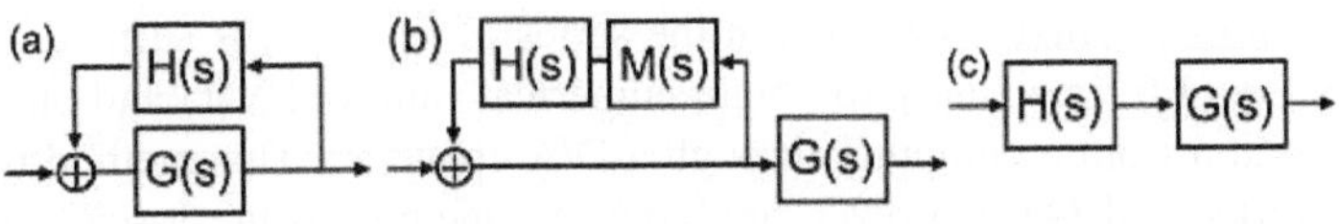

Fig. 1.3 Diagrams of feedback and feedforward active damping for scanner vibrations. **(a)** Conventional feedback Q-control, **(b)** feedback Q-control with mock scanner, **(c)** feedforward control. $G(s)$, $H(s)$, and $M(s)$ represent a resonant system to control, an operator for control, and a mock resonant system, respectively.

of two resonators connected in series, the composite transfer function $C(s)$ of the total system is expressed as

$$C(s) = \frac{1}{1 - M_1(s)H_1(s)} \times \frac{1}{1 - M_2(s)H_2(s)} \times G_1(s)G_2(s)\,. \tag{1.4}$$

Since $M_1(s)$ and $M_2(s)$ are the same as (or similar to) $G_1(s)$ and $G_2(s)$, respectively, Eq. (1.4) represents a target system consisting of damped resonators connected in series. Note that we can use $M_1(s)$ and $M_2(s)$ separately, unlike $G_1(s)$ and $G_2(s)$.

When the scanner consists of elemental resonators connected in parallel, active damping becomes more difficult. In this case, we use a mock z-scanner comprised of two low-pass filters connected in parallel, which correspond to the first and second resonant components of the z-scanner. It cannot damp the second resonant component significantly, but better than when using a single low-pass filter (12).

1.4.2 Feedforward Damping of x-Scanner

The feedforward control type of active damping (Fig. 1.3c) is based on inverse compensation (i.e., $H(s) \sim 1/G(s)$). Generally, inverse compensation-based damping has an advantage in that we can extend the scanner bandwidth. This damping method is much easier to apply to the x-scanner than to the z-scanner, because for the former, the scan waves are known beforehand and are periodic (hence, the frequencies used are discrete, i.e., integral multiples multiples of the fundamental frequency). The waveforms of the x-scan are isosceles triangles characterized by amplitude X_0 and fundamental angular frequency ω_0. Their Fourier transform is given by

$$F(\omega) = 2\pi X_0 \left[\frac{1}{2}\delta(\omega) - \frac{2}{\pi^2} \sum_{k=-\infty}^{+\infty} \frac{1}{k^2} \delta(\omega - k\omega_0) \right] \quad (k : \text{odd})\,. \tag{1.5}$$

To move the x-scanner in the isosceles triangle waveforms, the input signal $X(t)$ sent to the x-scanner characterized by the transfer function $G(s)$ is given by the inverse Fourier transform of $F(\omega)/G(i\omega)$, which is expressed as

$$X(t) = \frac{X_0}{2} - \frac{4X_0}{\pi^2} \sum_{k=1}^{+\infty} \frac{1}{k^2} \frac{1}{G(ik\omega_0)} \cos(k\omega_0 t) \quad (k : \text{odd})\,. \tag{1.6}$$

In practice, the sum of the first $\sim$10 terms in the series of Eq. (1.6) is sufficient. We can calculate Eq. (1.6) in advance to obtain numerical values of $X(t)$ and output them in succession from a computer through a D/A converter. This method is convenient and works well (4). However, applying this method to the y-scanner is not practical because the period of 'y-return' (scanning towards the origin after acquiring a frame) is much shorter than the total period of the y-scan (i.e., image acquisition time), and hence, we have to use too many terms in the series expansion of the y-scan waveform. We can simply use a round waveform for the y-return. Rounding the vertexes of the triangle waveforms is also useful for the x-scanner, although the deviation of the scan trajectory from the triangle waveform becomes large.

1.4.3 Practice of Active Damping of Scanner Vibrations

Here, we describe the practice of active damping applied to the scanner mentioned above. The z-piezoelectric actuator has a resonant frequency of ~450 kHz under free oscillation. The gain and phase spectra of the mechanical response of the z-scanner are shown with the lines marked with (a) in Fig. 1.4. The z-scanner exhibited large resonant peaks at 440 kHz and 550 kHz. The resonant frequency of 440 kHz is similar to that of the free oscillation of the piezoelectric actuator. Judging from the phase spectrum, the two resonators are connected in series. We used two mock z-scanners connected in series. The resulting gain and phase spectra are shown with the lines marked with (b) in Fig. 1.4. The peak at 440 kHz was almost completely removed and the frequency that gave a 90° phase delay reached 250 kHz (4). We could reduce the gain at 440 kHz but did not do so to minimize the phase delay. Since the imaging frequency (i.e., feedback frequency) does not go beyond ~150 kHz, the resonator of 440 kHz with a gain slightly larger than 1 is not excited during imaging.

Next, we show the effect of feedforward type of damping applied to the x-scanner contained in the most recent scanner mentioned above. The x-scanner has a major resonance at ~60 kHz and a relatively clean resonant pattern. When the line scan was performed at 3.3 kHz without damping, its displacement exhibited vibrations (Fig. 1.5a). When it was driven in a waveform calculated using Eq. (1.6) with the

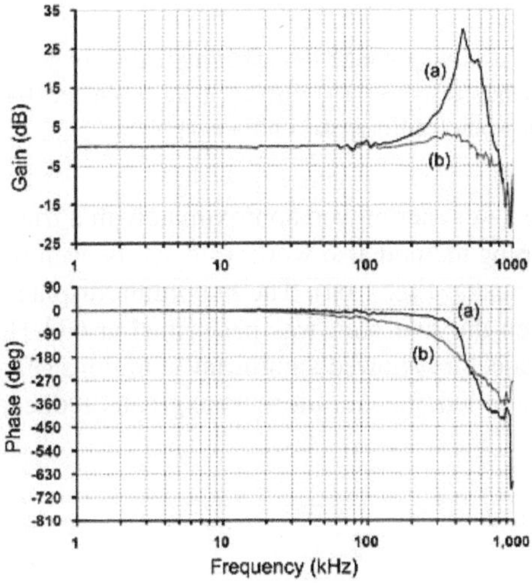

Fig. 1.4 Frequency response of z-scanner and effect on the response of active damping by feedback Q-control with mock z-scanner: (**a**) without damping, (**b**) with damping

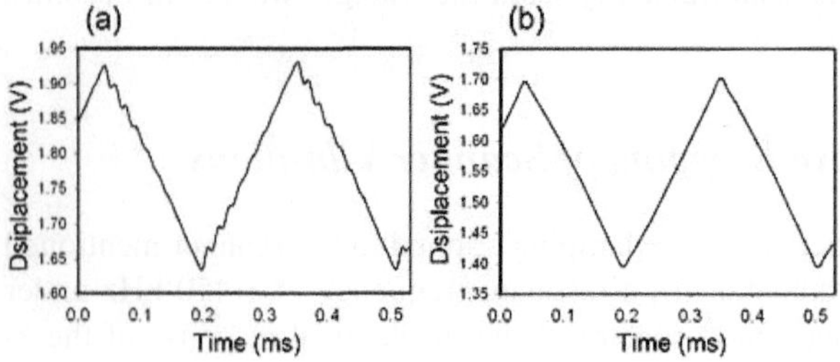

Fig. 1.5 Displacement of x-scanner and effect of active damping by feedforward control. (**a**) without damping, (**b**) with damping.

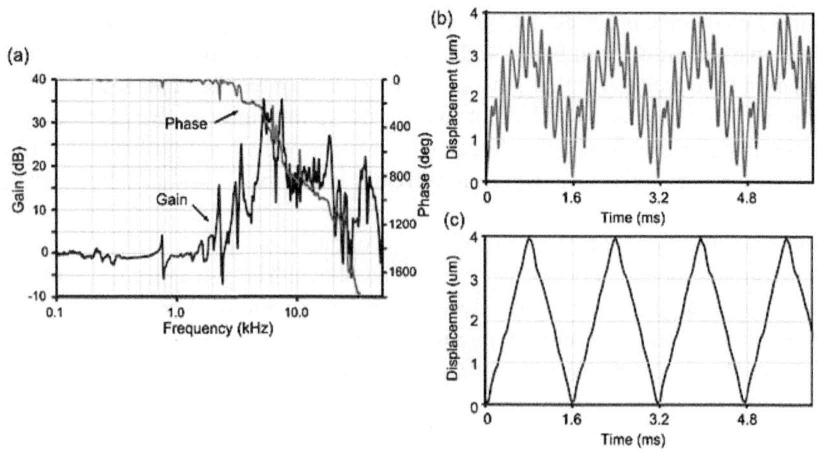

Fig. 1.6 Frequency response and displacement of wide-range x-scanner and effect of active damping by feedforward control. (**a**) Frequency response, (**b**) displacement when feedforward control is not applied, (**c**) displacement when feedforward control is applied.

maximum term $k = 17$ in the series, the x-scanner moved approximately in a triangle waveform (Fig. 1.5b) (4). This damping method also works effectively even for an x-scanner having multiple resonant peaks (Fig. 1.6a). The maximum displacement of this x-scanner is $\sim$5 μm. When the line scan was performed at 633 Hz with the displacement up to 4 μm, its trajectory showed significant vibrations (Fig. 1.6b). When feedforward damping was applied, the vibrations were greatly reduced (Fig. 1.6c).

1.5 Dynamic PID Control

In AFM, 'parachuting' means that a cantilever tip completely detaches from a sample surface at a steep down-hill region of the sample and does not re-land on the surface for a while. Parachuting more frequently occurs when the peak-to-peak

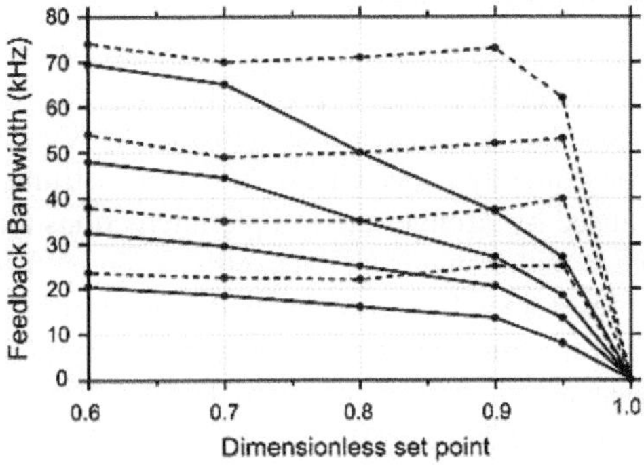

Fig. 1.7 Feedback bandwidth as a function of the set point and the effect of dynamic PID control. The dimensionless set point represents the cantilever's peak-to-peak amplitude set point normalized by the peak-to-peak free oscillation amplitude. *Solid line*, with conventional PID control; *dashed line*, with dynamic PID control.

amplitude set point (A_s) is set closer to the cantilever peak-to-peak free oscillation amplitude ($2A_0$). The error signal is saturated at $(A_s - 2A_0)$ during parachuting and the parachuting period is more prolonged when the saturated error signal is smaller. In fact, feedback bandwidth decreases with increasing set point due to parachuting (solid line, Fig. 1.7). However, for fragile biological samples, the tapping force exerted by an oscillating cantilever tip on the sample has to be minimized by using A_s very close to $2A_0$. To overcome this difficulty, we developed a dynamic PID controller in which the gain parameters are automatically and dynamically changed depending on the cantilever peak-to-peak oscillation amplitude (A) (13).

The sign of $(A - A_s)$ is a good indicator of whether the sample is being scanned at the down-hill region or the up-hill region; the plus and minus signs approximately correspond to down-hill and up-hill regions, respectively (note that A cannot be maintained exactly at A_s because of the delay of feedback control). In the dynamic PID controller, the gain parameters are increased when A exceeds A_s (or a level A_t slightly higher than A_s). The gain increment can be constant or proportional to $(A - A_s)$ or $(A_t - A_s)$. The dynamic PID controller is effective and the feedback bandwidth becomes independent of A_s as long as A_s is smaller than $\sim 0.95 \times 2A_0$, indicating that no parachuting occurs under the condition of $A_s < 0.95 \times 2A_0$ (dashed line, Fig. 1.7).

1.6 Drift Compensator

Using the dynamic PID controller, we can now use a set point A_s very close to the peak-to-peak free oscillation amplitude of a cantilever $2A_0$, without causing parachuting. However, the difference between them $(2A_0 - A_s)$ becomes very small when we use a small $2A_0$ for imaging biological samples. To not disturb the function of biomolecules, we usually have to use $2A_0 \sim 2-3\,\text{nm}$, so that the difference

($2A_0 - A_s$) becomes 0.2–0.3 nm under the condition of $A_s = 0.9 \times 2A_0$. This means that even a very small drift in A_0 affects imaging. A cantilever is usually excited by oscillation of a piezoactuator. The temperature of the piezoactuator is elevated during long runs, leading to a smaller oscillation and hence to a smaller A_0. Supposing that A_0 decreases by drift and consequently A decreases, the AFM instrument misinterprets this decrease in A as caused by stronger tip–sample interactions than expected. Therefore, the sample stage is withdrawn from the cantilever tip by feedback control, and eventually, the cantilever tip will be completely detached from the sample surface, resulting in inability of imaging. Thus, we need to precisely maintain the free oscillation amplitude constant, but there is no way to detect a change in the free oscillation amplitude during imaging.

This problem was previously challenged by Schiener *et al.* (14). They monitored the second-harmonic amplitude of the cantilever to detect the drift in A_0. The second-harmonic amplitude is sensitive to the tip–sample interaction, and therefore, the drift in A_0 is reflected in the second-harmonic amplitude averaged over a period longer than the image acquisition time. To compensate for drift in the cantilever excitation efficiency, we controlled the output gain of a wave generator connected to the excitation piezoactuator by measuring and averaging the second-harmonic amplitude (13). By this compensation, the cantilever free oscillation amplitude is maintained constant and thus very stable imaging is possible for a long time even with a small difference $(2A_0 - A_s) \sim 0.2\,\text{nm}$.

1.7 Bio-imaging

To demonstrate the high capability of our current high-speed AFM instrument, we here describe some imaging studies carried out on 2D crystals of bacteriorhodopsin, without getting into the biological significance.

1.7.1 Dynamic Equilibrium at the Bacteriorhodopsin Crystal Edge

The purple membrane (PM) exists in the plasma membrane of *Halobacterium halobium*, and its constituent protein, bacteriorhodopsin (bR), functions as a light-driven proton pump. In the PM, bR monomers are associated to form a trimeric oligomer and the trimers are arranged in a hexagonal lattice (15). In the 2D crystal of bR and any crystals in general, they are in dynamic equilibrium with the constituents at the interface between the crystal and the liquid phase. We visualized dynamic events at the interface in the PM (16).

The PM was adsorbed on a mica surface in a buffer solution (10 mM Tris-HCl (pH 8.0) and 300 mM KCl). Figure 1.8 shows typical AFM images taken at 3 frames/s. The bR trimers in the crystal are indicated by the thin-lined triangles. At 0.6 s, two bR trimers (red triangles) have newly bound to the crystal edge. Subsequently, two bR trimers (white triangles) have dissociated and another trimer (red triangle) has bound to the crystal edge at 2.1 s. One of the two dissociated trimers remains in the crystal area for ~0.9 s. In order to estimate the inter-trimer interaction energy, we analyzed the residence time of newly bound bR trimers at the crystal edge and its dependence on the number of interaction sites. For the analysis, we assumed that, within the 2D bR crystal, a trimer can interact with the surrounding trimers through six sites. Following this assumption, the number of interaction sites at the crystal edge is reduced, depending on the binding position. Successive AFM images as exemplified in Fig. 1.8 showed many binding and dissociation events in which bR trimers bound to different sites at the border between the crystal and non-crystal areas. These events can be classified into types 'I', 'II', and 'III' depending on the number of interaction sites involved (Roman numerals indicate the number of interaction bonds). Type II binding events are predominant (~74%), whereas type I (~6%) and type III (~20%) bindings are minor events. The lifetime of the type I bonds was too short to obtain clear images of the corresponding event, preventing reliable statistics.

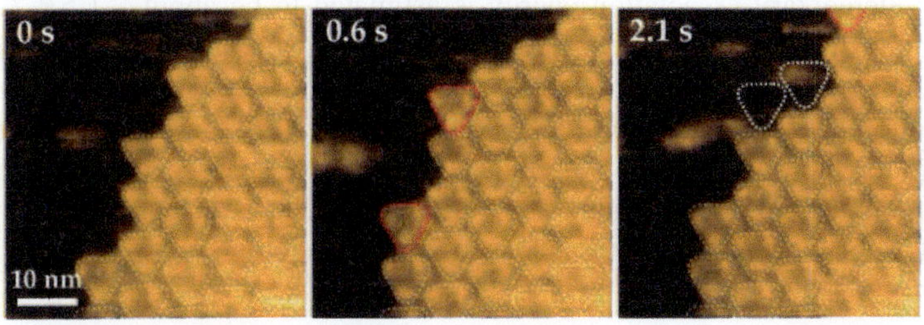

Fig. 1.8 Time-lapse high-magnification AFM images of PM at the interface between the crystalline and the non-crystalline area captured at 3 frames/s. The thin-lined triangles show bR trimers. The bR trimers encircled by the red dotted line at 0.6 s indicate newly bound bR trimers. The white dotted triangles at 2.1 s indicate the previously bound trimers.

Figure 1.9a shows a histogram of the lifetime for type II bonds which was measured using AFM images taken at 10 frames/s. The histogram can be well fitted by a single exponential (correlation coefficient, $r = 0.9$), from which the average lifetime τ_2 was estimated to be 0.19 ± 0.01 s. Figure 1.9b shows a histogram of the type III bond lifetime, from which the average lifetime τ_3 was estimated to be 0.85 ± 0.08 s. The longer lifetime of type III bonds compared with type II obviously arises from a relationship of $E_{III} < E_{II} < 0$, where E_{II} and E_{III} are the association energies responsible for type II and type III interactions, respectively. The average lifetime ratio, τ_2/τ_3, is given by

$$\tau_2/\tau_3 = \exp[-(E_{III} - E_{II})/(k_B T)] \ . \qquad (1.7)$$

Because the type II interaction contains two elementary bonds, whereas the type III interaction contains three, the energy difference $E_{III} - E_{II}$ corresponds to the association energy of the single elementary bond. From the ratio $\tau_2/\tau_3 = 0.22$ and

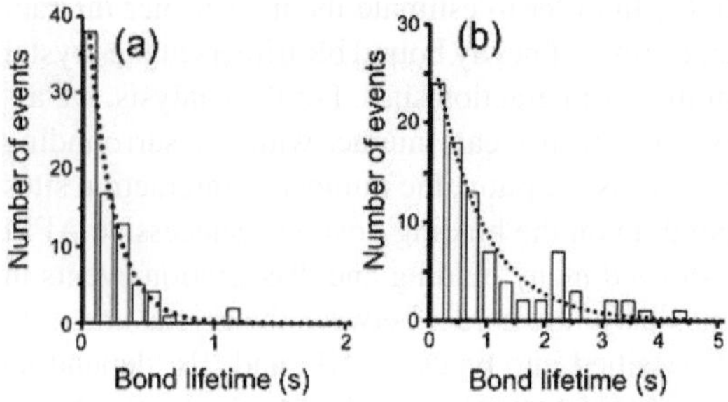

Fig. 1.9 (**a**) Histogram showing the lifetime distribution of type II binding events. The histogram was fitted by a single-exponential function (*dotted line*). (**b**) Histogram showing the lifetime distribution of type III binding events. The histogram was fitted by a single-exponential function (*dotted line*).

Eq. (1.7), this elementary association energy is estimated to be about $-1.5k_BT$, which corresponds to −0.9 kcal/mol at 300 K (16). This value is approximately consistent with that estimated by differential scanning calorimetry (17; 18).

The association energy of the type II bond is only $-3.0k_BT$. Nevertheless, the residence time for the type II bond was independent of the scan speed (60–105 µm/s corresponding to the imaging rate of 8–14 frames/s). This is a clear indication that very weak tip-sample interaction is achieved even at such high imaging rates.

1.7.2 Structural Changes of Bacteriorhodopsin under Light

Since bR is a light-driven proton pump, the detection of its structural change upon absorption of light has long been a goal. bR contains seven transmembrane α-helices (named A-G) enclosing the chromophore retinal (19; 20). Absorption of light by the retinal triggers a cyclic reaction (photocycle) that comprises a series of spectroscopically identified intermediates, designated as the J, K, L, M, N, and O states (21). The light-induced conformational changes in bR have been investigated by various methods (22; 23; 24; 25; 26), leading to a consensus that the proton channel at the cytoplasmic surface is opened by the tilting of helix F away from the protein center (25; 26). This movement of helix F leads to a rearrangement of the E-F loop, which results in large-scale conformation changes in the M and N intermediates. Recently, we directly captured the photo-induced conformational change at cytoplasmic surface of bR using high-speed AFM (27).

The photocycle of wild-type bR completes in $\sim$10 ms at pH 7.0 and therefore the photo-activated intermediate states are too brief to be clearly captured by high-speed AFM. Thus, we used the D96N mutant that has a longer photocycle ($\sim$10 s) at pH 7.0 but retains an ability of proton pumping. Figure 1.10a and b show successive images of D96N at the cytoplasmic surface captured at 1 frame/s. Under the un-photolyzed state (Fig. 1.10a), one can clearly see individual bR molecules forming well-ordered trimers in the two-dimensional hexagonal lattice. A trimer is encircled

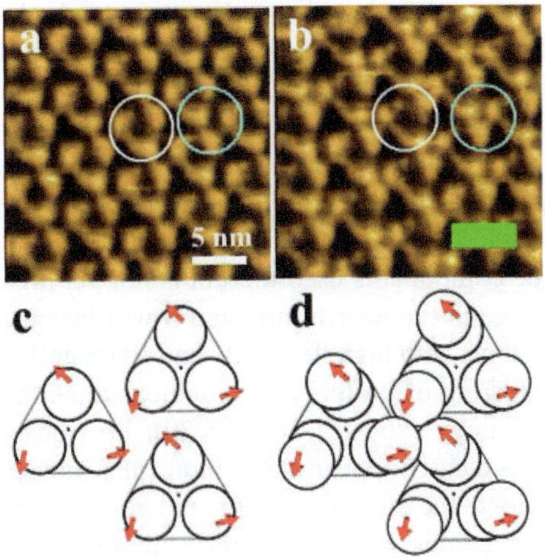

Fig. 1.10 High-speed AFM images of the cytoplasmic surface of D96N under dark (**a**) or illuminated (**b**) conditions. The green bar indicates light illumination. A trimer is highlighted by the white circles and a trefoil is highlighted by the light-blue circles. The direction of the conformational change in bR is shown by the red arrows (**c**, **d**). By the outward movement, bR monomers in a trefoil are brought into contact with each other (**d**).

by the white circle. Upon illumination with green light, bR drastically changes its structure as shown in Fig. 1.10b. This activated state with a unique structure returns to the prior unexcited state in a few seconds after light-off. This cyclic structural change is highly reproducible in repeated dark-illumination cycles. On the other hand, significant conformational changes of bR are not observed on the extracellular surface.

We analyzed the center of mass for each monomer during the dark-illumination cycles. Each activated bR monomer displaced outward from the center of trime by ∼0.7 nm on average (the direction of the movement is shown by arrows in Fig. 1.10c). As a result, three nearest-neighbor bR monomers, each of which belongs to a different adjacent trimer, are brought into contact with each other (Fig. 1.10d). Here, we use a new designation, 'trefoil', for the triad of the nearest-neighbor monomers (shown by the light-blue circles in Fig. 1.10a and b) to distinguish it from the original trimer. Remarkably, this transient assembly in a trefoil alters the decay kinetics of the activated state (27).

As demonstrated above, direct and dynamic observation of functioning protein molecules is a powerful new approach to studying conformational changes in proteins induced by external stimuli.

1.8 Conclusion

We believe that the techniques for increasing the imaging speed of AFM have already been almost matured. Further increasing the speed would be very difficult unless completely different approaches are introduced, such as self-sensing and self-actuation small cantilevers with a small spring constant. We have mostly employed analogue circuits to construct the high-speed AFM instrument. The analogue system lacks flexibility. For example, the analogue feedback Q-controller has to be newly constructed whenever a new scanner is developed. At present, digital systems such as FPGA and DSP are not fast enough. FPAA (Field Programmable Analogue Array), an analogue system corresponding to FPGA, is now becoming available (e.g., Anadigm) but the bandwidth of analogue components used therein is not yet high enough. However, this situation will change in the near future, and it will become possible to construct high-speed AFM systems with high flexibility and adaptability.

Recent studies of high-speed AFM imaging of proteins have demonstrated the usefulness of this new microscopy in biology (26; 27; 28; 29; 30). Previously known and unknown behaviors of proteins clearly appear in the molecular movies simultaneously. Even for known behaviors, the high-resolution movies provide corroborative 'visual evidence'. Unlike previous methods, high-resolution AFM imaging unselectively provides comprehensive information on the structure and dynamics of a functioning molecule. Thus, the high-speed AFM imaging of functioning biomolecules has the potential to transform the field of structural biology. In addition, high-speed AFM will provide a new approach to studying nanometer-scale dynamic phenomena occurring in liquids and at the solid-liquid interfaces, and therefore, will greatly contribute to creating new nanotechnology.

References

[1] Viani, M.B., Richter, M., Rief, M., Gaub, H.E., Plaxco, K.W., Cleland, A.N., Hansma, H.G., Hansma, P.K.: Fast imaging and fast force spectroscopy of single biopolymers with a new atomic force microscope designed for small cantilevers. Rev. Sci. Instrum. 70, 4300–4303 (1999)

[2] Fantner, G.E., Schitter, G., Kindt, J.H., Ivanov, T., Ivanova, K., Patel, R., Holten-Andrersen, N., Adams, J., Thurnera, P.J., Rangelowb, I.W., Hansma, P.K.: Components for high-speed atomic force microscopy. Ultramicroscopy 106, 881–887 (2006)

[3] Ando, T., Kodera, N., Takai, E., Maruyama, D., Saito, K., Toda, A.: High-speed atomic force microscope for studying biological macromolecules. Proc. Natl. Acad. Sci. USA 98, 12468–12472 (2001)

[4] Ando, T., Uchihashi, T., Fukuma, T.: High-speed atomic force microscopy for nano-visualization of dynamic biomolecular processes. Prog. Surf. Sci. 83, 337–437 (2008)

[5] Ando, T., Uchihashi, T., Kodera, N., Yamamoto, D., Taniguchi, M., Miyagi, A., Yamashita, H.: Review: High-speed atomic force microscopy for observing dynamic biomolecular processes. J. Mol. Recognit. 20, 448–458 (2007)

[6] Ando, T., Uchihashi, T., Kodera, N., Yamamoto, D., Taniguchi, M., Miyagi, A., Yamashita, H.: Invited Review: High-speed AFM and nano-visualization of biomolecular processes. Pflügers Archiv - Eur. J. Physiol. 456, 211–225 (2008)

[7] Hansma, P.K., Schitter, G., Fantner, G.F., Prater, C.: High-speed atomic force microscopy. Science 314, 601–602 (2006)

[8] Walters, D.A., Cleveland, J.P., Thomson, N.H., Hansma, P.K., Wendman, M.A., Gurley, G., Elings, V.: Short cantilevers for atomic force microscopy. Rev. Sci. Instrum. 67, 3583–3590 (1996)
[9] Kitazawa, M., Shiotani, K., Toda, A.: Batch fabrication of sharpened silicon nitride tips. Jpn. J. Appl. Phys. 42, 4844–4847 (2003)
[10] Ando, T., Uchihashi, T., Kodera, N., Miyagi, A., Nakakita, R., Yamashita, H., Matada, K.: High-speed AFM for studying the dynamic behavior of proteins molecules at work. Surf. Sci. Nanotechnol. 3, 384–392 (2005)
[11] Fukuma, T., Okazaki, Y., Kodera, N., Uchihashi, T., Ando, T.: High resonance frequency force microscope scanner using inertia balance support. Appl. Phys. Lett. 92, 243119 (2008)
[12] Kodera, N., Yamashita, H., Ando, T.: Active damping of the scanner for high-speed atomic force microscopy. Rev. Sci. Instrum. 76, 053708 (2005)
[13] Kodera, N., Sakashita, M., Ando, T.: Dynamic proportional-integral-differential controller for high-speed atomic force microscopy. Rev. Sci. Instrum. 77, 083704 (2006)
[14] Schiener, J., Witt, S., Stark, M., Guckenberger, R.: Stabilized atomic force microscopy imaging in liquids using second harmonic of cantilever motion for set-point control. Rev. Sci. Instrum. 75, 2564–2568 (2004)
[15] Henderson, R., Baldwin, J.M., Ceska, T.A., Zemlin, F., Beckmann, E., Downing, K.H.: Model for the structure of bacteriorhodopsin based on high-resolution electron cryomicroscopy. J. Mol. Biol. 213, 899–929 (1990)
[16] Yamashita, H., Voïtchovsky, K., Uchihashi, T., Contera, S.A., Ryan, J.F., Ando, T.: Dynamics of bacteriorhodopsin 2D crystal observed by high-speed atomic force microscopy. J. Struc. Biol. 167, 153–158 (2009)
[17] Jackson, M.B., Sturtevant, J.M.: Phase transitions of the purple membranes of Halobacterium halobium. Biochemistry 17, 911–915 (1978)
[18] Koltover, I., Raedler, J.O., Salditt, T., Rothschild, K.J., Safinya, C.R.: Phase behavior and interactions of the membrane-protein bacteriorhodopsin. Phys. Rev. Lett. 82, 3184–3187 (1999)
[19] Kimura, Y., Vassylyev, D.G., Miyazawa, A., Kidera, A., Matsushima, M., Mitsuoka, K., Murata, K., Hirai, T., Fujiyoshi, Y.: Surface of bacteriorhodopsin revealed by high-resolution electron crystallography. Nature 389, 206–211 (1997)
[20] Luecke, H., Schobert, B., Richter, H.T., Cartailler, J.P., Lanyi, J.K.: Structure of bacteriorhodopsin at 1.55 Å resolution. J. Mol. Biol. 291, 899–911 (1999)
[21] Lanyi, J.K.: Bacteriorhodopsin. Annu. Rev. Physiol. 66, 665–688 (2004)
[22] Dencher, N.A., Dresselhaus, D., Zaccai, G., Büldt, G.: Structural changes in bacteriorhodopsin during proton translocation revealed by neutron diffraction. Proc. Natl. Acad. Sci. USA 86, 7876–7879 (1989)
[23] Subramaniam, S., Gerstein, M., Oesterhelt, D., Henderson, R.: Electron diffraction analysis of structural changes in the photocycle of bacteriorhodopsin. EMBO J. 12, 1–8 (1993)
[24] Brown, L.S., Needleman, R., Lanyi, J.K.: Conformational change of the E-F interhelical loop in the M photointermediate of bacteriorhodopsin. J. Mol. Biol. 317, 471–478 (2002)
[25] Sass, H.J., Büld, G., Gessenich, R., Hehn, D., Neff, D., Schlesinger, R., Berendzen, J., Ormos, P.: Structural alterations for proton translocation in the M state of wild-type bacteriorhodopsin. Nature 406, 649–653 (2000)
[26] Subramaniam, S., Henderson, R.: Molecular mechanism of vectorial proton translocation by bacteriorhodopsin. Nature 406, 653–657 (2000)

[27] Shibata, M., Yamashita, H., Uchihashi, T., Kandori, H., Ando, T.: High-speed atomic force microscopy shows dynamic molecular processes in photo-activated bacteriorhodopsin. Nature Nanotech. 5, 208–212 (2010)
[28] Yamamoto, D., Nagura, N., Omote, S., Taniguchi, M., Ando, T.: Streptavidin 2D crystal substrates for visualizing biomolecular processes by atomic force microscopy. Biophys. J. 97, 2358–2367 (2009)
[29] Yamamoto, D., Uchihashi, T., Kodera, N., Ando, T.: Anisotropic diffusion of point defects in two-dimensional crystal of streptavidin observed by high-speed atomic force microscopy. Nanotechnology 19, 0384009 (2008)
[30] Miyagi, A., Tsunaka, Y., Uchihashi, T., Miyanagi, K., Hirose, S., Morikawa, K., Ando, T.: Visualization of intrinsically disordered regions of proteins by high-speed atomic force microscopy. Chem. Phys. Chem. 9, 1859–1866 (2008)

Chapter 2
High-Accuracy Atomic Force Microscope

David L. Trumper, Robert J. Hocken, Darya Amin-Shahidi,
Dean Ljubicic, and Jerald Overcash

Abstract. We have designed, built, and tested a high-accuracy atomic force microscope (HAFM) to be used for dimensional metrology. The HAFM is specialized for use in conjunction with our Sub-Atomic Measuring Machine (SAMM), serving as the surface measurement probe with 0.1 nm resolution over the SAMM travel range of 25 mm by 25 mm by 100 μm. In this configuration, all lateral scanning is provided by the SAMM, and so the HAFM is designed to move its probe with a single degree of freedom motion normal to the sample of interest. This sample-normal probe motion (Z-axis) is guided in the HAFM by symmetric monolithic flexures which are designed for high thermal stability and minimum lateral error motion. A piezoelectric stack drives the HAFM Z-axis over a range of 20 μm. The HAFM uses a commercially available self-sensing quartz-tuning-fork-based AFM probe, which is operated in constant-amplitude self-resonance via electronics described herein. In this configuration, the sample-probe separation is sensed via frequency shift of the probe resonance. The probe-sample separation is controlled using a discrete-time surface-tracking controller implemented on a field-programmable gate array (FPGA). The controller tracks the surface by actuating the piezo to maintain a constant self-resonance period. To avoid spurious mixing, the controller's sampling is made synchronous to the self-resonance oscillations. Three capacitive displacement sensors directly measure the surface tracking motion, providing a high-accuracy measurement of surface height. We have experimentally demonstrated surface tracking

David L. Trumper · Darya Amin-Shahidi · Dean Ljubicic
Mechanical Engineering Department,
Massachusetts Institute of Technology,
Cambridge, MA
e-mail: {trumper,d_a,ljubicic}@mit.edu

Robert J. Hocken · Jerald Overcash
Mechanical Engineering Department,
UNC-Charlotte,
Charlotte, NC
e-mail: {hocken,jloverca}@uncc.edu

E. Eleftheriou & S.O.R. Moheimani (Eds.): Cntrl. Tech. for Emerging Micro/Nanoscale Sys. LNCIS 413, pp. 17–46.
springerlink.com

control with 100 Hz unity-gain crossover frequency, 70 degrees phase margin, and 0.12-nm RMS noise in a 100-Hz measurement bandwidth. We have also used the HAFM to measure calibration gratings and the surface of a freshly cleaved sanded Mica sample. The HAFM probe has recently been installed on the SAMM stage; we show a preliminary image taken using this combination.

2.1 Introduction

The work presented here is directed to creating an instrument for nanometer-scale accuracy dimensional measurement of near-planar samples, with sub-atomic resolution. To achieve this goal we have undertaken a joint project between researchers at MIT and UNC-Charlotte. The system we have designed includes a magnetically-suspended 6 degree-of-freedom positioning stage, called the Sub-Atomic Measuring Machine (SAMM). This stage was designed in the doctoral thesis of Holmes (15), and has recently undergone extensive upgrades in measurement systems and control as described in (27). We have designed and tested several probing systems for use with the SAMM, starting with a scanning tunneling microscope probe (15), as well as a confocal microscope (34). Most recently, we have designed the HAFM atomic force microscope (AFM) probe described in detail below. The HAFM head is designed to interface with a kinematic mount on the SAMM. In this configuration, the SAMM stage performs the sample scan, and the HAFM head tracks the sample surface in the sample-normal direction.

The SAMM stage scans a 25 mm by 25 mm by 100 μm work volume with sub-nanometer resolution and 10-nm accuracy. Its moving stage is neutrally-buoyant in fluorosilicone oil, and is controlled in 6 degrees of freedom using 4 levitation linear motors.These linear motors use Halbach magnet arrays (12; 13) attached to the moving stage. The idea of combining oil flotation and magnetic suspension for a precision motion system was first explored in the short-range Angstrom stage (14). Following up on this idea, the long range stage (LORS) was developed in the doctoral thesis work of Holmes (15). The SAMM is the result of continuing improvement of the LORS stage (17; 16; 23; 18). Other researchers have also studied high accuracy measuring machines. Some examples of long-range scanning machines designed for metrology are (7; 8; 21; 9).

The LORS stage initially used a commercial STM. The STM probe was not sufficiently accurate, especially due to apparent multiple-tipping effects, and was later replaced by a metrological confocal microscope (34). Using the confocal microscope, the stage could achieve sub-nanometer resolution, 1 nm repeatability, and an estimated 10 nm accuracy. Recently, we have made efforts to design a metrological AFM for use with the SAMM stage.

An earlier design of a metrological AFM built for use with the SAMM is presented in the Master's thesis work of Stein (31). Accurate metrology configurations for this type of AFM were developed in the Master's thesis work of Mazzeo (25). This metrological AFM used a piezo-electric tube for scanning the probe in 3 degrees of freedom. Its AFM probe consisted of a tapered optical fiber glued to the side of a quartz tuning fork tine resonating at about 32 kHz. The probe was located at the center of a sectioned spherical target. The target location was measured using three capacitive displacement sensors measuring motion along three orthogonal radii of the sphere. Step height measurements with 4 nm RMS resolution could be made using this AFM. Building on this experience, the electromechanical configuration of the HAFM was designed and manufactured in the Master's thesis work of Ljubicic(22). The instrumentation and control of HAFM were later developed by Amin-Shahidi et al. (3) to enable surface tracking with sub-nanometer resolution. In this paper, we present the HAFM's design and experimental results.

A picture of the HAFM mounted on the Zerodur metrology frame of the SAMM is shown in Figure 2.1.

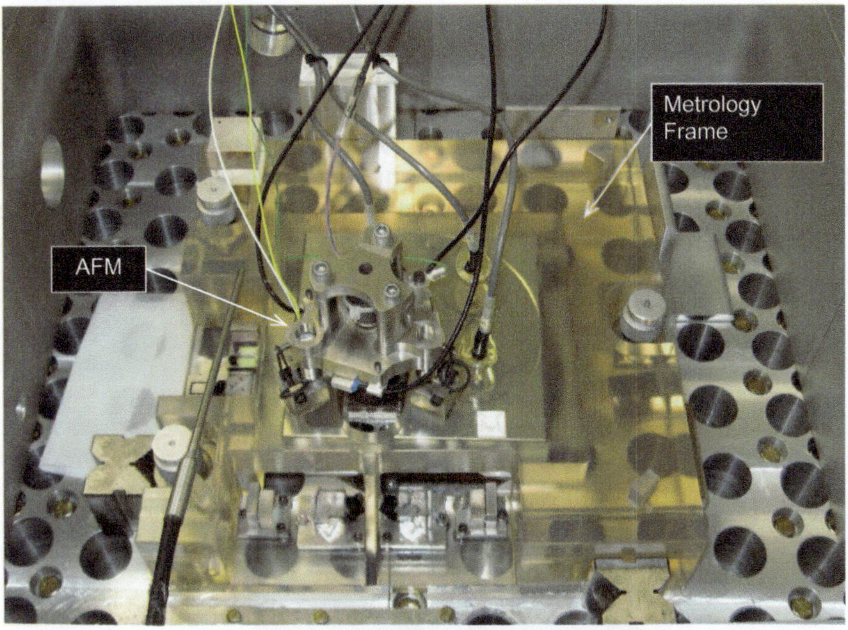

Fig. 2.1 HAFM head mounted on the SAMM's metrology frame in metrology laboratory at UNC-Charlotte

2.2 Mechanical Design

In this section, we present the mechanical design of the metrological AFM. We describe the overall configuration of the metrology system and the design of the SAMM and the AFM.

2.2.1 Overall Metrology System

Figure 2.2 shows schematically in cross-section the overall configuration of the SAMM/HAFM measuring machine. Figure 2.3 shows a three-dimensional view of the system configuration. As shown in these figures, the HAFM is fixed to the SAMM's metrology frame using three kinematic mounts. The moving stage of the HAFM extends down into the sample holder's cavity where the sample is placed. Both the sample holder and metrology frame are constructed of Zerodur low-expansion glass/ceramic material so as to give a dimensionally stable metrology loop. A borescope extending through the center provides a coarse image of the sample for probe alignment. The sample-holder's position is measured with respect to the SAMM's metrology frame using laser interferometers in the lateral directions (X, Y, and θ_z) and capacitive displacement sensors in the three vertical degrees of freedom (Z, θ_x, θ_y) . Three support struts, which extend between the metrology and the oil chambers, fix the sample holder to the moving platen. The platen is hollow so as to make the stage neutrally buoyant in oil and its position is controlled in 6 degrees of freedom using four levitation linear motors located at the bottom of the oil chamber. For imaging, the SAMM moves the sample in the lateral degrees of freedom, and provides vertical motion for long-term surface following, while the HAFM tracks the sample surface in the vertical direction (Z). More information on the earlier Angstrom stage can be found in (17; 16). More information on the SAMM can be found in (18; 34). More information on the levitation linear motors can be found in (19; 33; 24; 32).

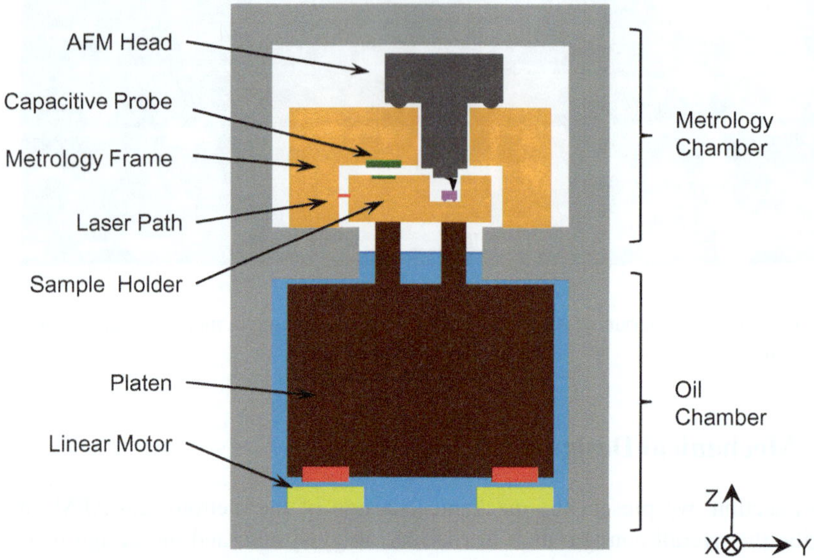

Fig. 2.2 Overall configuration of the metrological AFM including the HAFM and the SAMM

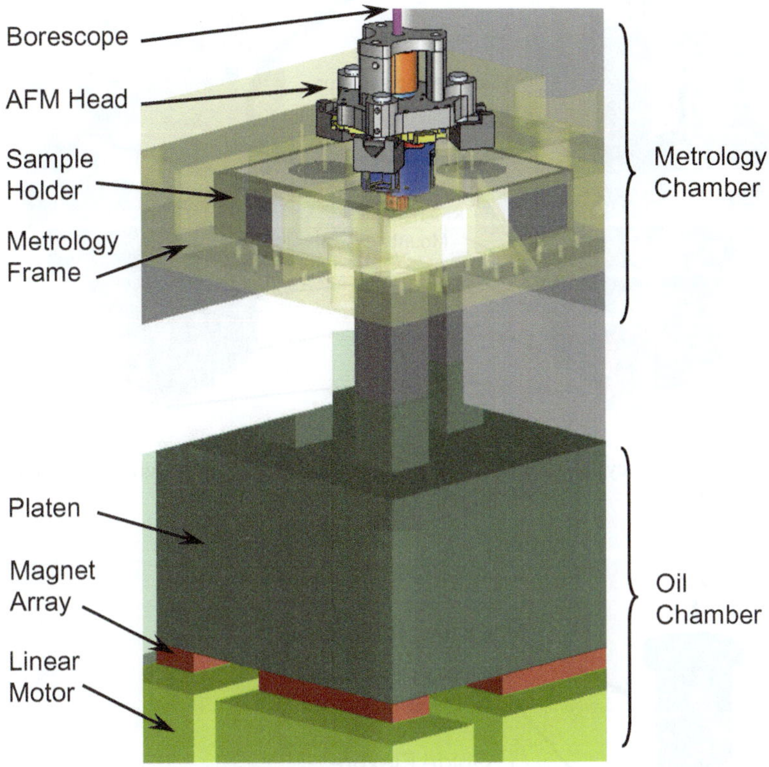

Fig. 2.3 Cross-sectional schematic model of the HAFM and SAMM combination

2.2.2 AFM Head

A more detailed view of the HAFM design is shown in Figure 2.4. The AFM probe is carried by the moving stage. The moving stage is constrained to only axial motion by the guide flexure. A piezoelectric stack, with a range of 20 μm, actuates the moving stage. A coupling flexure transmits the axial motion of the piezoelectric actuator to the moving stage. Three ADE8810 capacitive displacement sensors, with 50 μm range, measure the displacement of the moving stage with respect to the frame of the HAFM stator. This displacement of the AFM probe is calculated as the average of the three capacitive sensor readings, with better than 0.2 nm RMS resolution at 1000-Hz 3-dB measurement bandwidth. As shown in Figure 2.2, the force and metrology loops are made separate. The components within the metrology loop are manufactured from Invar low-expansion metal to minimize any error due to thermal deformation.

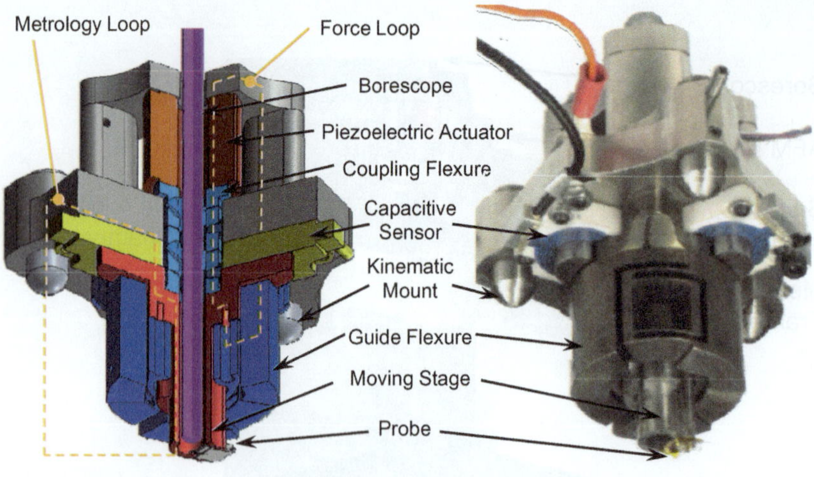

Fig. 2.4 High accuracy AFM head design: CAD crossectional view (left) completed head assembly (right)

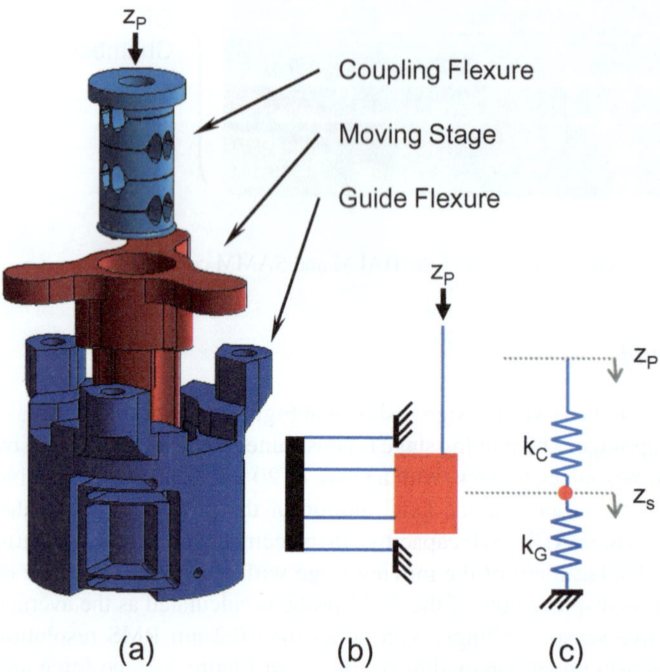

Fig. 2.5 Flexure design: (a) exploded CAD model; (b) flexure design; (c) static stiffness model

Figure 2.5 shows the flexure design, where z_p and z_s are defined as the displacement of the actuator and the moving stage respectively. The guide flexure, which is modeled as k_G, constrains the moving stage to axial motion only. It uses a pair of blades connected back-to-back in series to cancel out any parasitic radial error resulting from the axial motion. The coupling flexure, which is modeled as k_C, transmits the axial motion of the piezoelectric actuator to the moving stage. Complimentary to the guide flexure, the coupling flexure is only rigid in the axial direction. In this way, a minor portion of the piezoelectric actuator's error motion is transmitted to the moving stage. The displacement transmission ratio for any direction can be calculated as $T_R = k_C\ /\ (k_C + k_G)$. Table 2.1 lists the analytically derived stiffness values for both flexures and the resulting transmission ratio for different degrees of motion. Almost all of the axial motion is transmitted to the stage while the error motion is attenuated by at least 77 times. The HAFM head's design is described in more details in Master's thesis of Ljubicic (22).

Table 2.1 Analytically driven stiffness values of the guide and the coupling flexures and the resulting motion transmission ratio for different directions of motion

Direction of Motion	Guide Flexure Stiffness	Coupling Flexure Stiffness	Transmission Ratio
Z	1 N/μm	18 N/μm	95%
X, Y	56 N/μm	0.033 N/μm	0.06%
θ_Z	1.8 kN · m/rad	0.024 kN · m/rad	1.3%
θ_X, θ_Y	2.0 kN · m/rad	0.75 kN · m/rad	0.04%

2.3 Self-sensing Probe for Frequency Measuring AFM

The most common AFM configuration uses a microfabricated cantilever with sharp tip, and an optical lever sensing mechanism. However, the HAFM head needs to extend through a relatively narrow opening in the SAMM metrology frame to access the sample-holder cavity on the moving sample-holder. To achieve a compact and practical design, we use a self-sensing Akiyama probe, which eliminates the need for the optical lever sensing mechanism. We use the self-sensing probe in a frequency measuring configuration, with the probe controlled to oscillate at its natural frequency. The sample-probe spacing modifies the natural frequency, which is used for feedback control.

2.3.1 Self-sensing AFM Probe

Early tuning-fork-based self-sensing probes were developed for near field optical scanning microscopes (NSOM) as described in (28). Here a self-sensing design is advantageous, since an optical lever laser illumination beam could interfere with

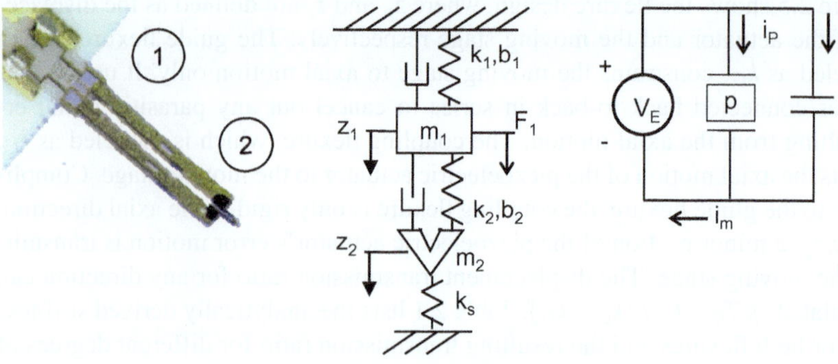

Fig. 2.6 Akiyama self-sensing probe (left), lumped parameter mechanical model (middle), and electrical circuit model (right). Akiyama probe is a product of NANOSENSORS^{TM}. Pictures of the probe are a courtesy of http://www.nanosensors.com, and are used with permission.

the near-field optical sensing mechanism. A typical self-sensing probe is created by attaching a tapered optical fiber to a quartz tuning fork tine. Commonly, the quartz tuning fork is extracted from the package used to provide the time-base in quartz watches. The HAFM uses a commercially-available Akiyama probe, which is a novel modification of the self-sensing probes. The Akiyama probe, shown in Figure 2.6, consists of a quartz tuning fork with an AFM cantilever symmetrically attached to both tines. The tuning fork's in-plane oscillation sets the cantilever tip in an out-of-plane tapping motion. The Akiyama probe is described in (1).

2.3.2 Electromechanical Model

The tip-sample atomic-force interactions are reflected back on the tuning fork dynamics and can be indirectly measured as the fork's electrical impedance. A lumped parameter model of the probe is shown in Figure 2.6, where m_1 and m_2 represent the tuning fork and the cantilever masses respectively. The connections are modeled as linear springs and dampers. The probe is modeled for free operation in air, where there is no tip-sample interaction. The electromechanical transformation of the piezoelectric tuning fork is modeled by $F_1 = pV$ and $i_p = p\dot{z}_1$, where F_1 is the force exerted by the piezo on mass m_1, p is the piezoelectric force constant in units of N/V, V is the voltage applied to the piezo, i_p is the current into the piezoelectric ideal transformer, and z_1 is the position of m_1.

We also model the piezoelectric actuator as including a shunt capacitance C. Near the first resonance, we can approximate the probe admittance transfer function from probe excitation voltage V_E to probe current I_P as measured by the transimpedance amplifier as

$$\frac{I_p(s)}{V_E(s)} = [\frac{p^2 s}{(m_1+m_2)s^2+b_1 s+k_1}+Cs]\frac{1}{\tau s+1}. \qquad (2.1)$$

Here the low-pass filter with the time-constant τ is a simplified model of the transimpedance amplifier dynamics. We also assume that the stiffness k_2 is high, so that at the first resonance, the masses m_1 and m_2 move as a rigid body. We model the probe in air where the stiffness k_s is equal to zero.

Figure 2.7 shows the experimentally obtained frequency response of an Akiyama probe with an overlaid fit based on Equation (2.1). On the basis of this experimental measurement, we fitted the parameters of Equation (2.1). The fitted parameters are listed below; note that the parameters of the electromechanical transfer function can only be determined as a ratio with respect to p^2.

$(m_1 + m_2)/p^2 = 1.372 \times 10^4$ [H]
$b_1/p^2 = 2.867 \times 10^6$ [Ω]
$k_1/p^2 = 1.265 \times 10^5 [\mathrm{F}^{-1}]$
$C = 0.99$ [pF]
$\tau = 2.59 \times 10^{-6}$ [s]

A pair of complex zeros following the resonance peak exists due to the current shunt path created by the parasitic capacitance C. The resulting asymmetric resonance peak does not correspond to the mechanical resonance and is not suitable for frequency measuring microscopy. As suggested in (26), we compensate for the parasitic current by injecting its inverse at the current measurement junction. Figure 2.7 shows the expected frequency response of a perfectly compensated probe derived by setting the parasitic capacitance, C, of the fitted model to zero.

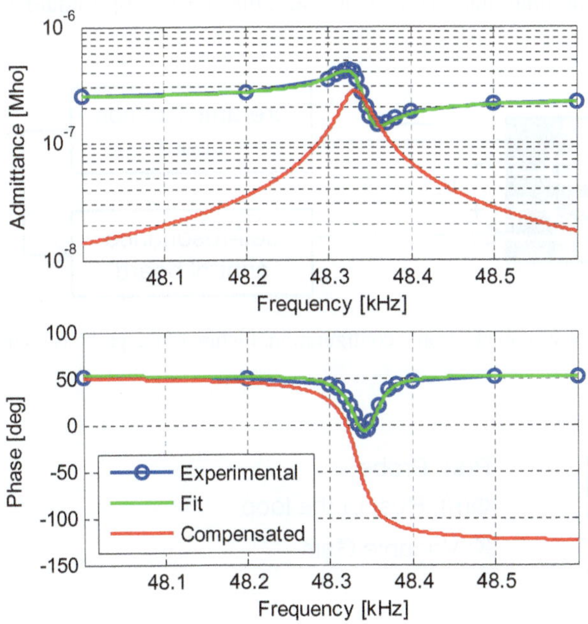

Fig. 2.7 Admittance frequency response for an Akiyama probe under test. Experimentally captured response, overlaid fitted model, and expected compensated response.

2.3.3 *Frequency Measuring AFM*

Given its tuning fork design, the Akiyama probe has a high quality-factor (500–1500 in air) which improves its sensitivity. However, a high-quality factor oscillator dissipates little energy per cycle, so any transient disappears relatively slowly. References (2; 4; 10; 30) have overcome this limitation by actively changing the probe's excitation frequency. They all use an excitation signal whose frequency shifts with the probe's shifting frequency response. Effectively, the excitation frequency is varied to keep a reference phase delay across the probe. In this configuration, the frequency of excitation changes with variations in the tip-sample distance. Therefore, the excitation frequency can be used as feedback to control the tip-sample gap. A good review of the frequency-measuring AFM methods is provided in (11).

References (10; 30) control the excitation frequency in closed-loop using a direct measurement of the phase difference across the probe. Alternatively, (2; 4) operate the probe in constant amplitude self-resonance. For HAFM, we use the latter operating mode. Figure 2.8 shows the overall configuration of our self-resonance control system electronics. Figure 2.9 shows the block diagram representation of our self-resonance control system. To establish self-resonance, the current passing through the probe is converted and phase-shifted to be constructively fed back as a voltage excitation. The oscillation amplitude is actively controlled in closed loop by varying the feedback gain.

Figure 2.10 shows a transfer function representation of the self-resonating loop. Being in constant-amplitude oscillation, the closed loop system has a pair of complex poles, $s = \pm j\omega_R$, on the imaginary axis. As a result, the closed loop frequency

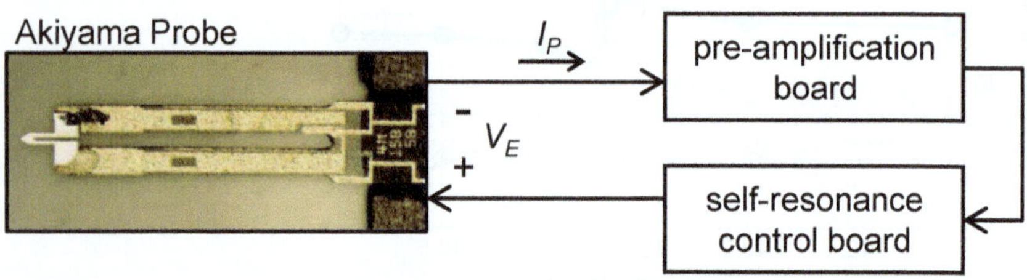

Fig. 2.8 Self resonance control system electronic configuration. Picture of the probe is courtesy of NANOSENSORS™.

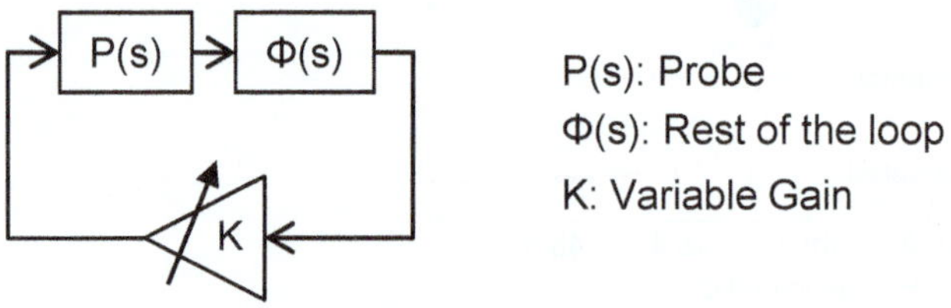

Fig. 2.9 Self resonance control system block diagram

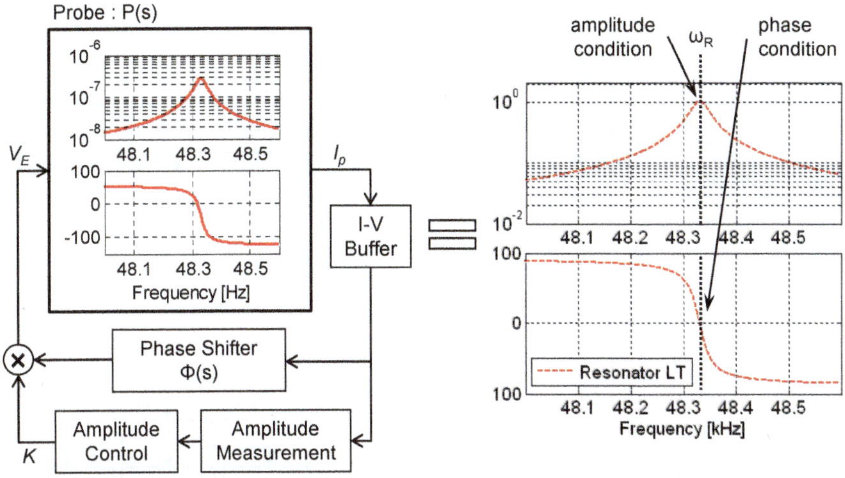

Fig. 2.10 Implementing amplitude controlled self-resonance by actively satisfying the self-resonance amplitude and phase conditions

response has an infinite gain for $\omega = \omega_R$. Based on this, we can derive amplitude and phase conditions

$$CL(s = j\omega_R) = \frac{P(s)\Phi(s)K}{-P(s)\Phi(s)K+1}|_{s=j\omega_R} = \infty \tag{2.2}$$

$$\Rightarrow P(s)\Phi(s)|_{s=j\omega_R} = 1/K \tag{2.3}$$

$$|P(s)\Phi(s)|_{s=j\omega_R} = 1/K \quad \text{Amplitude Condition} \tag{2.4}$$

$$\angle P(s)|_{s=j\omega_R} = -\angle\Phi(s)|_{s=j\omega_R} \quad \text{Phase Condition,} \tag{2.5}$$

that must be satisfied for constant amplitude self-resonance. The amplitude condition is satisfied by gain adjustments at the multiplier node via the output of the amplitude controller. This adjustment process is depicted from a root-locus viewpoint in Figure 2.11. The real zero-pole pair in Figure 2.11 are the singularities of the all-pass filter. The transimpedance buffer dynamics are not shown in this figure.

The phase condition is equivalent to phase control with the reference set by the phase lead of the rest of the loop, $\phi(s)$. The loop self-resonates at a frequency ω_R where the angle condition is met. Optimum sensing sensitivity is achieved at the natural resonance frequency, where the probe has minimum dynamic mechanical stiffness. We use an all-pass filter to tune the amount of phase shift $\phi(s)$ added external to the probe, so that the phase condition is met for $\omega_R = \omega_N$, where ω_N is the natural resonance frequency of the Akiyama probe. A good description of such oscillator amplitude control systems is presented in (29), and studied also in (5) and the Master's thesis of Cattell (6).

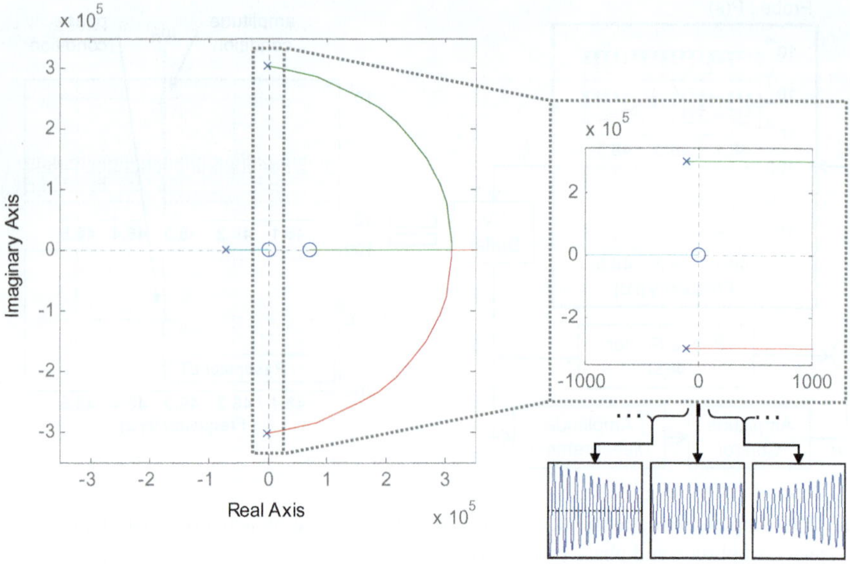

Fig. 2.11 Self-resonance control viewed from a root-locus point of view: as shown by the root locus of the self-resonance loop, by varying the gain K, the amplitude controller shifts the resonating complex pole pair between the right and the left half planes to increase or decrease the self-resonance amplitude respectively

2.3.4 Probe Electronics Implementation

The probe electronics consists of two boards, which are adapted from electronic designs provided as part of the Akiyama support documentation (26). We have designed a *pre-amplification* board which is used to convert the probe's current signal to an amplified voltage signal, and to provide for nulling of the probe stray capacitance. The pre-amplification board also implements an effectively guarded connection for the probe currents so that stray cable capacitance does not degrade the measurement. The second board is the *self-resonance control* board which is used to implement controlled-amplitude self-resonance which tracks the probe's mechanical resonant frequency. The feedback theory for this self-resonance control is described in section 2.3.3.

The electronic schematic of the *pre-amplification* board is shown in Figure 2.12. The electronic schematic of the *self-resonance control* board is shown in two parts, in Figures 2.13 and 2.14. These boards are described in more detail below. The design files as well as printed circuit board (PCB) manufacturing files for the boards are available at `http://web.mit.edu/trumper/Public/HAFM_Ckt.zip`

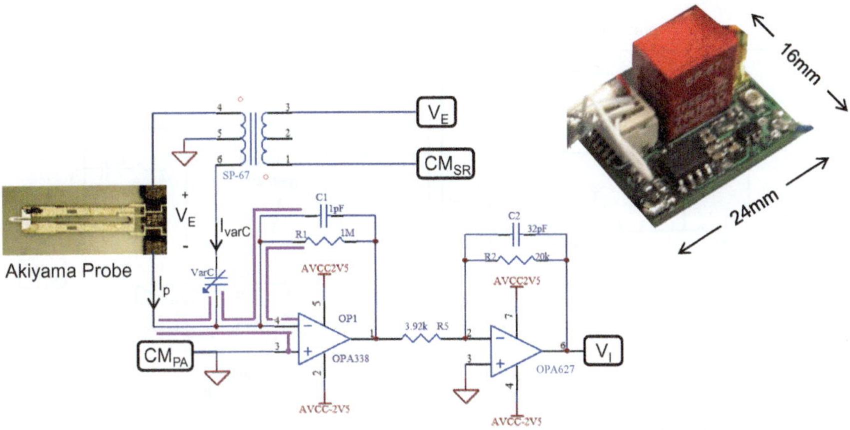

Fig. 2.12 Schematic design and photo of the manufactured pre-amplification board. Picture of the probe is courtesy of NANOSENSORS™.

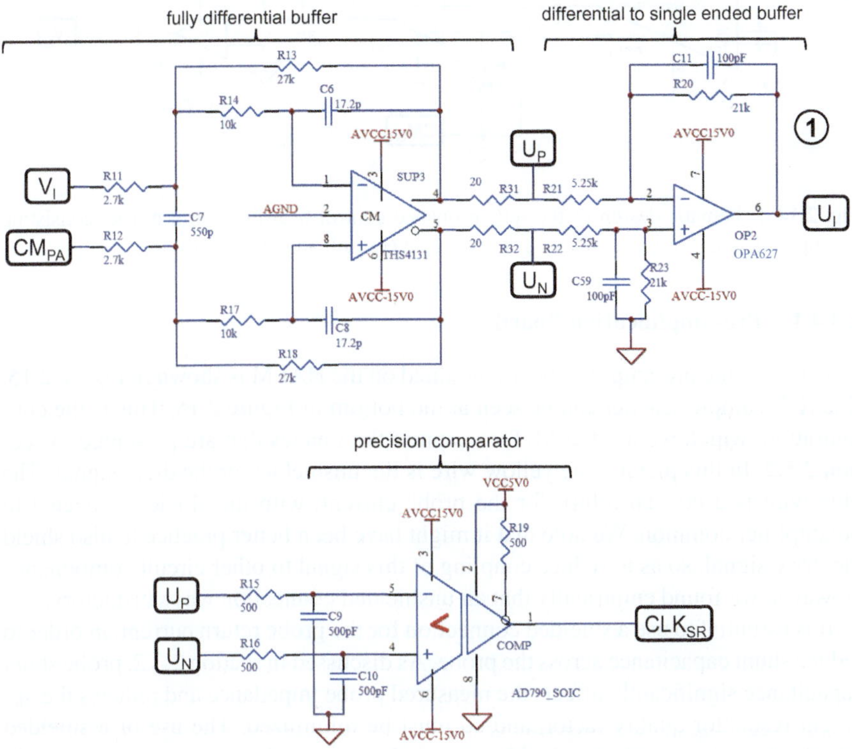

Fig. 2.13 Schematic design of the self-resonance control board's fully-differential input interface and precision comparator module used to digitize the self-resonance signal

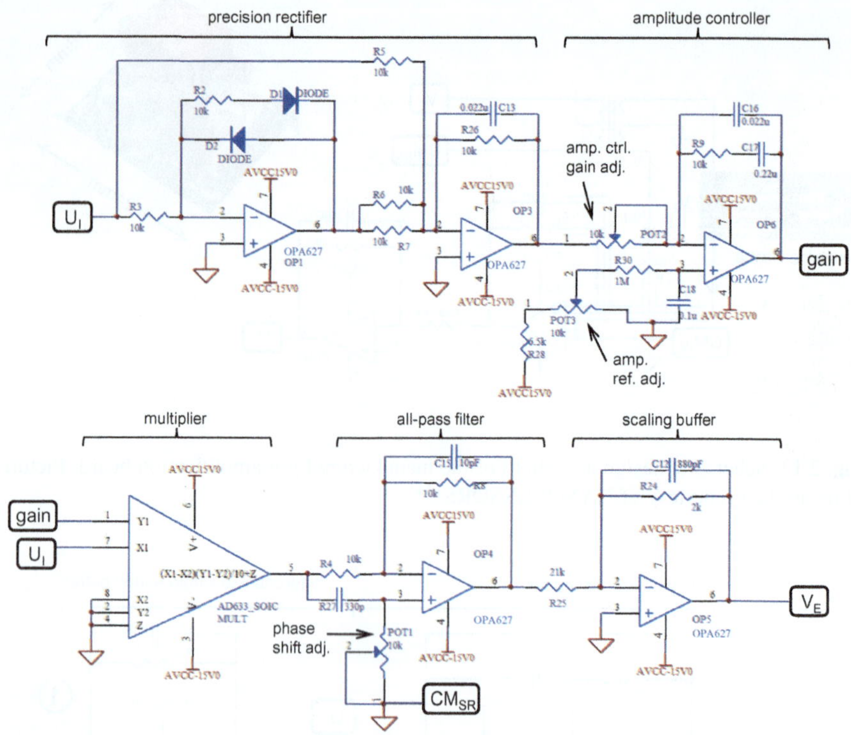

Fig. 2.14 Schematic design of the self-resonance control board's control blocks consisting of amplitude measurement and control, loop gain control, and phase shifting

2.3.4.1 Pre-amplification Board

A picture of the pre-amplifier board mounted on the HAFM is shown in Figure 2.15. The X-Y sample scanner can be seen at the bottom of Figure 2.15. This is the configuration, which we used at MIT, to capture the images that are presented in section 2.5.2. In this picture, the yellow wire is the unshielded probe drive signal. The blue wire is a coaxial return for the probe current, with the shield connected to preamplifier common. We note that it might have been better practice to also shield the drive signal, so as to reduce coupling of this signal to other circuit components. However, we found empirically that an unshielded connection was satisfactory.

It is essential to use a shielded connection for the probe return current, in order to reduce shunt capacitance across the probe. As discussed in section 2.3.2, probe shunt capacitance significantly affects the measured probe impedance and reduces the apparent resonator quality factor, and so must be minimized. The use of a shielded return connection, with the shield connected to preamplifier common, effectively serves as a guard connection for the probe return current. This is so because the inverting input of the OP338 transimpedance amplifier stage is at nearly zero voltage (virtual ground). Thus the capacitance between the center conductor and shield of

Fig. 2.15 Picture of the manufactured pre-amplification board

the return cable has effectively zero volts across it, and thus no shunt current. The use of such a guard connection was investigated in (31). The effect of stray cable capacitance and the use of such a guard is illustrated in Figure 2.16, as adapted from Stein's thesis.

2.3.4.2 Transimpedance Buffer

At the operating frequency of about 48 kHz, the transimpedance buffer shown in Figure 2.12, has a transimpedance of $R_1 = 1$ MΩ. That is, the output voltage of this stage at 48 kHz excitation is given to a good approximation by $V_o = -(I_p + I_{VarC})R_1$, where I_p is the probe current, and I_{VarC} is the current passing through the nulling capacitor. When properly adjusted, I_{VarC} cancels the probe stray shunt capacitance, since it is driven by the inverse voltage of the probe drive, via the

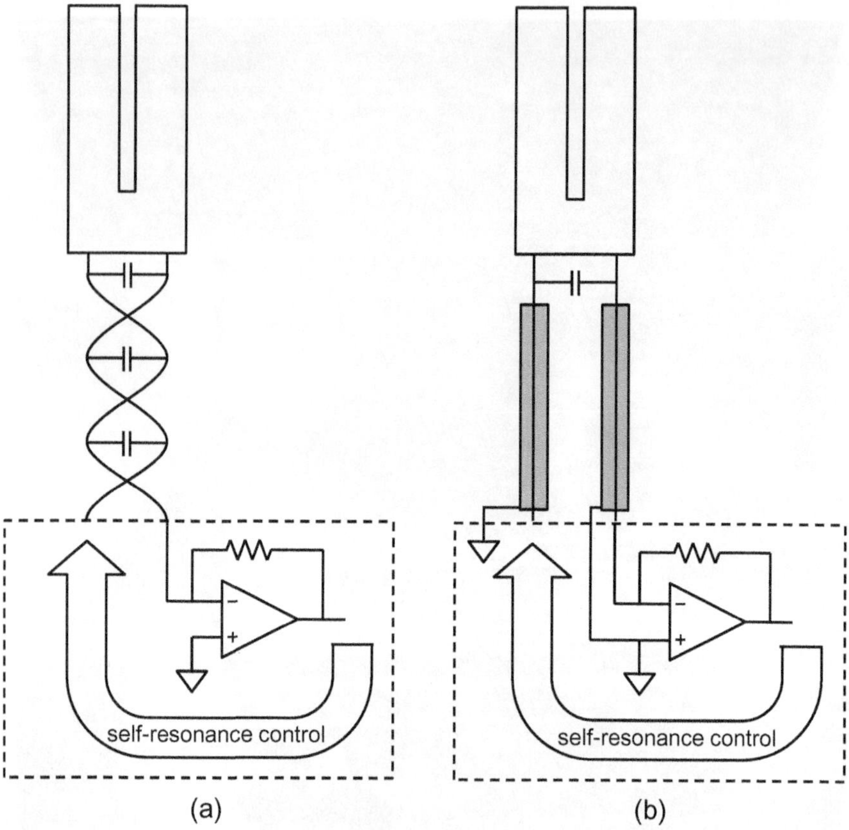

Fig. 2.16 Because the transimpedance amplifier input is a virtual ground, the use of a grounded shield connection (b) for the probe return current effectively guards out the cable capacitance, preventing degradation of the probe impedance measured quality factor. In earlier work, we had initially used a twisted pair (a), resulting in much higher probe shunt capacitance.

second leg of the center-tap transformer. The transformer also gives common-mode isolation of the probe drive from the other boards in the system. Also, as shown in the figure, the input connection from the probe which carries current I_p (yellow wire in Figure 2.15) is via a coaxial cable with shield connected to preamplifier board common, and the front end traces are shielded on the preamplifier circuit board as shown in Figure 2.12. This configuration was adapted from Akiyama probe support circuits shown on the NANOSENSORS™ website (26).

We must also consider the stability of the local feedback around the OP338 amplifier. That is, the loop around the OP338 includes the amplifier with its internal dominant pole dynamics, as well as the feedback transfer function from the amplifier output voltage to the inverting input voltage. The local feedback transfer function is

$$F(s) = \frac{R_1 C_1 s + 1}{R_1 (C_1 + C_p) s + 1}, \tag{2.6}$$

where C_p is the parasitic capacitance at the OP338 op-amp's inverting input terminal. The parasitic capacitance consists of the amplifier input capacitance as well as the circuit board connecting trace capacitance and coaxial cable stray capacitance to common.

The transfer function above has a DC gain of 1 for frequencies well below the pole of $\omega_p = 1/R_1(C_1 + C_p)$ and a gain of $C_1/(C_1 + C_p)$ for frequencies above the zero of $\omega_z = 1/R_1 C_1$. At the relatively high local loop crossover frequency for this amplifier, the feedback gain of $C_1/(C_1 + C_p)$ is attenuating. In order to maintain a relatively high crossover frequency, we chose the de-compensated OP338, which is not unity-gain stable.

With the selected the values of $C_1 = 1$ pF and $R_1 = 1\,\text{M}\Omega$ the local amplifier feedback loop's unity gain crossover is above the zero frequency ω_z. In this way, the feedback loop has enough phase margin, since the gain $C_1/(C_1 + C_p)$ has zero associated phase shift. The use of a decompensated amplifier increases the loop-gain at 48 kHz, which is the typical frequency of the probe's current signal. The selected OP338 op-amp also has low input bias current and good noise performance.

2.3.4.3 Self-resonance Control Board

The self-resonance control board is shown in two parts in Figures 2.13 and 2.14. Figure 2.13 shows the control board differential amplifier front end and precision comparator. Figure 2.14 shows the control board precision rectifier, resonance amplitude controller, multiplier and all-pass filter.

The THS4131 differential amplifier front end is used to achieve common-mode isolation from the pre-amplifier board, so as to minimize noise pickup. The inputs to this amplifier are voltage V_I, which is the output from the preamplifier buffer, and CM_{PA}, which is the preamplifier board common. The output of this differential amplifier stage is passed via signals U_P and U_N to the precision comparator stage, which turns the probe sinusoidal current into the self-resonance square wave clock signal CLK_{SR}. The 20 Ω resistors R_{31} and R_{32} prevent loading on the outputs of the THS4131. The differential signal is converted by an OPA627 stage to a single-ended representation U_1, which is used to drive the precision rectifier which gives an oscillation amplitude measurement.

The oscillation amplitude signal serves as the feedback measurement for the oscillation amplitude control stage, which has an output signal *GAIN*. The signal *GAIN* is used as one of the inputs to the AD633 multiplier. The other input to the multiplier is U_1, the probe sinusoidal oscillation signal. Thus this connection uses the value of *GAIN* to scale the probe oscillation signal, as set by the controller. The scaled probe oscillation signal is input to the all-pass filter stage, which is used to adjust the oscillation loop phase shift to meet the phase criterion. The phase-shifted signal is scaled down by approximately 10:1 by the scaling buffer stage to give an appropriate level on the signal V_E which drives the preamplifier transformer primary relative to control board common CM_{SR}.

As described in more detail in (29), and studied also in (5) and the Master's thesis of Cattell (6), when in oscillation at constant amplitude, the probe system appears as an integrator to the oscillation amplitude controller. That is, incremental changes in controller output *GAIN* cause the probe oscillation signal amplitude to integrate these changes. Since the resonator amplitude parameter acts as an integrator of control action, the amplitude controller is designed to stabilize these envelope dynamics. The controller transfer function has a pole at the origin, and a left-half plane zero at $s = -1/R_9C_{17} = -455\,\mathrm{s}^{-1}$. The frequency response breakpoint associated with this zero is at 72 Hz, which is below the amplitude loop crossover frequency. Thus, at crossover, the controller adds little negative phase shift to the -90 degrees from the plant integrator. The rectifier uses a low-pass filter to estimate the amplitude of oscillation. To avoid phase loss from the rectifier, the controller crossover must stay below the rectifier's poll at $s = -1/R_{26}C_{13} = -4545\,\mathrm{s}^{-1}$. Capacitor C_{16} rolls off the controller gain above crossover. The controller gain is adjusted empirically with POT2 to set the desired amplitude loop dynamics. The resonance signal reference amplitude is set by POT3.

2.3.5 Sensing Curves

Operating in the amplitude-controlled self-resonance, the probe can be used for tip-sample gap sensing. In this mode, variations in the tip-sample distance change the self-resonance frequency. Figure 2.17 shows the experimentally captured sensing-curves for different probe oscillation amplitudes. The curves are obtained by varying the tip-sample gap slowly for several cycles and recording the self-resonance frequency. For the probe under test, the self-resonance frequency is 46,646 Hz when oscillating freely in air. The frequency increases up to 46,938 Hz as the probe approaches the sample.

The probe's position-to-frequency sensitivity is inversely proportional to the oscillation amplitude. However, there is a limit on the oscillation amplitude, below which the probe's oscillation becomes unstable. This limit for the probe under test was found to be 20 nA. It can be seen that the sensing curves become more and more distorted as the oscillation amplitude approaches this limit. The sensing curve shows a repeatable offset between the approach and the retract curves. The retract curve is always below the approach curve and contains a discontinuity, which becomes more significant as the oscillation amplitude is reduced. Perhaps this could be due to tip-sample adhesion. Based on the identified sensing curves, the minimum stable oscillation current amplitude for imaging is determined to be 24 nA, with a corresponding sensitivity of 0.78 Hz/nm. Table 2.2 shows the measured sensitivity of the probe under test for different oscillation amplitudes.

The sensing curves can be used to estimate the mechanical oscillation amplitude at the tip of the probe. The frequency starts to shift as the probe tip starts to tap

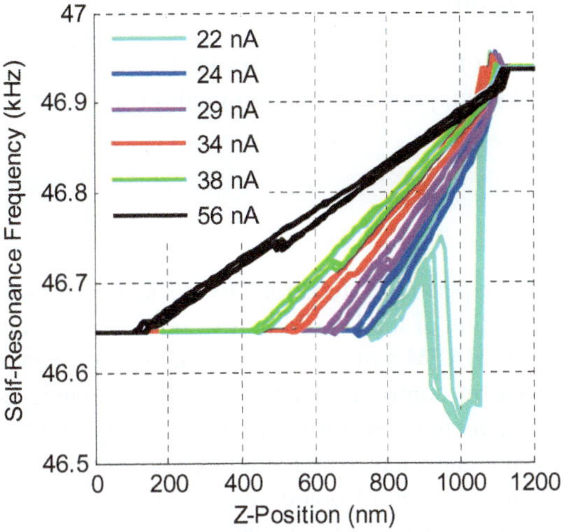

Fig. 2.17 Experimental sensing curves for an Akiyama probe under test at different oscillation current amplitudes. The free air self-resonance frequency is a fixed value of 46,646 Hz, and the frequency increases to a maximum value of about 46,938 Hz as the probe contacts the surface, at a Z-position of 1000 nm on the graph. Note that the position-to-frequency sensitivity increases with reduction in oscillation amplitude.

Table 2.2 Sensing sensitivity for different oscillation amplitudes for an Akiyama probe under test

Electrical Amplitude (nA)	Estimated Mechanical Amplitude (nm)	Sensitivity (Hz/nm)
22	376	0.777
24	442	0.661
29	527	0.554
34	616	0.474
38	702	0.416
56	1074	0.272

on the surface. The Z-position where maximum frequency is reached is almost the same for all oscillation amplitudes. As the oscillation amplitude is changed the only point that does not move is the center of oscillation. Therefore, we expect maximum frequency to be reached when the center of oscillations is at the sample surface. As a result, we can estimate the oscillation amplitude as the distance between the points where the frequency starts to shift to the point where it reaches a maximum.

2.4 Control and Instrumentation

In this section, we describe the control of the self-sensing AFM. We describe the overall control and instrumentation system, period estimation method, and surface tracking control.

2.4.1 System Overview

HAFM uses the probe's self-resonance period as feedback to track the surface. Figure 2.18 shows the overall configuration of the HAFM hardware. The Akiyama probe is attached to and moves with the HAFM's moving stage. The self-resonance control board sets the Akiyama probe in controlled-amplitude self-resonance as described in the previous section. For better noise performance, a small pre-amplification board compensates for the parasitic capacitance and amplifies the current signal close to the probe. The self-resonance control board converts the oscillation signal (V_I) into a digital self-resonance clock (CLK_{SR}), which is output to the real-time National Instrument controller. We use the real-time controller's FPGA board to estimate the self-resonance period (T_{SR}) and calculate control actions, which are sent to the piezo-driver. The piezoelectric actuator moves the probe to maintain a constant self-resonance period (T_{ref}), and consequently a constant tip-sample gap. As the controller tracks the surface, the real-time computer records the stage's surface-tracking motion measured by the three capacitive displacement probes.

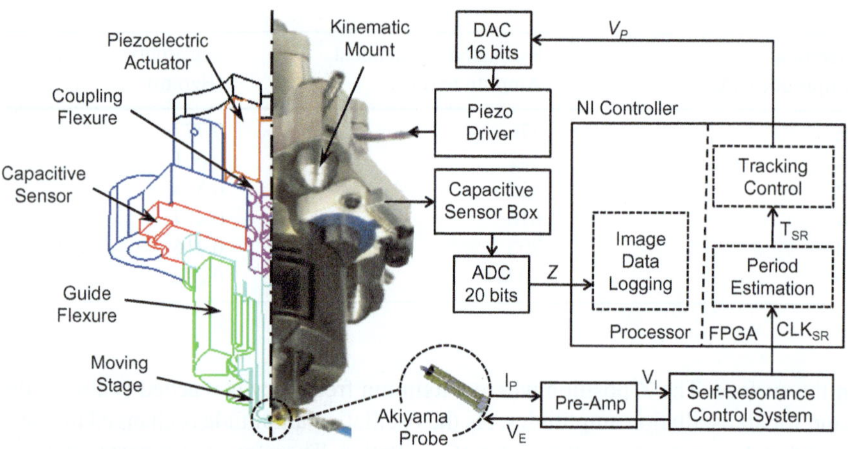

Fig. 2.18 Overall HAFM system design including the mechanical setup, sensors, actuators, and the controller. Picture of the probe is courtesy of NANOSENSORS™.

2.4.2 *Period Estimation*

A timer clocked at 200 MHz measures the time between the rising/falling edges of the resonance clock with 5ns resolution. A resolution of 5-ns is equivalent to about 10 Hz resolution when estimating the frequency of a 46-kHz signal. An easy way to achieve a better resolution is to measure the period of multiple cycles of a 46-kHz signal. This method results in a lower measurement update rate and does not use the information contained within the intermediate edges. A low update rate prevents effective low-pass filtering of the measurement noise. Another method is to use a moving average to estimate the period of a moving window consisting of multiple cycles. This method can use all of the edges, but adds phase lag to the measurement. As a better alternative, we measure the period between each consecutive edge and rely on the controller and the plant dynamics to low pass-filter the measurements. In this way, the measurement is updated at a fast rate of $2 \times 46 = 92$ kHz and has no phase lag when arriving at the controller.

2.4.3 *Tracking Control*

The tracking controller uses the self-sensing probe's feedback to follow the sample surface. Different aspects of the tracking controller are presented in this section.

2.4.3.1 Self-resonance Period Feedback

Converting the measured period to a frequency measurement requires a division, which is a non-linear operation. The division operation can mix the frequency content of the measurement noise, and prevents efficient filtering in the latter stages of our control loop. As a result, it is desirable to directly use the time measurement for control. Previously we showed that the self-resonance frequency changes linearly with the probe-sample gap. Given the small shift in frequency, we can assume, with good accuracy, the period variations versus the probe-sample gap to be linear as well.

2.4.3.2 Per-tap Controller Sampling

The control and timer algorithms are being clocked at different frequencies. We use a FIFO structure to transfer the period measurements between the two algorithms. The period measurement is updated at the edge detection rate. To avoid spurious mixing of the signal's frequency content, the controller must be sampled at an integer multiple of the measurement update frequency. However, the edge detection frequency is changing due to the variations in the self-resonance frequency as well as noise. To address this problem, we sample the controller at the edge rate of the resonance signal. In this way, the measurement update rate and the control sampling rate will be identical, and no mixing will occur. Given that the self-resonance frequency changes by less than 0.1%, the variations in the controller's sample-rate are negligible and do not cause nondeterministic behavior.

2.4.3.3 Loop Shaping Compensator Design

We designed a discrete-time surface tracking controller using the loop-shaping technique. Figure 2.19 shows the open-loop frequency response of the HAFM measured from the piezo-driver command (V_P) to the self-resonance period (T_{SR}). The two resonances at 1.6 kHz and 2.3 kHz are structural modes which are the bottle-neck on the bandwidth. The probe's sensing mechanism shows adequate bandwidth up to the structural resonances.

The compensated loop, shown in Figure 2.19, has a unity gain crossover frequency of 100 Hz with 65 degrees of phase margin. A simplified schematic diagram of the controller is shown in Figure 2.20. The controller uses an integrator compensator to create a clear unity gain cross over. A low pass filter with a 3 dB bandwidth of 1 kHz reduces the controller gain at frequencies higher than the closed-loop bandwidth. The two notch filters have knocked down the structural resonances. By masking the modal vibrations, the notch filters do not remove the vibrations but only prevent the controller from reacting to them. To stop such vibrations from appearing in the measurements, we apply a low-pass filter to the capacitive probe readings.

The discrete-time controller consists of the following blocks:

- The self-resonance signal's half-period is estimated as the number of ticks of a 200-MHz clock between consecutive edges of the self-resonance clock signal. An offset of 2164 ticks is removed from the half-period measurement to reduce the number size and avoid overflow at the later stages of the controller.
- An integrator is used to give a slope of -1 at the unity-gain cross-over point. the discrete transfer function of the integrator is given as below:

$$F_{integ}(z) = \frac{Y(z)}{U(z)} = \frac{z+1}{z-1} \tag{2.7}$$

- A low-pass filter is used to attenuate the controller's gain for frequencies well above its unity-gain cross-over frequency. The discrete transfer function of the low-pass filter is given as below:

$$F_{LPF}(z) = \frac{Y(z)}{U(z)} = \frac{2^{12}z}{2^8 z - 240} \tag{2.8}$$

- Two notch filters are used to mask the structural modes of the AFM head. The notch filters' center frequencies are at 1.6 kHz and 2.3 kHz and have 0.4 and 0.3 quality-factors respectively.

The discrete controller is implemented on an FPGA in 32-bit fixed point arithmetic. The controller is clocked via the approximately 96 kHz edge rate of the CLK_{SR} clock, i.e., via the squared-up probe oscillation signal. The transfer functions stated above are implemented at this clock rate. We note that this rate is very high relative to the loop crossover frequency of 100 Hz. However, the FPGA readily sustains this

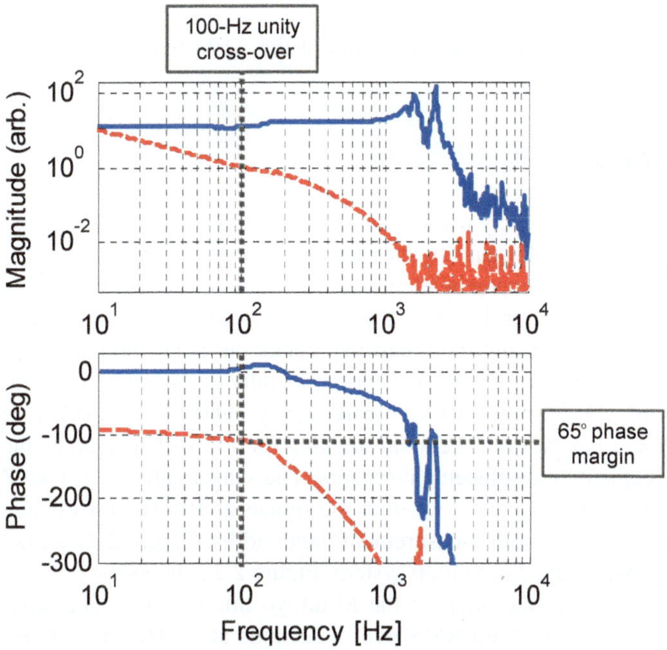

Fig. 2.19 Loop-shaping tracking controller design: the open-loop frequency response from the piezo's power amplifier command to the probe's self-resonance frequency, and the compensated loop frequency responses showing 100-Hz unity cross-over frequency and 65-degree phase margin

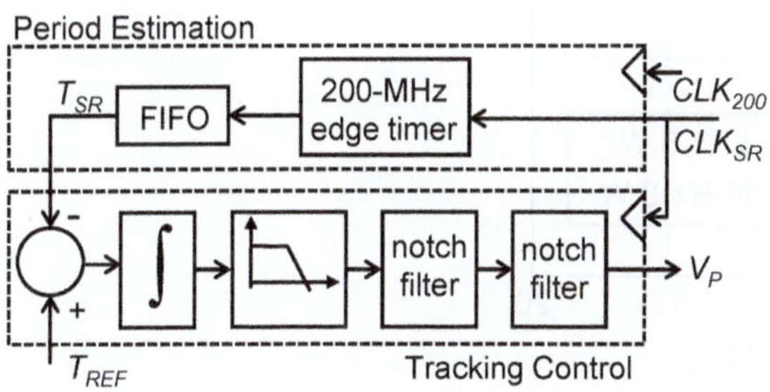

Fig. 2.20 Simplified diagram of control and period estimation blocks

rate. Further, using this synchronous clocking avoids noise frequency mixing which accompanies unsynchronized sampling, and the loop bandwidth is used to filter the measurement noise associated with the probe signal noise and with the finite counter time resolution of 5 ns.

2.5 Experimental Results

In this section, we present the experimental results of the AFM tests, which were conducted to confirm its operation.

2.5.1 *Single-Point Tests*

The HAFM was developed separately without an XY scanner. As a result, we started our tests with no in-plane sample scan. In such tests, the anvil of a Mitutoyo 110-series differential threaded micrometer head serves as the sample surface. We use the micrometer head to approach the probe. Once the sample is within the probe's range, we can lock the tip-sample gap and the controller tracks the axial motion of the micrometer anvil from that point forward. We experimentally demonstrated closed-loop point tracking with 100-Hz unity cross-over frequency and 0.12-nm RMS noise at 100-HZ measurement bandwidth. The axial tracking motion is displayed as the average of the three capacitive displacement sensors. The sensors are set at 1-kHZ bandwidth in hardware. For image data, we filter the signals at 100-Hz in software. The results are plotted here at both measurement bandwidths. Figure 2.21 shows the 100-nm step response of the closed loop system. Figure 2.22 shows the position noise for a stationary sample (the anvil of the Mitutoyo micrometer). The noise is calculated to be 0.24 and 0.12 nm RMS over 5 seconds at 1 kHz and 100 Hz measurement bandwidths respectively.

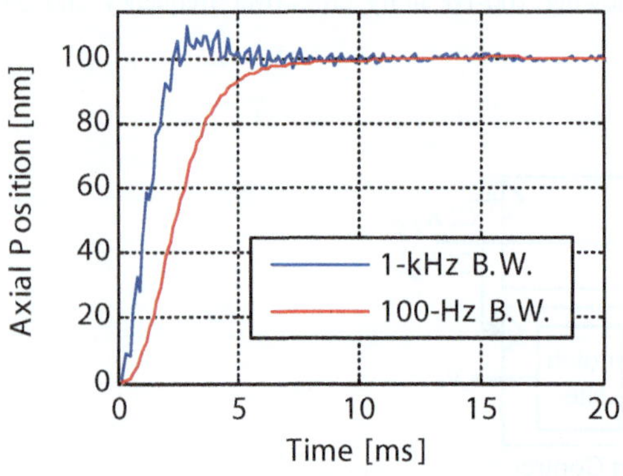

Fig. 2.21 HAFM's response to a step disturbance injected at the piezo-driver input V_P measured using the capacitive displacement sensors at 1000 and 100 Hz measurement bandwidths

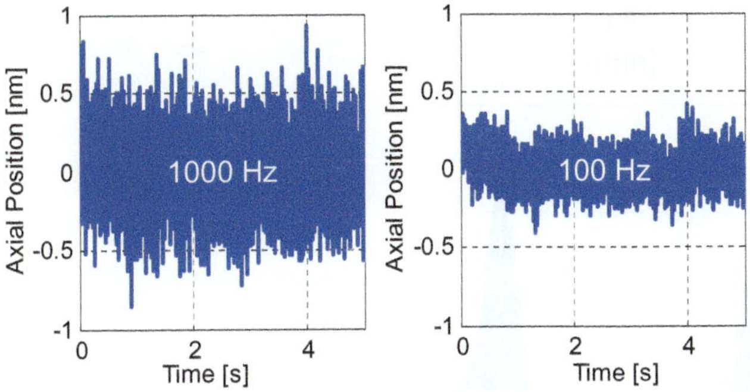

Fig. 2.22 Active tracking of a stationary point measured using the capacitive displacement sensors at 1000 Hz and 100 Hz measurement bandwidths

2.5.2 Imaging Tests at MIT

We tested the AFM for imaging by interfacing with a Veeco piezo-tube scanner. HAFM tracked the surface and the Veeco system was only used to scan the sample and record the image data generated by HAFM. Several samples, including sanded freshly cleaved mica and calibration gratings, were successfully imaged with the same bandwidth and noise level as in the initial no-scan test.

Figure 2.23 shows the image of a TGZ01 grating captured using HAFM at 5 μm/s scan speed. The grating is specified to have a peak-to-valley height of 25.5 ± 1 nm. Figure 2.24 shows the height distribution of the image. The peaks in the height distribution plot correspond to the flat faces. By measuring the distance between the two peaks, we estimate a peak-to-valley height of 25.6 for this grating which is very close to the grating calibration value. To investigate HAFM's performance more closely, we show extracted line profiles from other images of the same TGZ01

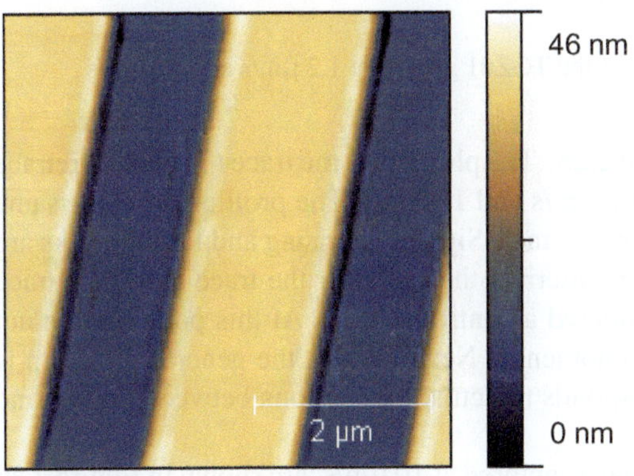

Fig. 2.23 Image of standard TGZ01 grating captured at 5 μm/s

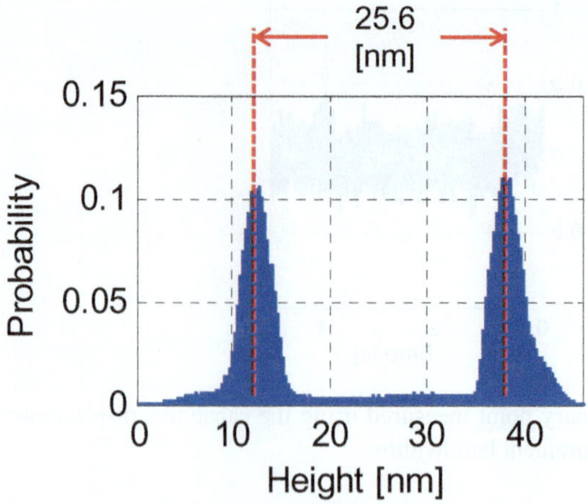

Fig. 2.24 Histogram of height across the image of TGZ01 grating shown in Figure 2.23

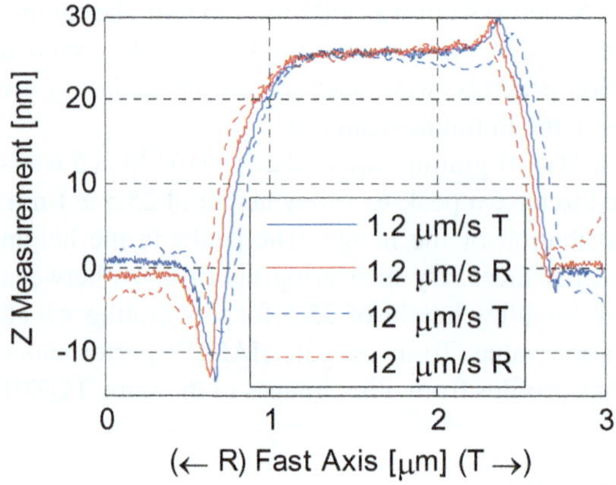

Fig. 2.25 Trace and retrace line scan of the TGZ01 grating at 1.2 μm/s and 12 μm/s

grating, which are shown in Figure 2.25. The plot shows the trace (T) and the retrace (R) scans at two scan speeds of 1.2 μm/s and 12 μm/s. The profiles show the same peak-valley distance and noise performance. Significant rising and falling peaks are visible at the edges. However, the pattern is the same for the trace and the retrace scans, and hence, cannot be considered a controller fault. At this point we are not certain about the cause of this phenomenon. Nevertheless, the generated pattern is found to be repeatable and corresponds to actual interactions between the tip and the edges.

Two other samples, which we have captured at 10 μm/s, are shown in Figure 2.26 and Figure 2.27.

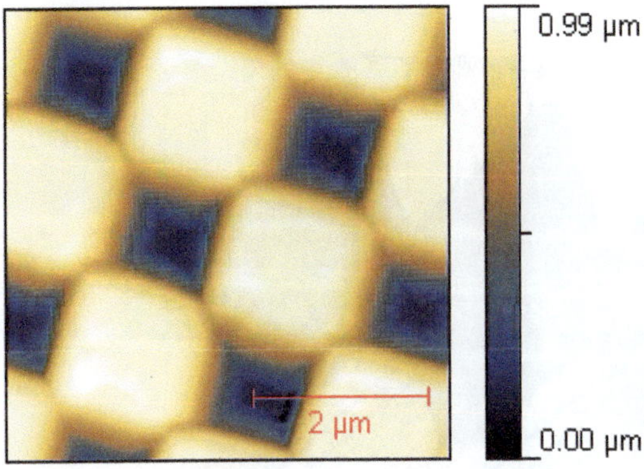

Fig. 2.26 Image of standard TGX01 grating captured at 10 μm/s

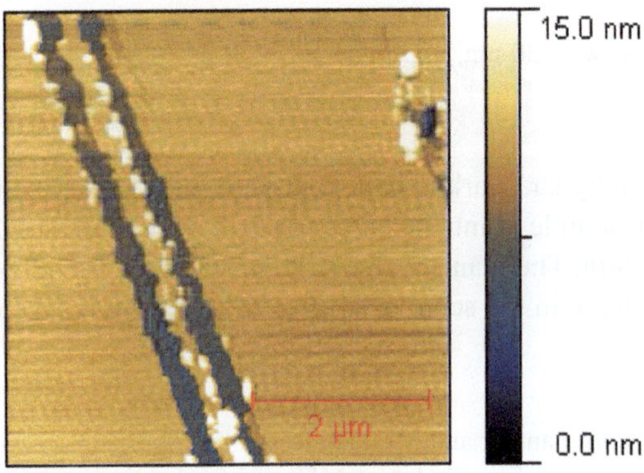

Fig. 2.27 Image of sanded freshly cleaved mica sample captured at 10 μm/s

Figure 2.26 shows a checkerboard TGX01 grating with about 0.9 μm height, which is typically used for lateral calibration of AFMs. The TGX01 and TGZ01 gratings are manufactured by Mikromasch. The sample, shown in Figure 2.27, is freshly cleaved sanded mica. Two parallel channels with 150 nm width and 4 nm depth are observed.

2.5.3 Imaging Test after Integration with SAMM

After confirming the HAFM's operation at MIT, we transferred it to the University of North Carolina at Charlotte to be integrated with the SAMM. We have taken a preliminary image using the SAMM stage operating in air in a temperature-controlled metrology lab (20 °C ± 0.1 °C). This image is shown in Figure 2.28. The sample in this image is a grating in photoresist on silicon, which was printed using an optical

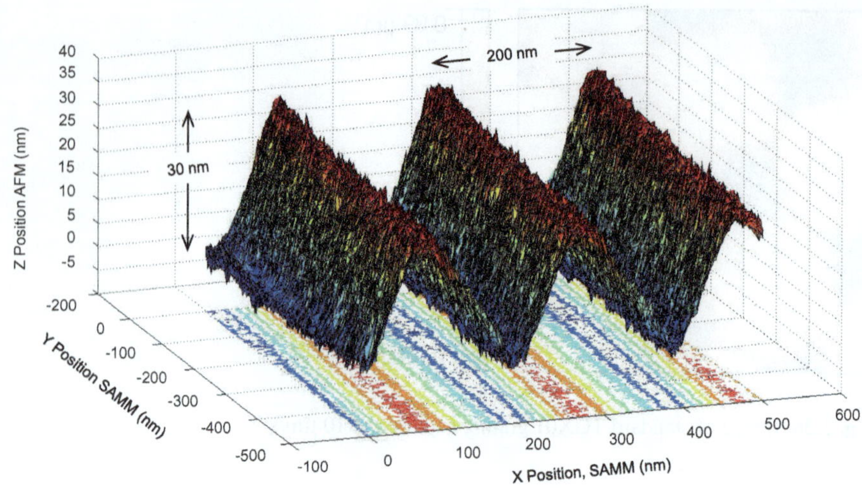

Fig. 2.28 Image of a sawtooth grating with 30-nm amplitude and 200-nm pitch captured using HAFM and SAMM

interference lithography tool created by Dr. Mark Schattenburg at MIT, as described in (20; 35). The image is 256×256 sample points taken over a 500 by 500 nm area. The fast scan-axis is in the X-direction. The scan speed is 195 nm/s, which is equivalent to 5 seconds per trace and retrace in the scan axis. The whole image is taken during an elapsed time of 11 minutes.

Acknowledgements. We thank Dr. Georg Fantner and Mr. Dan Burns for giving us access to their AFM scanner system in Prof. Youcef-Toumi's lab at MIT and for helping us with the HAFM imaging tests. We thank Prof. Patrick Moyer at UNC-Charlotte for guidance on building and using tuning-fork-based AFMs. The current work was supported by the U.S. National Science Foundation under contract DMII-0506898. Earlier elements of this work were supported by the U.S. National Science foundation under contracts DMII-0210543 and DMII-9414778.

References

[1] Akiyama, T., Staufer, U., de Rooji, N.: Symmetrically arranged quartz tuning fork with soft cantilever for intermittent contact mode atomic force microscopy. Rev. Sci. Instrum. 74, 112–117 (2003)

[2] Albrecht, T., Horne, D., Rugar, D.: Frequency modulation detection using high-q cantilevers for enhanced force microscope sensitivity. J. Appl. Phys. 69, 668–673 (1991)

[3] Amin-Shahidi, D., Ljubicic, D., Overcash, J., Hocken, R., Trumper, D.: High-accuracy atomic force microscope for dimensional metrology. In: Proceedings of ASPE Annual Meeting, Atlanta, GA (2010)

[4] Atia, W.A., Christopher, C.C.: A phase-locked shear-force microscope for distance regulation in near-field optical microscopy. Appl. Phys. Lett. 70, 405–407 (1997)
[5] Byl, M., Ludwick, S., Trumper, D.: A loop-shaping perspective for tuning adaptive feedforward controllers. Precis. Eng. J. 29(1), 27–40 (2005)
[6] Cattell, J.: Adaptive feedforward cancelation viewed from an oscillator amplitude control perspective. Master's thesis, Massachusetts Institute of Technology (2003)
[7] Dai, G., Pohlenz, F., Danzebrink, H., Xu, M., Hasche, K., Wilkening, G.: Metrological large range scanning probe microscope. Rev. Sci. Instrum. 75, 962–969 (2004)
[8] Dai, G., Wolff, H., Pohlenz, F., Danzebrink, H.U.: A metrological large range atomic force microscope with improved performance. Rev. Sci. Instrum. 80 (2009)
[9] Danzebrink, H.U., Koenders, L., Wilkening, G., Yacoot, A., Kunzmann, H.: Advances in scanning force microscopy for dimensional metrology. Annals CIRP 55, 841–878 (2006)
[10] Edwards, H., Taylor, L., Duncan, W.: Fast, high-resolution atomic force microscopy using a quartz tuning fork as actuator and sensor. J. Appl. Phys. 82, 980–984 (1997)
[11] Garcia, R., Perez, R.: Dynamic atomic force microscopy methods. Surf. Sci. Rep. 47, 197–301 (2002)
[12] Halbach, K.: Design of permanent multipole magnets with oriented rare earth cobalt material. Nuclear Instruments and Methods 169, 1–10 (1980)
[13] Halbach, K.: Physical and optical properties of rare earth cobalt magnets. Nuclear Instruments and Methods 187, 109–117 (1981)
[14] Holmes, M.: Analysis and design of a magnetically-suspended precision motion control stage. Master's thesis, University of North Carolina at Charlotte (1994)
[15] Holmes, M.: Long-range scanning stage. PhD thesis, University of North Carolina at Charlotte (1998)
[16] Holmes, M., Trumper, D.: Magnetic/fluid bearing stage for atomic-scale motion control. Precis. Eng. J. 18, 38–49 (1996)
[17] Holmes, M., Trumper, D., Hocken, R.: Atomic-scale precision motion control stage (the angstrom stage). CIRP Annals 44, 455–460 (1995)
[18] Holmes, M., Hocken, R., Trumper, D.: The long-range scanning stage: a novel platform for scanned-probe microscopy. Precis. Eng. J. 24 (2000)
[19] Kim, W., Trumper, D., Lang, J.: Modeling and vector control of a planar magnetic levitator. Precis. Eng. J. 4, 553–564 (1998)
[20] Konkola, P., Chen, C., Heilmann, R., Joo, C., Montoya, J., Chang, C.H., Schattenburg, M.: Nanometer-level repeatable metrology using the nanoruler. J. Vac. Sci. Technol. B 21, 3097–3101 (2003)
[21] Kramar, J.A.: Nanometer resolution metrology with the molecular measuring machine. Meas. Sci. Technol. 16, 2121–2128 (2005)
[22] Ljubicic, D.L.: Flexural based high accuracy atomic force microscope. Master's thesis, Massachusetts Institute of Technology (2008)
[23] Ludwick, S.: Modeling and control of six degree of freedom magnetic/fluidic motion control stage. Master's thesis, Massachusetts Institute of Technology (1996)
[24] Ludwick, S., Trumper, D., Holmes, M.: Six degree of freedom magnetic/fluidic motion control stage. IEEE Trans. Control Syst. Technol. 4, 553–564 (1996)
[25] Mazzeo, A.: Accurate capacitive metrology for atomic force microscopy. Master's thesis, Massachusetts Institute of Technology (2005)
[26] NANOSENSORS, Akiyama-Probe (A-Probe) technical guide. (2009), `http://www.akiyamaprobe.com`

[27] Overcash, J., Hocken, R., CG Stroup, J.: Noise reduction and disturbance rejection at the sub-nanometer level. In: Proceedings of ASPE Annual Meeting, Monterey, CA (2009)
[28] Paesler, M., Moyer, P.: Near-Field Optics: Theory, Instrumentation, and Applications. Wiley, Chichester (1996)
[29] Roberge, J.K.: Operational Amplifiers: Theory and Practice. Wiley, Chichester (1975)
[30] Rychen, J., Ihn, T., Studerus, P., Herrmann, A., Ensslin, K.: A low-temperature dynamic mode scanning force microscope operating in high magnetic fields. Rev. Sci. Instrum. 70, 2765–2768 (1999)
[31] Stein, A.: A metrological atomic force microscope. Master's thesis, Massachusetts Institute of Technology (2002)
[32] Trumper, D.: Levitation linear motors for precision portioning. IEEE J. Trans. Elec. Info. Syst. 10 (2006)
[33] Trumper, D., Kim, W.J., Williams, M.: Design and analysis framework for linear permanent magnet machines. IEEE Trans. Ind. Appl. 32, 371–379 (1996)
[34] Wang, C., Hocken, R., Trumper, D.: Dynamics and control of the uncc/mit sub-atomic measuring machine. CIRP Annals - Manufacturing Technology 50, 373–376 (2001)
[35] Zhao, Y., Trumper, D., Heilmann, R., Schattenburg, M.: Optimization and temperature mapping of an ultra-high thermal stability environmental enclosure. Precis. Eng. J. 34(1), 164–170 (2010)

Chapter 3
High-Speed, Ultra-High-Precision Nanopositioning: A Signal Transformation Approach

Ali Bazaei, Yuen K. Yong, S.O. Reza Moheimani, and Abu Sebastian

Abstract. Signal transformation is a novel strategy employed in feedback control to reduce the impact of measurement noise on positioning accuracy. This chapter addresses robustness issues of the method with respect to output disturbance and uncertainty in plant low frequency gain. The robustness problems can be solved by an inner loop with integral action before incorporating the signal transformation mappings. Feedback controllers are designed for two-dimensional positioning of a novel 12-electrode piezoelectric tube used for scanning probe microscopy. The closed-loop bandwidths are intentionally limited to set the standard deviation of the projected noise around 0.1 nm. For triangular waveform tracking and a general class of plants and compensators, necessary and almost sufficient conditions are derived for stability and convergence of tracking error. Effectiveness of the proposed method, regarding tracking and robust performances, is shown by simulations and experiments.

3.1 Introduction

Observation, control, and manipulation of matter at very small dimensions have attracted a great amount of attention in nanotechnology (1; 2). The invention of scanning probe microscopy (SPM) is one of the revolutionary events in nanoscience and nanotechnology (3; 4; 5).

Ali Bazaei · Yuen K. Yong · S.O. Reza Moheimani
School of Electrical Engineering and Computer Science,
The University of Newcastle Australia, Callaghan, NSW 2308, Australia
email:{Ali.Bazaei,Yuenkuan.Yong,
Reza.Moheimani}@newcastle.edu.au

Abu Sebastian
IBM Research – Zurich,
8803 Rüschlikon, Switzerland
e-mail: ase@zurich.ibm.com

E. Eleftheriou & S.O.R. Moheimani (Eds.): Cntrl. Tech. for Emerging Micro/Nanoscale Sys. LNCIS 413, pp. 47–65.
springerlink.com

Accurate tracking of a fast triangular waveform is one of the major challenges not only in SPM (6; 7; 8) but also in other scanner-based devices such as optical scanners and selective laser sintering (SLS) machines (9; 10). The performance of the piezoelectric tube scanner is often quantified by its positioning resolution (which is governed by measurement noise), tracking bandwidth and robustness to disturbances (5). There has been a significant effort to improve the tracking accuracy and speed of piezoelectric tube scanners using feedback control techniques. To track a fast triangular signal, high bandwidth closed-loop controllers have been implemented in many nanopositoning devices (7; 11; 12). However, the scanning speed is limited in feedback control systems due to hysteresis, thermal drift, sensor noise, uncertainty, and mechanical vibrations when piezoelectric tubes are used to follow non-smooth triangular trajectories (13). Capacitive and inductive sensors are commonly used in nanopositioning systems due to their capability of providing simple solution for non-contact, high-resolution measurement. These sensors typically have a noise density of 20 $\mathrm{pm}/\sqrt{\mathrm{Hz}}$ (14). For every hundredfold increment in the closed-loop system bandwidth, the position accuracy of a nanopositioning scanner will decrease tenfold. This potentially degrades the resolution of the scanner, hindering it from performing positioning tasks that require subnanometer resolution. Hence, feedback control methods with limited closed-loop bandwidth are of considerable importance.

Command pre-shaping methods can be considered as a possible way for vibration suppression in an already designed closed-loop control system, leaving the closed-loop bandwidth of the measurement noise unaffected (15; 16; 17; 18). However, these methods are not suitable for tracking of time-varying commands such as triangular waveforms or suffer from lack of robustness to plant uncertainties. Iterative learning control (ILC) can also be added as a feed-forward control action in a feedback system to improve the steady-state tracking error for repetitive references without altering the closed-loop bandwidth (19; 9). However, it may require a large number of iterations to converge. Feedback control methods such as repetitive control (RC) for tracking of periodic references introduce large closed-loop bandwidths, which may not be acceptable in the presence of measurement noise. Moreover, the trade-off between the tracking error and rejection of non-periodic disturbances in RC systems can cause problems when excessive cross coupling exist among the scanner axes (20; 21).

In (22), the concept of signal transformation was put forward as a novel approach for tracking of triangular waveforms in a nanopositioning system. The method showed significant closed-loop performance improvement compared with an ordinary feedback-control-system having a similar control bandwidth. However, the method is sensitive to DC gain variations and disturbances arising from cross coupling between the two axes.

This chapter addresses how signal transformation can be used along with traditional feedback control methods to improve tracking error in an atomic force microscope (AFM) scanner while keeping closed-loop measurement noise below a pre-specified level and providing stability and robustness to DC-gain variations and disturbances. A thorough stability analysis of the method in the absence of measurement noise and disturbances is also presented in the Appendix.

3.2 Objectives

To characterize the impact of measurement noise in feedback control systems, concept of projected noise is introduced. As shown in Fig. 3.1, consider a typical feedback control system designed to control a physical quantity x, which is measured by a sensor, that provides a measured signal $x_m := x + n$ for feedback, which is affected by measurement noise n. By the projected noise, we mean the direct effect of the measurement noise signal n on the actual controlled output x in the closed-loop feedback system. For linear systems, this effect can be quantified in terms of the noise signal n and the closed-loop transfer function from n to x. An objective in this paper is to evaluate the capability of signal transformation method in reducing projected measurement noise compared to ordinary feedback systems. To do this, we maintain the standard deviation of the projected measurement noise around 0.1nm at the actual displacement of x-axis. The other objective is to provide disturbance rejection capability and robustness when signal transformation is incorporated into the control systems.

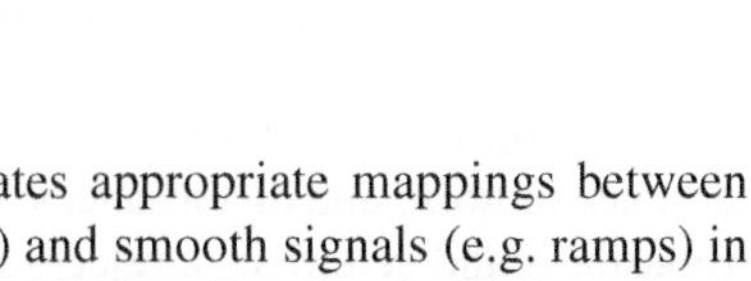

Fig. 3.1 Illustration of projected measurement noise in a typical feedback control system

3.3 Signal Transformation

The signal transformation approach incorporates appropriate mappings between non-smooth signals (e.g. triangular waveforms) and smooth signals (e.g. ramps) in a control system to improve the tracking error while keeping the closed-loop bandwidth low to limit the projected measurement noise (22). The signal transformation method for control of a SISO plant is described by the hybrid control system shown in Fig. 3.2, where Φ and Φ^{-1} refer to the signal transformation mappings, which in the case of triangular signal tracking use piecewise constant gains g_1 and g_2 as well as biases b_1 and b_2. The latter can be presented in the following form:

$$g_1 = g_2 = (-1)^k, \; b_2 = 2a_0 k, \; b_1 = -(-1)^k b_2, \tag{3.1}$$

where a_0 is the amplitude of the desired triangular waveform x_d, which has period $2T$, as shown in the left top insert in Fig. 3.2, and k is the index of the half period defined as

$$k(t) = \text{floor}\left(\frac{t}{T} + 0.5\right). \tag{3.2}$$

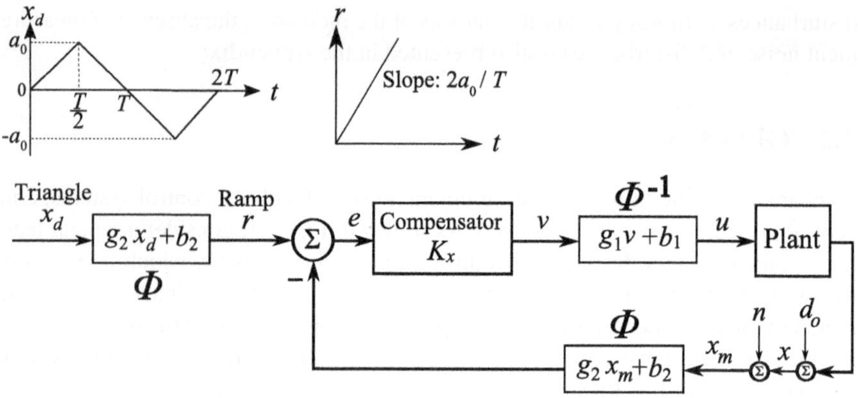

Fig. 3.2 Schematic diagram of signal transformation method for triangular waveform tracking

The signal transformation blocks, which use g_2 and b_2, can convert the non-smooth periodic triangular signal x_d into a smooth ramp signal denoted by r in the left top insert in Fig. 3.2. The signal transformation block between the plant and compensator does the reverse action, i.e. it can convert the smooth ramp signal into the non-smooth triangular signal. Consider an ideal situation, where the noise n and output disturbance d_o are zero, the plant is a unity gain transfer function, and its output is perfectly following the desired signal. In this case, the input/output signals at the compensator block will be smooth signals with no breaks or discontinuities, and the burden of providing appropriate non-smooth trajectories at the actuator, which demands a high control bandwidth in an ordinary feedback system, is carried by the signal transformation block. In this way, the compensator can be designed with a smaller closed-loop bandwidth in favor of rejecting the projected measurement noise without deteriorating the steady-state error. The Appendix addresses necessary and sufficient stability conditions of the signal transformation method in the absence of noise and disturbances. The signal transformation method, however, has robustness and disturbance rejection problems, which will be explained in Sec. 3.4.

3.4 Investigation of System Robustness

In this section, we use simulations to show that the signal transformation method mentioned in Sec. 3.3 can improve the tracking performance of feedback control systems with low closed-loop bandwidth, which translates into low projected noise. We also investigate the robustness of the method to DC gain variations and output disturbance using a model obtained for the x-axis of an AFM scanner. To do this, we use a model for the x-axis of the scanner, obtained after closing a damping loop through a low noise piezoelectric induced voltage to damp the first resonance of the tube. The piezoelectric induced voltage is obtained as in (23). The model has zeros at $230 \pm 6000i$, $-1180 \pm 876i$, and -2.1, and poles at $-1286 \pm 1992i$, $-1100 \pm 1497i$, and −2.3. A constant gain of $1/0.42$ was included at the input to force a unity DC gain for the plant in Fig. 3.2. The compensator in Fig. 3.2 is a double integrator plus an integrator in the following form:

$$K_x(s) = 2.3 \times \frac{50s + 250}{s^2}, \tag{3.3}$$

which provides a gain margin of 23.4 dB, a phase margin of 87°, and reduces the closed-loop bandwidth to 21 Hz. Such a low closed-loop bandwidth can keep the projected measurement noise around 0.1 nm, as shown in Sec. 3.6.1. For a 10-Hz triangular reference with amplitude $a_0 = 2$ V (10 μm peak-to-peak) and under different conditions, the resulting closed-loop steady-state displacement errors ($x_d - x$), scaled to voltage by sensitivity of capacitive displacement sensors (0.4 V/μm), are shown in Fig. 3.3. With unity DC gain and no disturbance, the signal transformation provides acceptable tracking (compare the thick solid line curve with a 4-V peak-to-peak triangular reference). However, when the plant DC gain is increased or is reduced twice (6 dB), which is much less than the gain margin, the error increases unacceptably, as shown in Fig. 3.3. This shows a lack of robustness against variations in the plant DC gain. With unity plant DC gain, the steady-state error of the system to the triangular reference along with a unity amplitude constant output disturbance d_o has also been included in Fig. 3.3 (dotted line), which shows an undesirable disturbance rejection performance. In this example, the 21 Hz bandwidth with signal transformation cannot provide acceptable disturbance rejection, while it is suitable for tracking in the absence of disturbances. To appreciate the benefit of signal transformation, we have replaced the signal transformation blocks by unity gains. The resulting ordinary feedback system with unity plant DC gain and no disturbances has the response labeled "No signal transformation" in Fig. 3.3, which shows that the 21 Hz bandwidth without signal transformation is not sufficient for acceptable tracking of a 10 Hz triangular reference.

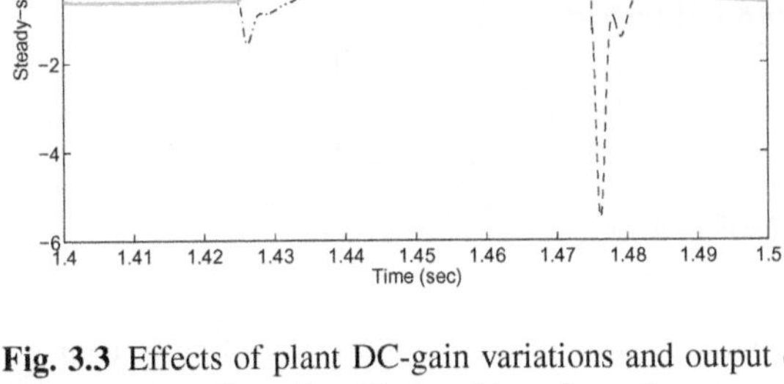

Fig. 3.3 Effects of plant DC-gain variations and output disturbance on closed-loop steady-state response of x-axis with signal transformation

3.5 Incorporating Robustness in Signal Transformation

In this section, we incorporate an intermediate feedback loop prior to the signal transformation blocks, as shown in Fig. 3.4, to improve the robustness properties mentioned in Sec. 3.4. In Fig. 3.4, the signal transformation mappings denoted by Φ and Φ^{-1} are as before, and d_{ox}, n_x, v_{cx}, v_{px}, and u_x stand for output disturbance, measurement noise, capacitive sensor output, piezoelectric induced voltage, and piezoelectric actuation voltage of the x-axis, respectively. A low-pass filter $F(s) = \left(1 + \frac{s}{1000}\right)^{-1}$ was used to reduce the effect of measurement noise. The intermediate and outer compensators were selected as

$$K_i(s) = \frac{166.667}{s} \tag{3.4}$$

$$K_{ii}(s) = \frac{50s + 250}{s^2} \quad . \tag{3.5}$$

The compensators $K_i(s)$ and $K_{ii}(s)$ were selected such that the overall transfer function from sensor noise n_x to the real displacement output signal x has a low bandwidth of 21 Hz, similar as in Sec. 3.4. This transfer function can be described as

$$T_{xn}(s) := \frac{x(s)}{n_x(s)} = -\frac{K_i(s)P_x(s)[F(s) + K_{ii}(s)]}{1 + K_i(s)P_x(s)[F(s) + K_{ii}(s)]}, \tag{3.6}$$

where $P_x(s)$ is the transfer function of the damped system of the x-axis. The robustification loop by itself (excluding the outer loop) provides a unity DC gain from u to x with a gain margin of 27.7 dB, phase margin of 90°, and bandwidth of 13 Hz. The overall system has a gain margin of 42 dB, phase margin of 55°, and bandwidth of 11.6 Hz for the forward transfer function from the reference to the displacement output. The simulation results shown in Fig. 3.5 correspond to closed-loop response of the proposed method where the same triangular reference signal and disturbance as in Sec. 3.4 are used and the plant DC gain in the legend refers to the DC gain of the damped system in Fig. 3.4. Clearly, the steady-state tracking error remains acceptable in the presence of DC-gain variations of the plant and output disturbance,

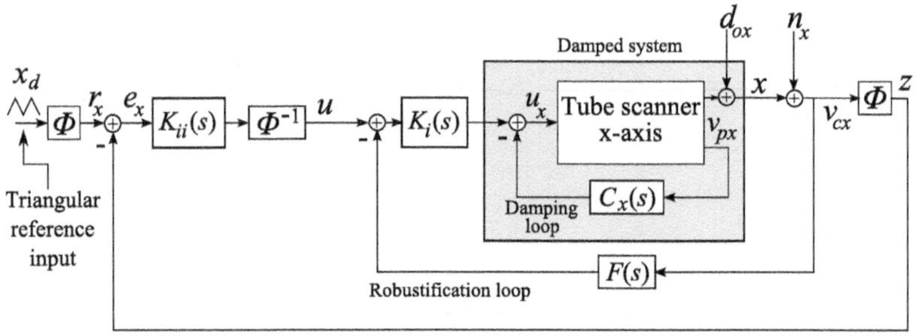

Fig. 3.4 Schematic diagram of signal transformation method with a robustification loop

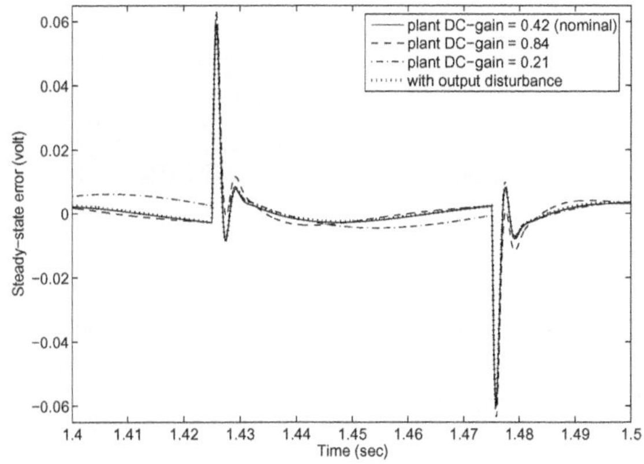

Fig. 3.5 Effects of plant DC-gain variations and output disturbance on closed-loop steady-state response of x-axis with robustified signal transformation

which shows that the robustification loop can improve the robustness of the signal transformation method without deteriorating its benefits (low tracking error with low bandwidth).

3.6 Experimental Results

In this section, the signal transformation method with the proposed robustifying scheme is performed on the x-axis of the actual scanner for further examination. The external electrode of the piezoelectric tube scanner is segmented into 12 equal sections, and the inner electrode is a continuous electrode which is grounded. One end of the tube is fixed. The free end serves as a stage over which a sample can be placed and its horizontal deflections are measured by two capacitive sensors. Figure 3.6 illustrates the wiring of the tube for actuation and sensing in the x-axis alone, where $\pm\hat{u}_x$ and $\hat{v}_{px}$ are the actuation and piezoelectric induced voltages after and before amplification, respectively. The same wiring is applied to the y-axis, but is not illustrated for the sake of clarity. A dSPACE-1103 rapid prototyping system was

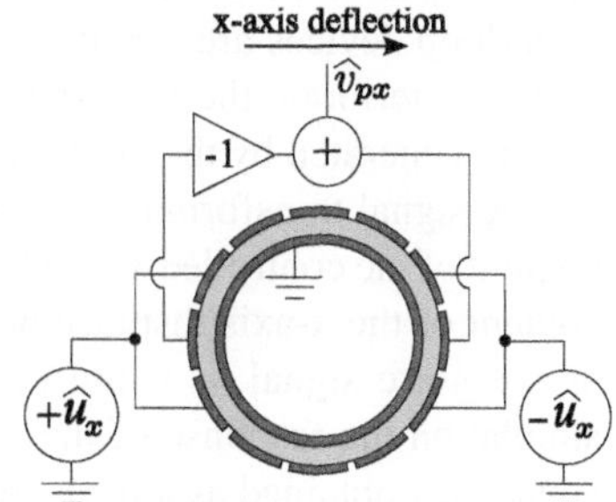

Fig. 3.6 Simultaneous piezoelectric actuation and sensing for x-axis of the tube scanner

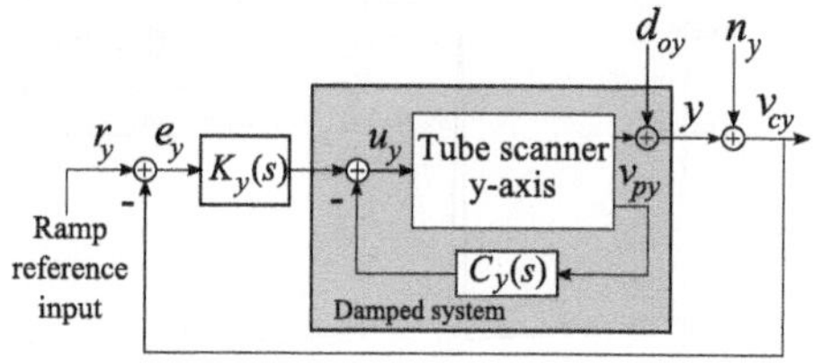

Fig. 3.7 Schematic diagram for control of y-axis

used to implement the x- and y-axis feedback controllers in real time. The z-axis displacement was controlled using the AFM's software and circuitry. The damped y-axis is controlled by an ordinary integral control as shown in Fig. 3.7, where the compensator

$$K_y(s) = \frac{1500}{s} \tag{3.7}$$

provides a gain margin 8.2 dB and a phase margin 86°. The y-axis reference signal r_y is a ramp signal whose slope is 512 times less than that of the x-axis triangular reference. The x-axis controllers and the triangular reference signal are as in Sec. 3.5. The overall noise transfer function $T_{xn}(s)$ for the x-axis controller has a bandwidth of 21 Hz as before. A calibration grating (MikroMasch TGQ1) with a 3 μm period, 1.5 μm square side and 20 nm height was used for imaging. A contact mode ContAl cantilever probe with a resonance frequency of 13 kHz was used to perform the scan. To evaluate the scanning performance of the controllers, a 9.8 Hz triangular reference signal was applied to the x-axis and the aforementioned synchronized ramp signal was applied to the y-axis of the piezoelectric tube scanner to generate a 10 μm × 10 μm image (with 256×256 scan lines). Figure 3.8(a) shows the scanned image and the tracking performance of the x-axis displacement with signal transformation of the piezoelectric tube scanner. The RMS error of the tracking signal is 80 nm.

3.6.1 *Tracking Performance and Noise*

The resolution of the piezoelectric tube is often governed by the sensor noise due to the noise being fed back to the actuator in closed-loop systems. This makes the open-loop architecture a more attractive solution than the closed-loop one. However, open-loop devices are sensitive to nonlinear effects such as drift and creep. These effects deteriorate the tracking performance and subsequently degrade the image quality generated by the devices.

The signal transformation method presented in this paper ensures that the noise content of the controlled x-position signal x, in Fig. 3.4, is low. To estimate the noise content of the x-axis displacement, the capacitive sensor output was first recorded as the noise signal n_x while the piezoelectric tube remained stationary. Then by simulation the response of the noise transfer function $T_{xn}(s)$ to the recorded noise signal was obtained as a measure of noise projected into the actual x-position. The

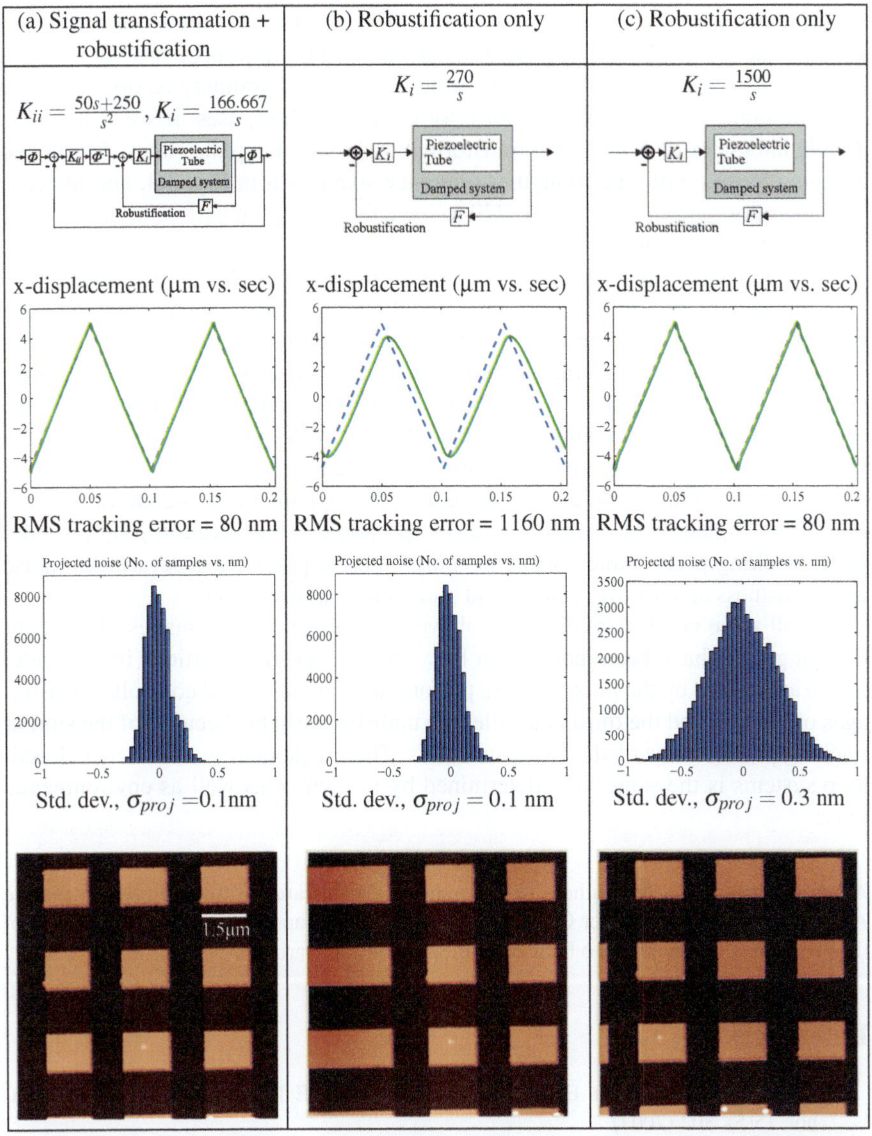

Fig. 3.8 Comparison of the projected noise, tracking error and AFM image (10 μm × 10 μm) of the three closed-loop systems. (a) Closed-loop system with signal transformation and robustification loop. (b) Closed-loop system without signal transformation. The integrator gain was tuned to 270; therefore the projected noise is similar to that of system (a). (c) Closed-loop system without signal transformation. The integrator gain was tuned to 1500; therefore the tracking performance is similar to that of system (a). σ_{proj} is the standard deviation of the projected noise. The tracking performance in the x-axis of each closed-loop system is illustrated. The triangular reference signal (dashed line) and output displacement (solid line) are also plotted.

resulting histogram is shown in Fig. 3.8(a). The standard deviations of the sensor noise n_x and the projected noise are 3.075 nm and 0.11 nm, respectively. To evaluate the efficacy of signal transformation, we now consider an ordinary feedback system with the same level of projected noise for comparison purposes. If we remove the double integrator and the signal transformation blocks, keep the robustifiction and damping loops, apply the triangular reference signal at u in Fig. 3.4, and increase the integrator gain in Eq. (3.4) to 270 to have the same standard deviation of 0.11 nm for the projected noise, we obtain the resulting steady-state tracking error of the x-axis as shown in Fig. 3.8(b), where the root-mean-square (RMS) tracking error is 14.5 times more than that of the signal-transformation method. This tracking error is mostly contributed by the low closed-loop bandwidth of the system.

Alternatively, if in the latter system, which has no signal transformation, we increase the integrator gain of the x-channel to 1500 to keep the RMS value of the resulting steady-state tracking error equal to that with signal transformation, the standard deviation of the projected noise will increase to 0.3 nm, as shown in Fig. 3.8(c), which is almost three times more than that obtained with signal transformation. Thus, signal transformation with the proposed robustification loop provides better tracking performance, while keeping both the projected measurement noise and robustness against disturbance and parameter variations low.

For all three controllers designed above, the corresponding images of the calibration grating have been included in Fig. 3.8. The severe distortions in the center image are caused by the poor tracking performance of the second controller. The images of the first and the third controller are undistinguishable because of the similar tracking performance of the two controllers. The image quality of the two closed-loop systems is the same; it is determined by the sensor as well as environmental noise.

Acknowledgements. The authors wish to thank IBM Research – Zurich Research and the ARC Centre of Excellence for Complex Dynamic Systems and Control for their support of this work. Special thanks go to Evangelos Eleftheriou and Haris Pozidis for their support.

References

[1] Devasia, S., Eleftheriou, E., Moheimani, S.O.R.: IEEE Trans. Control Systems Technol. 15(5), 802 (2007)
[2] Sugimoto, Y., Pou, P., Custance, O., Jelinek, P., Abe, M., Pérez, R., Morita, S.: Science 322, 413 (2008)
[3] Bhushan, B.: Nanotribology and Nanomechanics. An Introduction. Springer, Heidelberg (2005)
[4] Kalinin, S., Gruverman, A.: Scanning Probe Microscopy: Electrical and Electromechanical Phenomena at the Nanoscale. Springer, Heidelberg (2006)
[5] Salapaka, S., Salapaka, M.: IEEE Control Systems Mag. 28(2), 65 (2008)
[6] Lining, S., Changhai, R., Weibin, R., Liguo, C., Minxiu, K.: J. Micromechan. Microeng. 14(11), 1439 (2004)

[7] Mahmood, I.A., Moheimani, S.O.R., Liu, K.: IEEE Trans. Nanotechnology 8(1), 55 (2009)
[8] Payam, A., Yazdanpanah, M., Fathipour, M.: In: Proceedings 2009 4th IEEE Int'l Conf. on Nano/Micro Engineered and Molecular Systems, pp. 717–721 (2009)
[9] Yen, J.Y., Yeh, Y.C., Peng, Y.H., Lee, J.F.: Mechatronics 19(1), 65 (2009)
[10] Zhiqiang, D., Zude, Z., Wu, A., Youping, C.: The Int'l J. Adv. Manufacturing Technol. 32(11-12), 1211 (2007)
[11] Salapaka, S., Sebastian, A., Cleveland, J.P., Salapaka, M.V.: Rev. Sci. Instrum. 73(9), 3232 (2002)
[12] Yong, Y.K., Aphale, S., Moheimani, S.O.R.: IEEE Trans. Nanotechnology 8(1), 46 (2009)
[13] Moheimani, S.O.R.: Rev. Sci. Instrum. 79(7), 071101 (2008)
[14] Fleming, A.J., Wills, A., Moheimani, S.O.R.: IEEE Trans. Control Systems Technol. 16(6), 1265 (2008)
[15] Smith, O.J.M.: Feedback Control Systems. McGraw-Hill, New York (1958)
[16] Singer, N.C., Seering, W.P.: J. Dynamic Syst., Measurement, and Control 112(1), 76 (1990)
[17] Pennacchi, P.: Shock and Vibration 11(3/4), 377 (2004)
[18] Kogiso, K., Hirata, K.: Robotics and Autonomous Syst. 57(3), 289 (2009)
[19] Tien, S., Zou, Q., Devasia, S.: IEEE Trans. Control Systems Technol. 13(6), 921 (2005), doi:10.1109/TCST.2005.854334
[20] Pipeleers, G., Demeulenaere, B., Al-Bender, F., De Schutter, J., Swevers, J.: IEEE Trans. Control Systems Technol. 17(4), 970 (2009), doi:10.1109/TCST.2009.2014358
[21] Aridogan, U., Shan, Y., Leang, K.K.: ASME J. Dynamic Syst., Measurement, and Control 131(6), 061103 (2009)
[22] Sebastian, A., Moheimani, S.O.R.: Rev. Sci. Instrum. 80, 076101 (2009)
[23] Moheimani, S.O.R., Yong, Y.K.: Rev. Sci. Instrum. 79(7), 073702 (2008)

Appendix

Here the stability of the proposed signal transformation method in Fig. 3.2 is investigated in the absence of noise and disturbances.

Stability Analysis

We assume that the plant and compensator dynamics are of degrees n_p and n_c and are described by linear-time-invariant state-space matrix sets $[A_p, B_p, C_p]$ and $[A_c, B_c, C_c]$ with X_p and X_c referring to the corresponding state vectors, respectively. The feedthrough matrices have been assumed zero. To start the analysis, we merge the plant and its adjacent signal transformation blocks into a unified state-space model, called *equivalent plant*. Hence, we wish to determine under what circumstances the simple control system shown in Fig. 3.9 is equivalent to the original hybrid control system in Fig. 3.2, i.e., with the same ramp signal $r(t)$ in both control systems, the time histories of variables e, v, X_c, and y in the equivalent system shown in Fig. 3.9 are the same as the corresponding variables in the original system shown in Fig. 3.2.

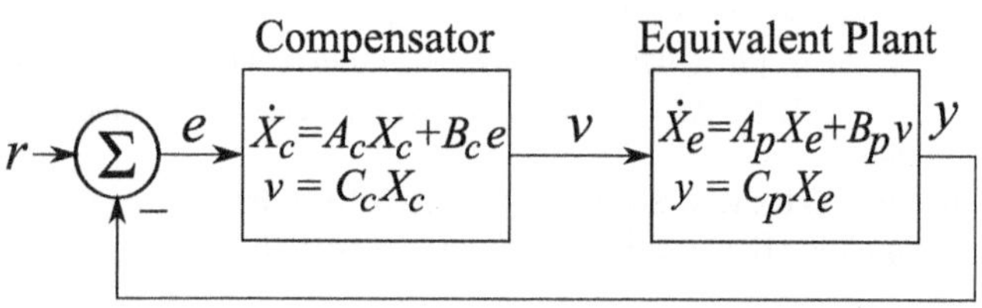

Fig. 3.9 Schematic diagram of the equivalent control system

The following theorem provides the conditions for the foregoing equivalence.

Theorem 3.1. *In a time interval* $t \in (kT - \frac{T}{2}, kT + \frac{T}{2})$, *the hybrid control system in Fig. 3.2 is equivalent to the control system in Fig. 3.9, and state* X_e *of the equivalent plant is related to the plant state by*

$$X_e := \frac{1}{g_1}(X_p + F), \; F := A_p^{-1} B_p b_1, \tag{3.8}$$

provided that the gains and biases are constants (in the time interval) satisfying the following relationships

$$g_1 g_2 = 1, \; b_2 - g_2 C_p A_p^{-1} B_p b_1 = 0, \tag{3.9}$$

and the equivalent state vector at the start of the time interval is initialized according to (3.8).

Proof. The signal transformations in Fig. 3.2 are described in the following forms.

$$u = g_1 v + b_1 \tag{3.10}$$

$$y = g_2 x + b_2. \tag{3.11}$$

Consider the time interval $t \in (iT - T, iT)$. Since the gain and bias signals are constant in this interval, the following state-space model is readily obtained using the plant state dynamics and Eqs. (3.10) and (3.11), if the plant has a no poles at the origin:

$$\begin{cases} \dot{X}_e = A_p X_e + B_p v \\ y \ = g_1 g_2 C_p X_e + (b_2 + g_2 \delta_0 b_1) \end{cases}, \tag{3.12}$$

where $\delta_0 = -C_p A_p^{-1} B_p$ is the DC gain of the plant. It is clear from Eq. (3.12) that we can replace the blocks between nodes v and y in Fig. 3.2 with the equivalent plant, as described by Fig. 3.9 and Eq. (3.8), and the control systems are equivalent if the conditions mentioned in Theorem 3.1 are satisfied.

Conditions (3.9) are satisfied with the selected gains and biases in Eqs. (3.1) if the plant has a unity DC gain ($\delta_0 = 1$). If the plant has a transfer function of the form:

$$P_{ol}(s) := \frac{x(s)}{u(s)} = \frac{\delta_0 + \delta_1 s + \cdots + \delta_{n_p-1} s^{n_p-1}}{1 + \varepsilon_1 s + \cdots + \varepsilon_{n_p} s^{n_p}}, \tag{3.13}$$

its state-space realization can be written by the following canonical form:

$$\begin{aligned} A_p &= \left[0_{(n_p-1)\times 1}, I_{n_p-1}; \frac{-1}{\varepsilon_{n_p}}, \frac{-\varepsilon_1}{\varepsilon_{n_p}}, \ldots, \frac{\varepsilon_{n_p-1}}{-\varepsilon_{n_p}} \right], \\ B_p &= \left[0_{(n_p-1)\times 1}; \frac{1}{\varepsilon_{n_p}} \right], C_p = \left[\delta_0, \cdots, \delta_{n_p-1} \right]. \end{aligned} \tag{3.14}$$

The overall state vector X of the equivalent closed-loop system, defined as

$$X := \begin{bmatrix} X_e \\ X_c \end{bmatrix}, \tag{3.15}$$

obeys the following state-space equation:

$$\begin{cases} \dot{X} = AX + Br \\ y \ = CX \end{cases}, \tag{3.16}$$

where

$$A := \begin{bmatrix} A_p & B_p C_c \\ -B_c C_p & A_c \end{bmatrix}, B := \begin{bmatrix} 0_{n_p \times 1} \\ B_c \end{bmatrix},$$

$$C := \left[C_p \ 0_{1\times n_c} \right]. \tag{3.17}$$

The equivalent plant state X_e must be initialized by (3.8) at the start of each half period, which requires knowledge of plant state X_p. To use the equivalent control system as a stand-alone machinery for analysis, appropriate formulas are necessary to update the equivalent state at the switching moments $t = kT - \frac{T}{2}$. The following theorem gives the updating relationships at the switching moments.

Theorem 3.2. *With the triangular reference signal shown in Fig. 3.2, signal transformation parameters (3.1), and unity DC gain for the plant, the overall state vector of the equivalent control system just before a switching moment obeys the recursive formula:*

$$X_1^- := X(T^-/2) = E^{1/2}X_0 + H + A^{-1}Ba_0, \tag{3.18}$$

$$X_{k+1}^- := X(t)|_{t=kT^- + \frac{T}{2}} = \hat{A}X_k^- + J(k-0.5) + H,\ k = 1,2,3,\ldots, \tag{3.19}$$

and the state just after a switching moment is updated using its value just before the switching moment as

$$X_k^+ := X\left(kT^+ - \frac{T}{2}\right) = \hat{I}X_k^- + L(k-0.5),\ k = 1,2,3,\ldots, \tag{3.20}$$

where L is the constant $(n_p + n_c) \times 1$ vector:

$$L := [4a_0, 0, \ldots, 0]^T, \tag{3.21}$$

and

$$\hat{I} := \begin{bmatrix} -I_{n_p} & 0 \\ 0 & I_{n_c} \end{bmatrix},\ X_0 := \begin{bmatrix} X_p(0) \\ X_c(0) \end{bmatrix}, \tag{3.22}$$

$$E := e^{AT},\ H := 2\left[\frac{1}{T}(E-I)A^{-1} - I\right]A^{-1}Ba_0, \tag{3.23}$$

$$\hat{A} := E\hat{I},\ J := 2(E-I)A^{-1}Ba_0 + EL\,. \tag{3.24}$$

Proof. For brevity, only a sketch of the proof is presented here. In the original control system shown in Fig. 3.2, the gains and biases have discontinuous changes at the switching times $t = kT - \frac{T}{2}$ $(k = 1,2,3,\ldots)$, which makes the signals y and u discontinuous. However, the states and outputs of the plant and compensator (X_c, X_p, v, x) are continuous due to zero feedthrough matrices and the inherent integration actions in the compensator and plant state equations. Hence, the equivalent plant state X_e has discontinuities at the switching times because of g_1 and b_1 (see Eq. (3.8)). Thus, to maintain the equivalence of the simple control system shown in Fig. 3.9 with the original control system over time intervals longer than a half period, we have to intentionally incorporate appropriate jumps in the equivalent plant state X_e at each switching time, which can be described in the following form using Eq. (3.8):

$$\Delta X_{e_k} := X_{e_k}^+ - X_{e_k}^- = \left(\frac{1}{g_{1_k}^+} - \frac{1}{g_{1_k}^-}\right)X_{p_k} + A_p^{-1}B_p\left(\frac{b_{1_k}^+}{g_{1_k}^+} - \frac{b_{1_k}^-}{g_{1_k}^-}\right) \tag{3.25}$$

where the lowest subscript k for each variable refers its value at the switching moment $t = kT - \frac{T}{2}$ ($k = 1,2,3,\ldots$), and the minus and plus superscripts refer to the values of the corresponding variable at infinitesimal times just before and after the switching moment indicated by the lowest subscript, respectively, as defined in the following forms:

$$X_{p_k} := X_p\left(kT - \frac{T}{2}\right),\ X_{e_k}^{+} := X_e\left(kT^{+} - \frac{T}{2}\right),\ X_{e_k}^{-} := X_e\left(kT^{-} - \frac{T}{2}\right),\ldots \quad (3.26)$$

If we use Eq. (3.8) to replace X_{p_k} by $g_{1_k}^{-}X_{e_k}^{-} - F_k^{-}$ in Eq. (3.25), the equivalent plant state just after the switching moment can be described in terms of its value just before the switching moment as:

$$X_{e_k}^{+} = \frac{1}{g_{1_k}^{+}}\left[g_{1_k}^{-}X_{e_k}^{-} + A_p^{-1}B_p(b_{1_k}^{+} - b_{1_k}^{-})\right]. \quad (3.27)$$

The inverse of the plant state matrix is in the following form:

$$A_p^{-1} = \begin{bmatrix} -\varepsilon_1 & -\varepsilon_2 & -\varepsilon_3 & \cdots & -\varepsilon_{n_p-1} & -\varepsilon_{n_p} \\ 1 & 0 & 0 & \cdots & \cdots & 0 \\ 0 & 1 & 0 & \cdots & \cdots & 0 \\ \vdots & \ddots & \ddots & \ddots & \cdots & \vdots \\ 0 & \cdots & 0 & 1 & 0 & 0 \\ 0 & \cdots & \cdots & 0 & 1 & 0 \end{bmatrix} \quad (3.28)$$

Hence, $A_p^{-1}B_p$ is $[-1,0,\ldots,0]^T$. Using (3.27) and the signal transformation gains and biases selected in (3.1) for the triangular reference, the equivalent plant state at the switching moments can be updated based on the following relationship:

$$X_{e_k}^{+} = -X_{e_k}^{-} + L(k - 0.5),\ \text{for}\ k = 1,2,3,\ldots \quad (3.29)$$

where L is a $n_p \times 1$ vector defined similar to (3.21). Using (3.29) and the fact that the compensator state is continuous at the switching moments, the overall state X just after the switching moment is easily obtained in the form of (3.20). Using the commutativity of multiplication of A^{-1} and e^{At}, the solution of the state X from state-equation (3.16) in the time interval $t \in (kT - \frac{T}{2}, kT + \frac{T}{2})$ with ramp input $r = \frac{2a_0 t}{T}$, in terms of the state just after the switching moment $t = kT - \frac{T}{2}$, can be written in the following form.

$$X(t) = e^{At'}X_k^{+} + \{(e^{At'} - I)[a_0(2k-1)I + A^{-1}\alpha] - \alpha t' I\}A^{-1}B, \quad (3.30)$$

where

$$t' := t - \left(kT - \frac{T}{2}\right),\ \alpha := \frac{2a_0}{T},\ X_k^{+} := X(t)|_{t=(kT-\frac{T}{2})^{+}}. \quad (3.31)$$

Equation (3.19) defines a discrete-time LTI dynamic system with $\hat{A}$ as the state matrix, $[J,H]$ as the input matrix, and $[k-0.5,1]^T$ as the input vector whose first element is a discrete-time ramp signal. Hence, a necessary condition for the closed-loop system to be free from exponentially unstable modes is that all eigenvalues of $\hat{A}$ are inside the unit disk. This condition is also a sufficient one because the state at the arbitrary time $t = kT - \frac{T}{2} + t'$ depends on X_k^- through Eqs. (3.20) and (3.30) and variable t' is limited to $t' \in (0,T)$, which shows that if X_k^- does not have any exponentially unstable mode, neither does $X(t)$. In the more general case, where the desired signal x_d is an arbitrary bounded signal but the signal transformation parameters are kept as before with unity DC gain for the plant, Eq. (3.20) will not change but Eqs. (3.30), (3.18), and (3.19) can be represented in the following forms:

$$X(t'+kT-\frac{T}{2}) = e^{At'}X_k^+ + 2(e^{At'}-I)A^{-1}Ba_0k + W(k,t'), \tag{3.32}$$

$$X_1^- = EX_0 + \int_0^{\frac{T}{2}} e^{A(\frac{T}{2}-t)}Bx_d(t)dt, \tag{3.33}$$

$$X_{k+1}^- = \hat{A}X_k^- + Jk + W(k,T) - 0.5EL,\; k = 1,2,3,\ldots, \tag{3.34}$$

where

$$W(k,t') = \int_0^{t'} e^{A(t'-\tau)}x_d\left(\tau + kT - \frac{T}{2}\right)d\tau B(-1)^k. \tag{3.35}$$

Since vector $W(k,t')$ is bounded, because of the boundedness of x_d, the aforementioned condition about the absence of exponentially unstable modes is not restricted to the triangular waveform and is also valid for arbitrarily bounded reference inputs.

Theorem 3.3. *Assuming unity DC gain for the plant and signal transformation parameters (3.1), the hybrid control system is free from exponentially unstable modes if and only if the eigenvalues of matrix* $\hat{A}$*, defined in (3.24), are inside the unit circle.*

Note that the hybrid control system may have exponentially stable responses, whereas the closed-loop state matrix A, defined in (3.17), may have unstable eigenvalues, which means under some circumstances, incorporation of signal transformation into an ordinary unstable feedback system may stabilize the closed-loop responses.

Steady-State Behavior with Triangular Reference

The signal transformation converts the original triangular reference into a ramp signal. However, it is not desirable for the plant states to grow linearly with time. The following theorem provides conditions under which the states of the plant remain bounded in steady-state conditions.

Theorem 3.4. *Assuming the triangular reference input in Fig. 3.2, unity DC gain for the plant, signal transformation parameters (3.1), and eigenvalues of matrix $\hat{A}$ inside the unit circle, the plant state in the hybrid control system will remain bounded if and only if either of the following conditions is satisfied:*

$$\delta_c := -CA^{-1}B = 1 \tag{3.36}$$

$$P(t') := [I_{n_p}, 0]e^{At'}\hat{I}(I-\hat{A})^{-1}L + 0.5L = 0,\ \forall t' \in (0,T). \tag{3.37}$$

Proof. Successive use of Eq. (3.19) leads to the following equation:

$$X_{k+1}^{-} = \hat{A}^k X_1^{-} + \sum_{l=0}^{k-1} \hat{A}^l[(k-l)J + H - 0.5J],\ k = 1,2,3,\ldots, \tag{3.38}$$

which represents the state value just before a generic switching moment in terms of its value just before the first switching moment. We can simplify solution (3.38) if matrix $(I-\hat{A})$ is invertible. Given that eigenvalues of $\hat{A}$ have magnitudes less than 1 this condition is met. Under such assumption, the following equalities hold:

$$\sum_{l=0}^{k-1} (k-l)\hat{A}^l = [\hat{A}^{k+1} - (k+1)\hat{A} + kI](I-\hat{A})^{-2} \tag{3.39}$$

$$\sum_{l=0}^{k-1} \hat{A}^l = (I-\hat{A}^k)(I-\hat{A})^{-1}. \tag{3.40}$$

Substituting the right-hand sides of Eqs. (3.39), (3.40), and (3.18) into Eq. (3.38), we obtain the closed-form formula:

$$X_{k+1}^{-} = \hat{A}^k E X_1^{-} + [\hat{A}^{k+1} - (k+1)\hat{A} + kI](I-\hat{A})^{-2}J + (I-\hat{A}^k)(I-\hat{A})^{-1}(H-0.5J), \tag{3.41}$$

which is valid for $k = 1,2,3,\ldots$, and represents the overall state just before the switching moment $t = kT + \frac{T}{2}$ in terms of the state just before the first switching moment initial state. For a stable closed-loop system, where the eigenvalues of matrix $\hat{A}$ are within the unit circle, $\lim_{k\to\infty} \hat{A}^k$ is zero and (3.41) is reduced to the following relationship:

$$\lim_{k\to\infty} X_{k+1}^{-} = (I-\hat{A})^{-1}[J(k+0.5) + H - (I-\hat{A})^{-1}J] \tag{3.42}$$

Using (3.42) with k replaced by $k-1$, (3.20), and (3.30), the steady-state expression for $X(t)$ is obtained as

$$\lim_{k\to\infty} X\left(kT - \frac{T}{2} + t'\right) = Q(t')k + U(t'),\ t' \in (0,T), \tag{3.43}$$

where

$$Q(t') = e^{At'}[\hat{I}(I-\hat{A})^{-1}J + L + 2A^{-1}Ba_0] - 2A^{-1}Ba_0 \tag{3.44}$$

$$U(t') = e^{At'}[\hat{I}(I-\hat{A})^{-1}(H-0.5J) - \hat{I}(I-\hat{A})^{-2}J - 0.5L] + [(e^{At'}-I)(A^{-1}-\frac{T}{2}I) - t'I]A^{-1}B\alpha. \tag{3.45}$$

Since the DC gain of the equivalent plant in Fig. 3.9 is unity, and the DC gain from input v to state vector X_e is $-A_p^{-1}B_p = [1,0,\ldots,0]^T$, the DC gain of the closed-loop system from input r to the overall state X, considering no signal transformation, is

$$-A^{-1}B = \begin{bmatrix} -A_p^{-1}B_p\delta_c \\ V \end{bmatrix} = \begin{bmatrix} \delta_c \\ 0_{(n_p-1)\times 1} \\ V \end{bmatrix}, \tag{3.46}$$

where vector V describes the closed-loop DC gain from r to the compensator state X_c, and δ_c is the closed-loop DC gain from input r to output y. Using (3.24), (3.22), (3.46), and the fact that $\hat{I}^{-1} = \hat{I}$, the coefficient of k in (3.43) can be simplified to

$$Q(t') = e^{At'}\hat{I}(I-\hat{A})^{-1}L(\delta_c - 1) - 2A^{-1}Ba_0. \tag{3.47}$$

Using (3.8), (3.1), (3.15), and (3.43), one can show that the plant state in steady-state can be described as

$$\lim_{k\to\infty} X_p\left(kT - \frac{T}{2} + t'\right) = g_1\left\{[I_{n_p},0]Q(t') + 2A_p^{-1}B_p a_0\right\}k + g_1[I_{n_p},0]U(t'). \tag{3.48}$$

Using equality $-A_p^{-1}B_p = [1,0,\ldots,0]^T$, (3.47), and (3.46) in (3.48), the coefficient of k in the steady-state solution of the plant state vector can be expressed in the following form:

$$(-1)^k P(t')(\delta_c - 1), \tag{3.49}$$

which reveals that the plant state tends to a bounded value if and only if either the closed-loop DC gain δ_c is unity, or all the elements in the time-dependent vector $P(t')$, defined in (3.37), are identically zero.

Since condition (3.37) is almost impossible to occur, condition (3.36) is almost a necessary condition for boundedness of the plant state. The plant does not have any pole at the origin because of its unity DC gain. Hence, the only way for the closed-loop system to have a unity DC gain is that the compensator has at least one pole at the origin. Thus, a *sufficient and almost necessary* condition for boundedness of the plant state is that the compensator has at least one pole at the origin.

In the more general case of an arbitrary bounded reference signal x_d, using (3.34), the constant vector $H - 0.5J$ in (3.38) should be replaced by the bounded vector $W(k-l,T) - 0.5EL$. In this case, the last term in the right-hand side of (3.41) should be replaced by $\sum_{l=0}^{k-1}\hat{A}^l[W(k-l,T) - 0.5EL]$, which will not grow with k, because the state matrix $\hat{A}$ in the discrete-time LTI system (3.34) is stable. In this way, all of the terms in the right-hand sides of Eqs. (3.42), (3.43), and (3.48), which grow with k, remain unchanged. Hence, the aforementioned condition about the boundedness of the plant state is not restricted to the triangular desired waveform and is valid for any arbitrary bounded reference signal x_d as well.

Theorem 3.5. *Assuming unity DC gain for the plant, signal transformation parameters (3.1), and eigenvalues of matrix* $\hat{A}$ *within the unit circle, a sufficient and almost necessary condition for the plant state to remain bounded in the hybrid control system is that the compensator has at least one pole at the origin.*

Chapter 4
2DOF Control Design for Nanopositioning

Chibum Lee, Gayathri Mohan, and Srinivasa Salapaka

Abstract. This chapter will focus on control systems theoretic analysis and synthesis that significantly expand the range of performance specifications and positioning capabilities of scanning probe microscopes (SPMs). We will present a systems theory framework to study fundamental limitations on the performance of one and two degree-of-freedom (DOF) control designs for nanopositioning systems. In particular, this chapter will present, compare, and analyze optimal-control methods for designing two-degree-of-freedom (2DOF) control laws for nanopositioning. The different methods are motivated by various practical scenarios and difficulty in achieving simultaneously multiple performance objectives of resolution, bandwidth, and robustness by tuning-based or shaping of open-loop-plants based designs. The analysis shows that the primary role of feedback is providing robustness to the closed-loop device whereas the feedforward component is mainly effective in overcoming fundamental algebraic constraints that limit the feedback-only designs. Experimental results indicate substantial improvements (over 200% in bandwidth) when compared to optimal feedback-only controllers.

4.1 Introduction

Positioning of one component with respect to the other with high precision is a pivotal requirement in many micro/nano-scale studies and applications. For instance ultra-high precision positioning systems are crucial to auto focus systems in optics (26), disk spin stands and vibration cancellation (16; 34; 9) in disk drives, wafer and mask positioning in microelectronics (39; 40; 37), piezo hammers in precision mechanics (19), cell penetration and micro dispensing devices in medicine and biology (24), and to the large class of studies and applications enabled by scanning probe microscopes (SPMs). Recently, there is an added impetus on design of nanopositioning systems since they form the bottleneck in terms of speed and accuracy of most

Chibum Lee · Gayathri Mohan · Srinivasa Salapaka
University of Illinois, Urbana Champaign
e-mail: {clee62,gmohan2,salapaka}@illinois.edu

E. Eleftheriou & S.O.R. Moheimani (Eds.): Cntrl. Tech. for Emerging Micro/Nanoscale Sys. LNCIS 413, pp. 67–82.
springerlink.com

devices for nano-investigation, especially in SPMs. For example, there is considerable interest in designing positioning systems for atomic force microscopes (AFMs), since the positioning resolution and tracking bandwidth of positioning systems is typically few orders less than the imaging resolution and bandwidth that microcantilever probe provides. Besides high precision positioning, most nano-scientific studies and applications impose severe demands on the tracking bandwidth and reliability in terms of repeatability of experiments. High tracking bandwidth is required since many studies, especially in biology and material science, require assaying matter with nanoscale precision over areas with characteristic lengths that are typically three orders or more. Repeatability of experiments is essential for validation of the underlying studies. This requirement translates to robustness of positioning systems to modeling uncertainties and operating conditions. Devices that are insensitive to (robust to) diverse operating conditions give repeatable measurements, and are hence reliable.

Typical nanopositioning systems comprise of a flexure stage that provides frictionless motion through elastic deformation, an actuator, typically made from piezoelectric material that provides the required force to deform the flexure stage and/or sensing system along with the control system. The main challenges to the design of robust broadband nanopositioning systems come from flexure-stage dynamics that limit the bandwidth of the positioning stage, from nonlinear effects of piezoactuation such as hysteresis and creep that are difficult to model, and from sensor noise management issues in control feedback that can potentially hamper the tracking-resolution of the device.

The existing literature on designs for improving nanopositioning systems can largely be characterized as those that aim at redesigning the flexure stage mechanisms and those that propose new control methods for existing stages (28; 6). The most common redesign approach is to design smaller and stiffer flexure stages which result in higher resonance frequencies, and therefore provide better tracking-bandwidths (30; 17). However, the downside of this approach is that the resulting smaller, stiffer devices have reduced traversal ranges. The control approaches for better positioning performance can be broadly classified into feedforward and feedback control. A comprehensive review of these designs and their success in increasing positioning resolution is presented in (28; 6). Many of these designs attenuate the hysteresis and creep effects, common in piezo-actuated positioning systems, by carefully modeling these phenomena and compensating them through model-inversion based control designs (13; 36; 11; 4). However, the piezoelectric nonlinearities are sensitive even to small changes in operating conditions such as, ambient temperature, residual polarization in piezo-actuators, and the reference point about which the flexure stage is calibrated; thus making it difficult to estimate accurately their effects and compensate for them. Therefore, many of these designs result in devices that are sensitive to experimental conditions and do not guarantee repeatability of experiments.

Feedback-based control design offers significant compensation against the nonlinear effects of the piezoelectric actuator without explicit incorporation of their models (27; 28). With a careful feedback design, the nonlinear effects of piezo-actuation can be robustly attenuated even when the measurements are noisy; that is, the net gains in resolution from the suppression of unmodeled piezoelectric effects can be made significantly larger when compared to the negative effects of feeding back noisy measurements (5). Some of these designs have been reviewed in (28). The feedback-only control designs show considerable improvements in positioning resolution, tracking bandwidth, and reliability of positioning systems, however are limited by certain algebraic structure and hardware constraints. These improvements as well as limitations can be explained by analyzing the control block diagram schematic of a typical feedback-only nanopositioning system shown in Figure 4.1. In terms of this diagram, the tracking error is given by $e = Sr + Tn$ where $r, n, S = 1/(1+GK)$, and $T = GK/(1+GK)$ represent the reference, the measurement noise, the sensitivity and the complementary sensitivity transfer functions. Since the complementary sensitivity transfer function T determines the effect of the measurement noise on the tracking error, high-resolution positioning objective requires the control design K to be such that the magnitude $|T|$ is small over the range of frequencies where noise is dominant; more specifically, small values of the roll-off frequency (ω_T) of T guarantee smaller bandwidth of T and therefore good resolution. On the other hand, designing the control K to achieve small values of $|S|$ over large frequency ranges results in a large tracking bandwidth. However the control design K is constrained by the feedback-only structure of the block diagram 4.1(a). For instance, the algebraic limitation $S(j\omega) + T(j\omega) = 1$ at all frequencies ω implies that no control design can achieve small values for the sensitivity and complementary sensitivity simultaneously at the same frequency. A control framework discussed in (21) establishes and quantifies the trade-offs between the performance objectives, and suggests feedback controller designs when feasible.

The Bode integral law (3; 10) poses another fundamental constraint on achieving the performance objectives in a nanopositioning system. According to this law the integral $\int_0^\infty \log|S(j\omega)|d\omega = 0$, for any stable system, relative degree greater than or equal to two (the relative degree condition is generally true for most positioning system models) (35). In a typical positioning system, in fact this limitation holds over a finite frequency range. This restriction implies that any control design that achieves small sensitivity function S (for small tracking error) at a certain frequency range necessarily implies that it will result in large values of S (high-tracking

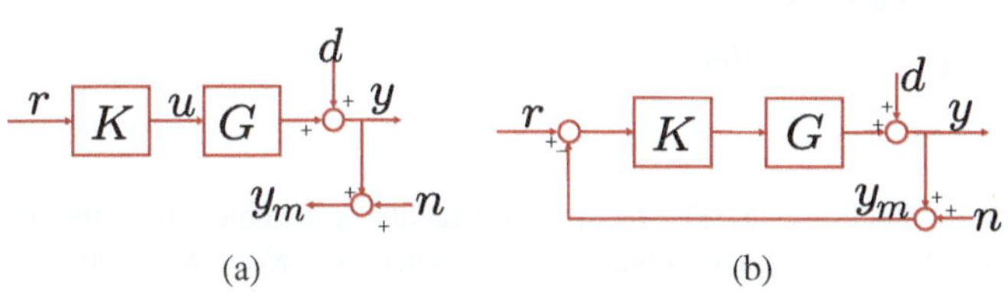

Fig. 4.1 Block diagram schematics for nanopositioning systems: The controller acts only on the reference input signal in open-loop nanopositioning systems shown in (a) while it has access to the difference between the reference and the position y in the closed-loop (feedback-only) positioning system as shown in (b)

error as well as low robustness to modeling uncertainties) at some other frequency ranges. Another important fundamental limitation on control design with respect to the nanopositioning systems is that the complementary sensitivity bandwidth (ω_T) is greater than the bandwidth ω_b of the sensitivity function. Since ω_b signifies tracking bandwidth and ω_T noise-rejection bandwidth, this limitation works against the performance objectives. Feedback controller designs in (31; 27; 33; 2) demonstrate robustness to modeling uncertainties using H_∞ control techniques, and also address the problem of simultaneously achieving the performance objectives. In (7), a multi-input multi-output control design approach is shown that aims to decouple the different positioning directions, and establish a trade-off between the bandwidth and robustness needs.

Some control schemes that combine both the feedforward and feedback techniques are referred to as 2DOF designs. In (14), iterative learning control (ILC) is coupled with feedback control to demonstrate improvements over purely feedback based designs. An inversion based feedforward controller, to compensate for hysteresis effects, along with a feedback controller for pole placement are used in (1) to achieve better tracking.

4.2 Two Degree-of-Freedom Design

The 2DOF control design is analogous to a generalized version of feedback control where the reference inputs and the measurements are processed independently by the controller. The 2DOF controller can take several structures as depicted in Figure 4.2.

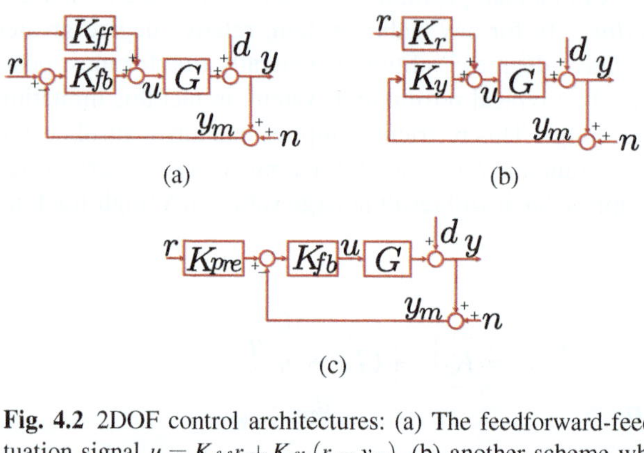

Fig. 4.2 2DOF control architectures: (a) The feedforward-feedback scheme where the actuation signal $u = K_{ff}r + K_{fb}(r - y_m)$, (b) another scheme where $u = K_r r + K_y y_m$, and (c) prefilter architecture where $u = K_{fb}(K_{pre}r - y_m)$. The schemes (a) and (b) are equivalent as control designs in that one can be retrieved exactly in terms of the other. Practical implementable designs for controllers in (a) and (b) can easily be derived from control design in (c), however the vice-versa may require certain factorization procedures.

We primarily consider the 2DOF design shown in Figure 4.2(a) where the control input $u = K_{ff}r + K_{fb}(r - y)$. In this design, the transfer function from d to y that characterizes the robustness to modeling uncertainties is determined by the sensitivity function $S = (I + GK_{fb})^{-1}$, and the transfer function from n to y that characterizes resolution is determined by the complementary sensitivity function $T = (I + GK_{fb})^{-1}GK_{fb}$. Although these characteristics are similar to feedback-only design, the transfer function from r to y and from n to y are distinct and can be designed independently in 2DOF design. The 2DOF design achieves better trade-offs between performance objectives owing to this increased freedom in design. The transfer functions from r to y and r to e are denoted respectively by $S_{er} = S(T - GK_{ff})$ and $T_{yr} = SG(K_{ff} + K_{fb})$. The closed-loop signals namely position, tracking error and control signal, in terms of the new transfer functions S_{er} and T_{yr} are given by, $y = T_{yr}r - Tn + Sd$, $e = S_{er}r + Tn - Sd$, $u = S(K_{ff} + K_{fb})r - SK_{fb}n - SK_{fb}d$ respectively.

Good tracking, a primary performance objective in nanopositioning problems, can be achieved by making the tracking error ($e = S_{er}r + Tn - Sd$) small, which is effected by designing S_{er}, T and S to be small in frequency ranges where the signals r, n and d are dominant respectively. Designing $|T|$ to have small roll-off frequency and high roll-off rates ensures good resolution, keeping $|S_{er}|$ small over large frequency ranges helps achieve large bandwidth and designing the peak magnitude value of S to be close to 1 improves robustness to modeling uncertainties. Here, the norm $||S||_\infty$ gives a measure of the robustness and higher its value lower will be the robustness.

Even though 2DOF control designs provide for larger space of achievable performance specifications over feedback-only designs, they also are constrained by fundamental limitations, both practical and algebraic. Since robustness to modeling uncertainties, and positioning resolution in these designs are still decided by feedback-only transfer functions (S and T respectively), constraints such as Bode-integral law also limit the achievable performance in 2DOF designs.

4.2.1 Feedforward Control Design for Fixed Feedback System

Commercially available nanopositioning systems are typically fitted with pre-designed feedback controllers. These feedback controllers K_{fb} are often not accessible for modification or re-design by users. However, in such cases it is possible to append the system with a feedforward controller in the form of a prefilter on the reference signal (see Figure 4.3). This feedforward component K_{pre} can be designed such that the closed-loop positioning transfer function matches a desired transfer function T_{ref}. The target transfer function T_{ref} is chosen such that it reflects the desired performance objectives such as good resolution and high tracking bandwidth. Additionally, desired transient characteristics including settling times and overshoots, can be attained by choosing an appropriate T_{ref}. Now since the closed-loop transfer function is driven to behave like T_{ref}, it inherits the transient characteristics too.

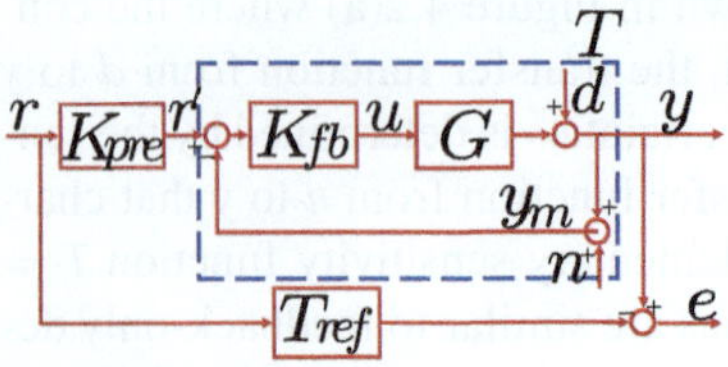

Fig. 4.3 Model matching through prefilter problem

The closed-loop transfer function of the positioning system is given by TK_{pre}. The transfer functions T_{ref} and T are typically stable and proper. A mismatch transfer function $E = T_{ref} - TK_{pre}$ is defined that captures the error between the actual and desired transfer functions. Subsequently, the feedforward component K_{pre} is deduced by posing a minimization problem on the mismatch transfer function E (in the $||.||_\infty$ sense). Small values of $\|E\|_\infty$ guarantees small values for mismatch error signal (see Figure 4.3) given by

$$e = T_{ref}r - y = (T_{ref} - TK_{pre})r. \tag{4.1}$$

This minimization problem at hand becomes trivial if T is a minimum phase transfer function, that is if it has only stable zeros. In this case, the solution K_{pre} can be deduced as $T^{-1}T_{ref}$. However, typical nanopositioning systems are flexure-based with non-collocated actuators and sensors, which give rise to non-minimum phase zeros of T. In such cases, an alternate interpretation of the model matching problem is considered. The model matching problem is analogous to minimizing the parameter γ such that $\|T_{ref} - TK_{pre}\|_\infty \leq \gamma$. If the optimal solution exists the minimum is given by $\gamma = \gamma_{opt}$ for some stable K_{pre}. A further re-formulation of this problem is to find a stable K_{pre} such that $\|E_\gamma\|_\infty \leq 1$, where $E_\gamma = \frac{1}{\gamma}(T_{ref} - TK_{pre})$ for $\gamma > 0$. For stable K_{pre}, the quantity E_γ satisfies the interpolating conditions $E_\gamma(z_i) = \frac{1}{\gamma}T_{ref}(z_i)$ for every non-minimum phase zero z_i of the scanner G. The feedforward component K_{pre} can then be obtained by solving this optimal-control problem by using the Nevanlinna Pick interpolation theory (8).

In Figure 4.4, a 2DOF design comprising a pre-designed feedback controller (designed using $\mathscr{H}_\infty$ optimal framework) and a prefilter (designed using the Nevanlinna-Pick solution) is presented. Its performance is compared to that of the pre-designed feedback-only controller. Since both the 2DOF and feedback-only designs use the same feedback components the resolution and robustness to modeling uncertainties are identical for both cases. Therefore the comparison is made by means of the corresponding sensitivity transfer functions S and S_{er} (refer Figure 4.4), which represent the transfer function from the reference signal to the tracking error. The 2DOF controller shows a significant improvement (330%) over the feedback-only design with respect to tracking bandwidth (see (21) for details).

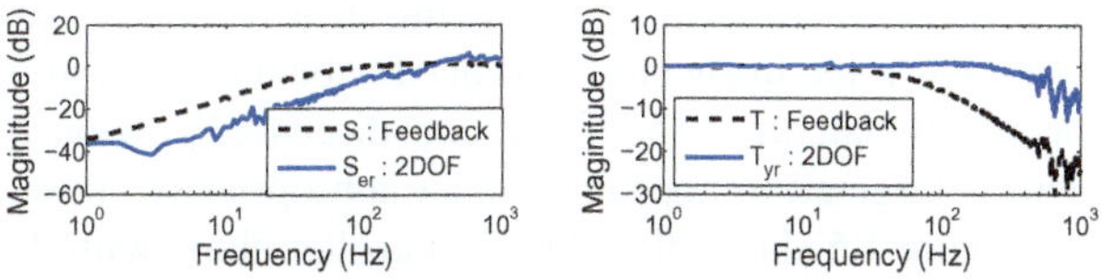

(a) (b)

Fig. 4.4 Comparison of experimentally obtained magnitude of $S(s)$ and $T(s)$ from $\mathscr{H}_\infty$ feedback-only control design (dashed) with $S_{er}(s)$ and $T_{yr}(s)$ from prefilter model matching 2DOF control (solid). The feedforward controller designed using prefilter model matching design achieves over 330% improvement in the tracking bandwidth of the closed-loop design. The robustness and resolution are determined by the feedback components S and T, and therefore remain the same for the two cases.

4.2.2 Improving Robustness to Operating Conditions for Given Feedback System

The pre-designed feedback-only controllers present in commercial scanners are developed for satisfactory resolution and bandwidth when operated near nominal operating conditions. The flexure based scanners are extremely sensitive to changes in operating conditions such as ambient temperature, humidity and vibrational noise. Therefore, when there are deviations from the nominal operating conditions, the performance of the feedback-only controller K_s tends to deteriorate. The robustness to such uncertainties can be improved by constructing an additional feedback controller over the existing one. One such scheme that uses the Glover-McFarlane method (12; 23) is illustrated in (33) and demonstrates substantial increase in robustness.

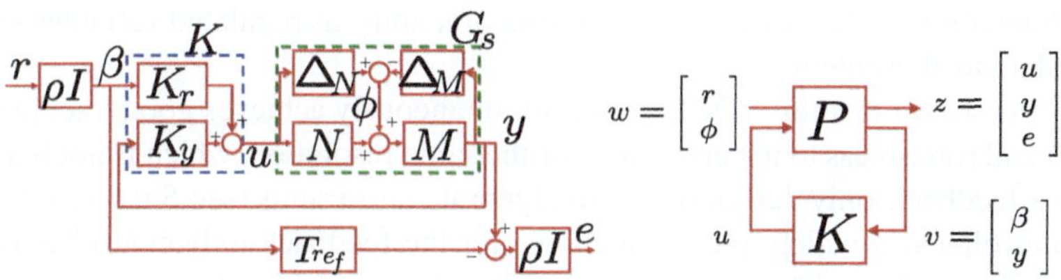

(a) (b)

Fig. 4.5 Control design for model matching and robustness to modeling errors: (a) The signals u, y and e that represent the control signal, the noise component in the position signal, and the mismatch error are chosen as regulated variables z. The transfer function from the reference signal r to these regulated variables z reflects the performance objectives of bounded control signals, noise attenuation, as well as the model matching. The transfer function from the effects of modeling error ϕ to z represents the effect of modeling errors (unmodeled dynamics) on performance objectives. (b) To achieve robust performance, a control design $K = [K_r\ K_y]$ which minimizes the $\mathscr{H}_\infty$-norm of the transfer function from w to z is sought through the optimal control problem.

This section discusses a 2DOF control design comprising of a feedback controller designed for robustness and a feedforward controller for better bandwidth. The 2DOF design adopts the model matching (15; 22) scheme discussed previously where the closed-loop system is targeted to approximate an appropriately chosen model transfer function T_{ref}. Figure 4.5 shows the 2DOF controller along with the plant (scanner) G. The nominal plant model is represented by G_s and the robustness objective implies that the controller based on the optimal solution should be stabilizing for plant models that are in the *neighborhood* of G_s. Using co-prime factorization (38) the nominal plant can be written in terms of functions M and N as $G_s = M^{-1}N$. The vector function $[\Delta_M \ \Delta_N]$ accounts for the unmodeled dynamics in the system. The robustness condition on the plant can now be translated to the set described by G_p,

$$\{G_p | G_p = (M - \Delta_M)^{-1}(N + \Delta_N), \text{ where } \|[\Delta_M \ \Delta_N]\|_\infty \leq \gamma^{-1}\}, \tag{4.2}$$

where γ specifies a bound on the uncertainty. An appropriate optimal control problem, in this scenario, seeks a stabilizing controller that is robust to modeling uncertainties in the system and minimizes the mismatch function (difference between the transfer function from r to y and the target transfer function T_{ref}) under a preferred norm. The target transfer function T_{ref} (similar to Section 4.2.1) is chosen to satisfy response characteristics that are desired for the closed-loop transfer function.

The optimal control problem posed minimizes the transfer function $\Phi(K)$ from signals r and ϕ (an exogenous disturbance signal that captures the effects of uncertainties, see Figure 4.5) to the closed-loop variables $[u^T \ y^T \ e^T]^T$ under the $\mathscr{H}_\infty$ norm (21). A weighting factor ρ maybe used to decide the relative emphasis on the robustification aspect over model matching during optimization. Here, minimizing the norm of $\Phi(K)$ ensures that the effect of ϕ on the mismatch (error) signal e is minimized. As a consequence, the mismatch signal becomes insensitive to modeling uncertainties. Also the constraints on the saturation limit of control and the problem of noise attenuation are handled by the inclusion of u and y as regulated variables in the optimal control problem.

In (21) it is shown that this 2DOF design simultaneously achieves good tracking bandwidth and robustness to uncertainties, for the same resolution, which is not feasible with a feedback-only design owing to algebraic constraints (see Section 4.1). The PII controller is a widely prevalent choice for the feedback-only controller in nanopositioning systems. This can be attributed to the ease of implementation and ability of the PII controller to track ramp signals (representative of triangular raster scans). Figure 4.6 compares the performance of the feedback-only controller (PII) and the 2DOF controller in terms of the sensitivity and complementary sensitivity functions of the corresponding closed-loop systems. It is shown that there is over 64% improvement in bandwidth (S_{er} over S). The improvement in robustness is illustrated through comparison of $S(j\omega)$, the value of $||S||_\infty$ is 1.52 for the 1DOF and 1.21 for the 2DOF design. As mentioned earlier, the smaller the value of $||S||_\infty$, higher the robustness.

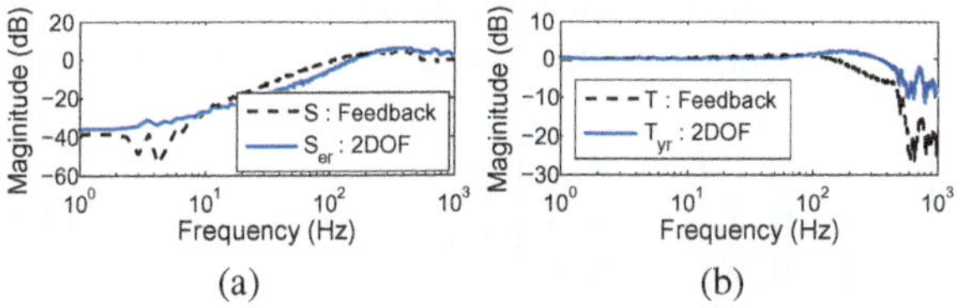

Fig. 4.6 Comparison of experimentally obtained magnitude of $S(s)$ and $T(s)$ from PII feedback-only control design (dashed) with $S_{er}(s)$ and $T_{yr}(s)$ from 2DOF optimal robust model matching control (solid). The 2DOF optimal robust model matching controller achieves over 64% improvement in the tracking bandwidth.

4.2.3 Simultaneous Feedback and Feedforward Control Design

For the 2DOF design considered in this section, an optimal control problem is cast that simultaneously decides the feedforward and feedback control. It is different from the previous formulations in the sense that the relative duties of feedforward and feedback components are not explicitly specified. The objectives are focused on robust stability, disturbance rejection and noise attenuation. Also, the closed-loop response is designed to mimic a target response $T_{ref}r$ similar to previous sections. The chosen controller configuration is illustrated in Figure 4.7. In order to satisfy the desired objectives, the regulated signals are chosen to be

$$z_m = T_{ref}r - y, \qquad z_s = W_s(r - n - y), \qquad z_t = W_t y,$$

where z_m, z_s, and z_t represent the deviation from target response, the weighted tracking error including noise, and the weighted system output. The objectives therefore require these regulated variables to be small. For higher flexibility in design the transfer functions S and T are weighted by functions W_s and W_t. Following a

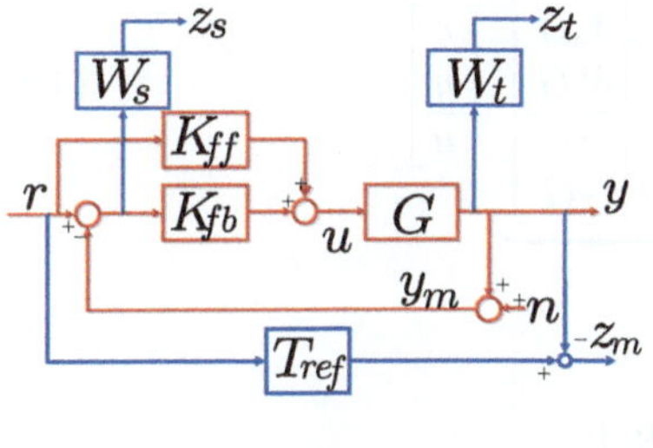

Fig. 4.7 Control design for 2DOF stacked sensitivity with model matching design. The signals z_m, z_s, and z_t represent the deviation from target response, the weighted tracking error with noise, and the weighted system output. The weights W_s and W_t are chosen to reflect the design specifications of robust stability, disturbance rejection and noise attenuation. The target response function T_{ref} is chosen to specify the performance of close loop response to reference signal.

procedure similar to Section 4.2.2, the matrix transfer function from the exogenous inputs, $w = [r^T \;\; n^T]^T$ to the regulated output signals, $z = [z_m^T \;\; z_s^T \;\; z_t^T]^T$ is formulated as

$$\begin{bmatrix} z_m \\ z_s \\ z_t \end{bmatrix} = \underbrace{\begin{bmatrix} T_{ref} - T_{yr} & T \\ W_s S_{er} & -W_s S \\ W_t T_{yr} & -W_t T \end{bmatrix}}_{=\Phi(K)} \begin{bmatrix} r \\ n \end{bmatrix}. \tag{4.3}$$

Now it is possible to pose an optimal control problem to minimize this matrix transfer function under the $\mathcal{H}_\infty$ norm. However, the minimization of this matrix transfer function $\Phi(K)$ also includes minimization of terms such as T, $W_s S_{er}$ and $W_t T_{yr}$ that are unrelated to the specified design objectives. In addition, the algebraic constraints discussed previously ($S + T = I$ and $S_{er} + T_{yr} = I$) impose severe limitations on the feasible controller subspace for the minimization problem. Therefore, instead of casting in terms of the regulated output signals, a different multi-objective approach where the goal is to minimize the transfer functions $T_{ref} - T_{yr}$, $W_s S$ and $W_t T$ is considered. The problem is translated to an optimal control problem with LMI conditions (29).

The problem is reduced to minimizing the transfer functions from r to z_m and from n to $[z_s^T \;\; z_t^T]^T$ and written as

$$\min_{K \in \mathcal{K}} \rho \left\| T_{ref} - T_{yr} \right\|_{\alpha_1} + \left\| \begin{matrix} W_s S \\ W_t T \end{matrix} \right\|_{\alpha_2}, \tag{4.4}$$

where $K = [K_{ff} \;\; K_{fb}]$, $\mathcal{K}$ is the set of stabilizing controllers and the parameter ρ is chosen to decide the relative importance between model matching and robust performance and $\|\cdot\|_{\alpha_i, i \in \{1,2\}}$ can either be the 2-norm or the ∞-norm.

The multi-objective problem can be realized in the generalized control framework as shown in Figure 4.8. The generalized matrix function P in this formulation is given by,

$$\begin{bmatrix} z_m \\ z_s \\ z_t \\ \hline r \\ r - n - y \end{bmatrix} = \underbrace{\left[\begin{array}{cc|c} T_{ref} & 0 & -G \\ W_s & -W_s & W_s G \\ 0 & 0 & W_t G \\ \hline I & 0 & 0 \\ I & -I & -G \end{array}\right]}_{=P} \begin{bmatrix} r \\ n \\ \hline u \end{bmatrix}, \tag{4.5}$$

and is denoted in the state space by,

$$P \equiv \left[\begin{array}{c|cc} A & B_w & B \\ \hline C_z & D_{zw} & D_z \\ C & D_w & 0 \end{array}\right]. \tag{4.6}$$

By adopting this scheme, the redundancies in minimization present initially are overcome. If the controller is assumed to take the form

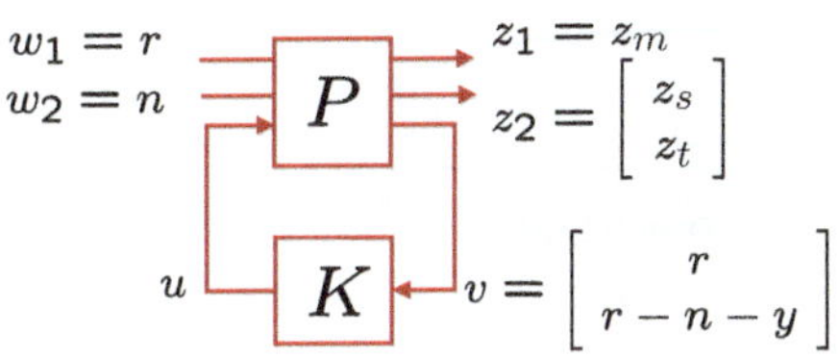

Fig. 4.8 General framework for 2DOF control in Eq. (4.4): The multi-objective optimization problem is to design a controller that minimizes the sum of ρ times the transfer function from w_1 to z_1 and the transfer function from w_2 to z_2. This is in contrast to stacked sensitivity framework that minimizes the transfer function from $w = [w_1^T \; w_2^T]^T$ to $[z_1^T \; z_2^T]^T$.

$$K(s) = [K_{ff}(s) \;\; K_{fb}(s)] \equiv \left[\begin{array}{c|c} A_k & B_k \\ \hline C_k & D_k \end{array}\right], \tag{4.7}$$

then the final closed-loop transfer function can be deduced as,

$$\begin{aligned} \Phi &= \left[\begin{array}{cc|c} A + BD_kC & BC_k & B_w + BD_kD_w \\ B_kC & A_k & B_kD_w \\ \hline C_z + D_zD_kC & D_zC_k & D_{zw} + D_zD_kDw \end{array}\right] \\ &\equiv \left[\begin{array}{c|c} \bar{A} & \bar{B} \\ \hline \bar{C} & \bar{D} \end{array}\right]. \end{aligned} \tag{4.8}$$

In Figure 4.8, the transfer functions from w_i to z_i (for $i = 1,2$) reflect the performance objectives desired and can be re-written in terms of the transfer function Φ in Eq. (4.3) as $\Phi_i = L_i \Phi R_i$, where the matrices L_i and R_i are chosen appropriately. The optimization problem in Eq. (4.4) is recast as the following problem,

$$\min \quad \rho\gamma_1 + \gamma_2 \tag{4.9}$$

when subject to the LMIs corresponding to the required performance objectives for Φ_1 and Φ_2. For instance, we could pose $\mathscr{H}_\infty$ performance conditions on these transfer functions in terms of LMIs. Now, the solution can be obtained by using standard convex optimization techniques following which the controller is deduced from the optimal solution (see (20) for details).

Experimental results illustrate that this 2DOF control design leads to 216% improvement in bandwidth over a 1DOF stacked sensitivity optimal control design, for almost the same robustness and resolution (refer Figure 4.10). Figure 4.9(a) shows that S_{er} achieves good tracking over the 1DOF design with good robustness to modeling uncertainties (note the $\|S\|_\infty$ value). Figure 4.9(b) shows the roll-off frequency of T_{yr} as against T from the experiment.

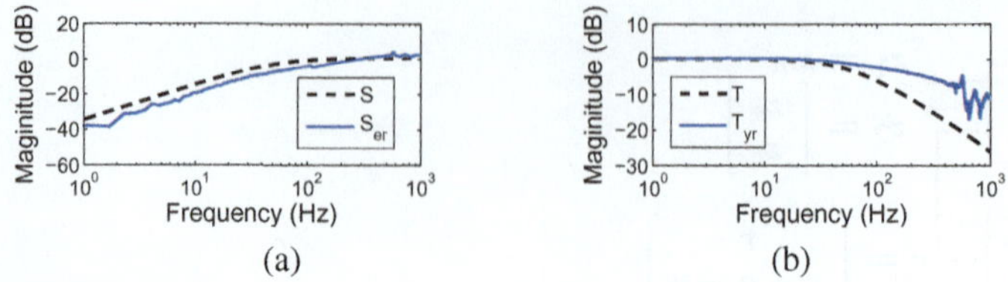

Fig. 4.9 Magnitude of $S_{er}(s)$ and $T_{yr}(s)$ (solid) obtained from experiment with S and T (dashed)

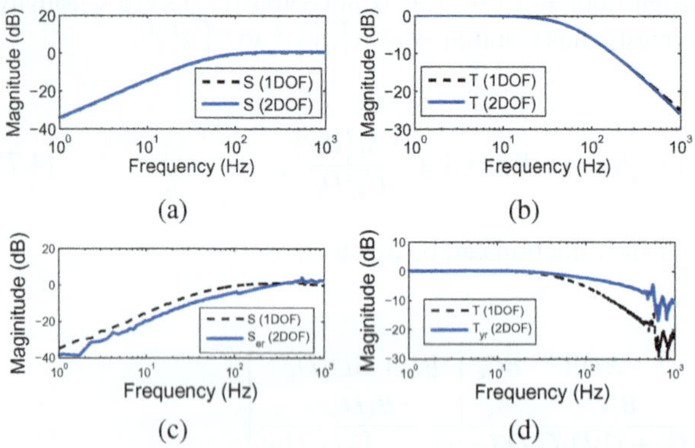

Fig. 4.10 Comparison of $\mathscr{H}_\infty$ feedback-only stacked synthesis and 2DOF multi-objective synthesis control design. (a),(b) $|S|$ and $|T|$ from 1DOF control design (dashed) with $|S|$ and $|T|$ from 2DOF control design (solid). (c),(d) $|S|$ and $|T|$ from 1DOF control design (dashed) with $|S_{er}|$ and $|T_{yr}|$ from 2DOF control design (solid) obtained from experiment. The 2DOF multi-objective synthesis control achieves about 216% improvement in the tracking bandwidth.

4.2.4 Role of Feedforward and Feedback Components

The sensitivity and the complementary sensitivity functions determine the positioning resolution and robustness to modeling uncertainties in both the 1DOF and 2DOF designs. Therefore the direct advantage of the 2DOF design is in its ability to achieve larger tracking bandwidths when compared to 1DOF designs. This advantage also translates to indirect benefits in terms of being able to achieve better trade-offs between the resolution, robustness, and bandwidth objectives. The fundamental limitations that apply to the control design in 1DOF design also apply to the design of the feedback component K_{fb} in the 2DOF design. For instance, the bode integral law described in Section 4.1 does apply to the sensitivity function in 2DOF design and therefore poses a constraint on the design of K_{fb}. The feedforward component on the other hand is not constrained by such limitations, but at the same time does not affect the resolution or the robustness objectives. Therefore to simultaneously achieve the multiple performance objectives, the feedback component plays a primary role as it

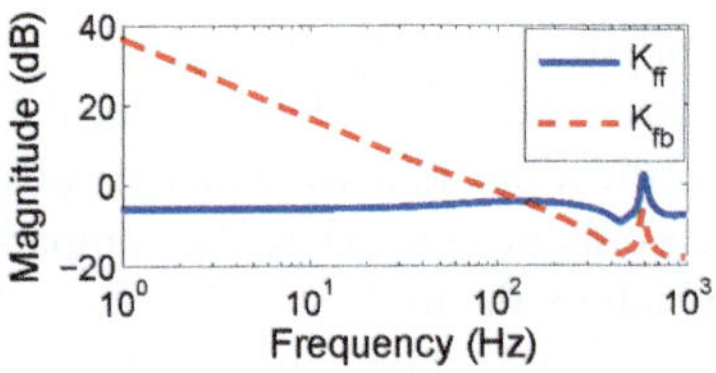

Fig. 4.11 The optimization problem automatically separates the roles of the feedforward and feedback control. The feedback component K_{fb} is dominant in the low frequency region and the feedforward component K_{ff} shows dominance in the higher frequency range.

decides both the resolution and the robustness characteristics in addition to affecting the bandwidth. In this sense, the feedforward component plays a secondary role and is hardly effective in the frequency ranges where feedback component is large (that is where the sensitivity function is small). In fact, the feedforward component K_{ff} becomes prominent only in the frequency ranges where the sensitivity function S can no longer be made small. This division of roles is corroborated by the solutions to the optimal control formulations where both the feedback and feedforward controllers are treated as decision variables with no distinguishing role assigned to them a priori (see Figure 4.11). Therefore the advantage of the feedforward component is primarily in increasing the bandwidth from the corner frequency of S to about the resonance frequencies of the flexure stages. To make the system bandwidth larger than the flexure resonance typically requires a huge control effort and the saturation limit on control becomes a serious limitation.

An analysis of the 2DOF designs show that they achieve specifications that are impossible for the 1DOF designs, that is, some of the fundamental limitations on the 1DOF design do not apply to the 2DOF designs. As stated in Section 4.1, high values of tracking bandwidth ω_{BW} and small values of roll-off frequency (ω_T) of the complementary sensitivity function T are desired for good tracking and noise attenuation. However, in feedback-only controller design this is not achievable, since ω_{BW} cannot be made higher than ω_T owing to fundamental limitations (35). The 2DOF control design on the other hand, is not bound by such restrictions, as can be seen from the experimental illustrations in the previous sections (in optimal prefilter model matching control based on $\mathcal{H}_\infty$ controller ω_{BW} of 214.5 Hz and ω_T of 60.1 Hz while in 1DOF $\mathcal{H}_\infty$ control design the bandwidth is 49.4 Hz with the same roll-off frequency). Additionally, it is shown using the 2DOF designs in this chapter, it is possible to improve the bandwidth and robustness to modeling uncertainties for a given resolution while fundamental limitations pose restrictions on the feedback-only design to achieve this trade-off.

4.3 Summary

In this chapter the role and design of 2DOF control methods for nanopositioning systems are discussed. It is shown that the trade off imposed by fundamental limitations in 1DOF can be relaxed through 2DOF control. The control designs when implemented on an AFM give substantial improvements in bandwidth (as high as 330%) for the same resolution and robustness. The experiments demonstrate that 2DOF design presented here achieved design specifications that were analytically (and therefore practically) impossible for feedback-only designs. Here the comparisons have been made to feedback-only designs which by themselves have obtained significant improvements over commercial devices (27). It is shown that the modern control theoretic framework provides an apt platform to quantify and analyze the design objectives of robustness, bandwidth and resolution as well as to design control laws to achieve them. Even though this chapter is focused on nanopositioning, the tools provided in this chapter are applicable to various applications in nanotechnology (25; 32). For instance, these tools have been applied in a variety of contexts in scanning probe microscopy (18). The underlying framework allows for quantification of desired objectives, analysis of fundamental limitations, computing inherent trade-offs between performance objectives, and designing and solving optimal control problems to achieve best feasible performance.

References

[1] Aphale, S.S., Devasia, S., Moheimani, S.O.R.: High-bandwidth control of a piezoelectric nanopositioning stage in the presence of plant uncertainties. Nanotechnology 19(12), 125–503 (2008)

[2] Bashash, S., Jalili, N.: Robust adaptive control of coupled parallel piezo-flexural nanopositioning stages. IEEE/ASME Transactions on Mechatronics 14(1), 11–20 (2009)

[3] Bode, H.: Network Analysis and Feedback Amplifier Design. Van Nostrand Reinhold, New York (1945)

[4] Croft, D., Shedd, G., Devasia, S.: Creep, hysteresis and vibration compensation for piezoactuators: Atomic force microscopy application. In: Proceedings of the Amercian Control Conference, Chicago, IL, pp. 2123–2128 (2000)

[5] Daniele, A., Salapaka, S., Salapaka, M.V., Dahleh, M.: Piezoelectric scanners for atomic force microscopes: design of lateral sensors, identification and control. In: Proceedings of the American Control Conference, San Diego, CA, pp. 253–257 (1999)

[6] Devasia, S., Eleftheriou, E., Moheimani, S.O.R.: A survey of control issues in nanopositioning. IEEE Transactions of Control Systems Technology 15(5), 802–823 (2007)

[7] Dong, J., Salapaka, S., Ferreira, P.: Robust control of a parallel-kinematic nanopositioner. Journal of Dynamic Systems, Measurement, and Control 130, 041007 (2008)

[8] Doyle, J.C., Francis, B.A., Tannenbaum, A.R.: Feedback Control Theory. MacMillan, New York (1992)

[9] Du, C., Xie, L., Teoh, J.N., Guo, G.: An improved mixed $\mathscr{H}_2/\mathscr{H}_\infty$ control design for hard disk drives. IEEE Transactions on Control Systems Technology **13**(5), 832–839 (2005)

[10] Freudenberg, J.S., Looze, D.P.: Right half-plane poles and zeros and design tradeoffs in feedback systems. IEEE Transactions on Automatic Control 30(6), 555–565 (1985)

[11] Ge, P., Jouaneh, M.: Tracking control of a piezoceramic acuator. IEEE Transactions on Control Systems Technology 4(3), 209–216 (1996)

[12] Glover, K., McFarlane, D.: Robust stabilization of normalized coprime factor plant descriptions with $\mathscr{H}_\infty$-bounded uncertainty. IEEE Transactions on Automatic Control 34(8), 821–830 (1989)
[13] Hatch, A.G., Smith, R.C., De, T., Salapaka, M.V.: Construction and experimental implementation of a model-based inverse filter to attenuate hysteresis in ferroelectric transducers. IEEE Transactions on Control Systems Technology 14(6), 1058–1069 (2006)
[14] Helfrich, B.E., Lee, C., Bristow, D.A., Xiaohui, X., Dong, J., Alleyne, A.G., Salapaka, S.: Combined $\mathscr{H}_\infty$-feedback and iterative learning control design with application to nanopositioning systems. In: Proceedings of American Control Conference, Seattle, WA, pp. 3893–3900 (2008)
[15] Hoyle, D.J., Hyde, R.A., Limebeer, D.J.N.: An $\mathscr{H}_\infty$ approach to two degree of freedom design. In: Proceedings of the IEEE Conference on Decision and Control, pp. 1581–1585 (1991)
[16] Jianxu, M., Ang Jr., M.: High-bandwidth macro/microactuation for hard disk drive. In: Proceedings of the SPIE, vol. 4194, pp. 94–102 (2000)
[17] Leang, K., Fleming, A.: High-speed serial-kinematic AFM scanner: Design and drive considerations. In: Proceedings American Control Conference, pp. 3188–3193 (2008)
[18] Lee, C.: Control-systems based analysis and design methos for scanning probe microscopy. Ph.D. thesis, University of Illinois at Urbana-Champaign (2010)
[19] Lee, C., Lin, C., Hsia, C., Liaw, W.: New tools for structural testing: Piezoelectric impact hammers and acceleration rate sensors. Journal of Guidance, Control and Dynamics 21(5), 692–697 (1998)
[20] Lee, C., Salapaka, S., Voulgaris, P.: Two degree of freedom robust optimal control design using a linear matrix inequality optimization. In: 48th IEEE Conference on Decision and Control (December 2009)
[21] Lee, C., Salapaka, S.M.: Robust broadband nanopositioning: fundamental trade-offs, analysis, and design in a two-degree-of-freedom control framework. Nanotechnology 20(3), 035,501 (2009)
[22] Limebeer, D., Kasenally, E., Perkins, J.: On the design of robust two degree of freedom controllers. Automatica 29(1), 157–168 (1993)
[23] McFarlane, D., Glover, K.: A loop shaping design procedure using $\mathscr{H}_\infty$ synthesis. IEEE Transactions on Automatic Control 37(6), 759–769 (1992)
[24] Meldrum, D.R., Pence, W.H., Moody, S.E., Cunningham, D.L., Holl, M., Wiktor, P.J., Saini, M., Moore, M.P., Jang, L.S., Kidd, M., Fisher, C., Cookson, A.: Automated, integrated modules for fluid handling, thermal cycling and purification of DNA samples for high throughput sequencing and analysis. In: Proc. IEEE/ASME International Conference on Advanced Intelligent Mechatronics, vol. 2, pp. 1211–1219 (2001)
[25] Pantazi, A., Sebastian, A., et al.: Probe-based ultrahigh-density storage technology. IBM Journal of Research and Development 52(4/5), 493–511 (2008)
[26] Rihong, Z., Daocai, X., Zhixing, Y., Jinbang, C.: Research on systems for measurements of CCD parameters. In: Proceedings of the SPIE, vol. 3553, pp. 297–301 (1998)
[27] Salapaka, S., Sebastian, A., Cleveland, J.P., Salapaka, M.V.: High bandwidth nanopositioner: A robust control approach. Review of Scientific Instruments 73(9), 3232–3241 (2002)
[28] Salapaka, S.M., Salapaka, M.V.: Scanning probe microscopy. IEEE Control Systems Magazine 28(2), 65–83 (2008)
[29] Scherer, C., Gahinet, P., Chilali, M.: Multiobjective output-feedback control via LMI optimization. IEEE Transactions on Automatic Control 42(7), 896–911 (1997)

[30] Schitter, G., Astrom, K., DeMartini, B., Thurner, P., Turner, K., Hansma, P.: Design and modeling of a high-speed AFM-scanner. IEEE Transactions on Control Systems Technology 15(5), 906–915 (2007)
[31] Schitter, G., Menold, P., Knapp, H.F., Allgower, F., Stemmer, A.: High performance feedback for fast scanning atomic force microscopes. Review of Scientific Instruments 72(8), 3320–3327 (2001)
[32] Sebastian, A., Pantazi, A., et al.: Nanopositioning for probe storage. In: Proceedings of the American Control Conference, Portland, OR, pp. 4181–4186 (2005)
[33] Sebastian, A., Salapaka, S.: Design methodologies for robust nano-positioning. IEEE Transactions on Control Systems Technology 13(6), 868–876 (2005)
[34] Shim, D., Lee, H.S., Guo, L.: Mixed-objective optimization of a track-following controller using linear matrix inequalities. IEEE/ASME Transactions on Mechatronics 9(4), 636–643 (2004)
[35] Skogestad, S., Postlethwaite, I.: Multivariable Feedback Control, Analysis and Design, 2nd edn. John Wiley and Sons, West Sussex (2005)
[36] Smith, R.C., Hatch, A.G., De, T., Salapaka, M.V., del Rosario, R.C.H., Raye, J.K.: Model development for atomic force microscope stage mechanisms. SIAM Journal on Applied Mathematics 66(6), 1998–2026 (2006)
[37] Verma, S., Jong Kim, W., Shakir, H.: Multi-axis maglev nanopositioner for precision manufacturing and manipulation applications. IEEE Transactions on Industry Applications 41(5), 1159–1167 (2005)
[38] Vidyasagar, M.: Control System Synthesis: A Factorization Approach. MIT Press, Cambridge (1985)
[39] White, D., Wood, O.: Novel alignment system for imprint lithography. Journal of Vacuum Science & Technology B 18(6), 3552–3556 (2000)
[40] Zandvliet, M.J., Scherer, C.W., Hol, C.W.J., van de Wal, M.M.J.: Multi-objective $\mathscr{H}_\infty$ control applied to a wafer stage model. In: Proc. 43rd IEEE Conference on CDC Decision and Control, vol. 1, pp. 796–802 (2004)

Chapter 5
Improving the Imaging Speed of AFM with Modern Control Techniques

Stefan Kuiper and Georg Schitter

Abstract. In Atomic Force Microscopy (AFM), the dynamics and non-linearities of the positioning stage are major sources of image artifacts and distortion, especially when imaging at high-speed. This contribution discusses some recent developments to compensate for these adverse effects of the positioning stage dynamics in high-speed AFM by utilizing modern control methods. The improvements on both the lateral scanning motion and in controlling the tip-sample interaction force are demonstrated to allow significantly faster, and more accurate AFM imaging.

5.1 Introduction

Atomic Force Microscopy (AFM) (7) is an important tool in micro- and nanotechnology to obtain topographical images of samples with molecular or even atomic resolution. In AFM, the sample is probed by a very sharp tip, while scanning in a raster scan-pattern. To measure the interaction force between the tip and the sample, the measurement tip is mounted at the free end of a micro cantilever, of which the deflection is detected with sub-nanometer precision using an optical sensing system (2), as depicted in Figure 5.1. During imaging, the measured interaction force is controlled in a feedback loop, manipulating the vertical position of the sample and/or probing tip (cf. Fig. 5.1).

Stefan Kuiper
Delft Center for Systems and Control,
Delft University of Technology,
Mekelweg 2, 2628CD Delft, The Netherlands
e-mail: stefan.kuiper@tudelft.nl

Georg Schitter
Automation and Control Institute,
Vienna University of Technology,
Gusshausstrasse 27-29, 1040 Vienna, Austria
e-mail: schitter@acin.tuwien.ac.at

E. Eleftheriou & S.O.R. Moheimani (Eds.): Cntrl. Tech. for Emerging Micro/Nanoscale Sys. LNCIS 413, pp. 83–100.
springerlink.com

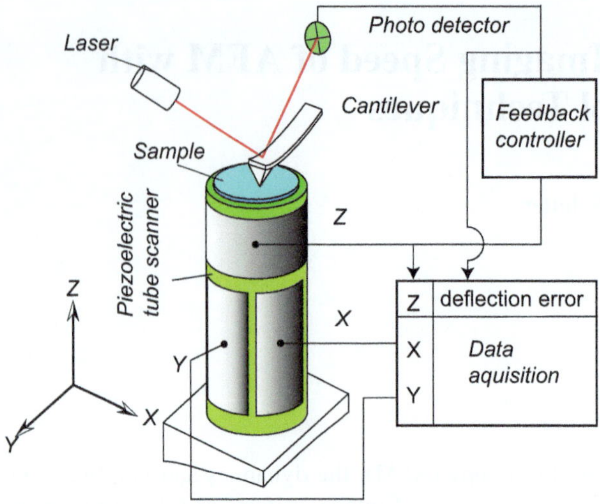

Fig. 5.1 Atomic force microscope with a piezoelectric tube scanner-based positioning stage

The operation of an AFM can be split into two main tasks: (i) to provide the lateral scanning motion between the tip and the sample, and (ii) to control the tip-sample interaction force while scanning. To perform both tasks a positioning stage is used to position the sample and/or measurement tip in all three spatial direction. To achieve the desired image resolution the positioning stage is required to have a precision in the nanometer or even sub-nanometer range.

Although the AFM is capable of providing high resolution topographical images in a wide range of applications and environmental conditions, one of its major drawbacks is its low imaging speed, taking on the order of several minutes per image for most commercially available AFM systems. The low imaging speed makes AFM imaging a very time consuming, and therefore costly process (15). As in AFM the image is obtained by probing the sample point by point, the main limitation on the imaging speed is the bandwidth of the positioning stage. In most AFM systems the positioning stages are based on piezoelectric actuation. Although these piezoelectric actuators provide high resolution, their accuracy is hindered by hysteresis, creep and stage dynamics, especially when operated at high-speed (11; 34; 1). To overcome these limitations considerable research has been done on improving the mechanical and mechatronic design of these piezo-based positioning stages (4; 27; 31), as well as applying modern control methods for both the lateral scanning motion (11; 28; 34; 36), and the vertical feedback loop (39; 32; 18). The ultimate goal is to enable capturing dynamical processes at molecular scale in real time by imaging at video-rate, which requires all the components of the AFM system to be redesigned and optimized in terms of bandwidth and imaging speed (4; 27; 15; 25).

This contribution discusses several AFM developments to allow faster and more accurate AFM imaging, utilizing modern control methods. Section 5.2 presents improved control of the lateral scanning motion, discussing a method to improve the tracking accuracy in an iterative manner over several scanning trails, and a method to provide active damping of the resonant modes of the positioning stage without requiring dedicated position sensors. Section 5.3 discusses dual-actuated control of the tip-sample interaction force, which allows a higher control bandwidth without sacrificing effective positioning range.

5.2 Improved Control of the Lateral Scanning Motion

In AFM imaging the sample or measurement tip is scanned laterally, i.e. in the image plane, in order to obtain a topography image of the sample surface. Although in some applications a sinusoidal scanning motion (25) is used, most often a triangular scanning motion is preferred because of the constant tip-sample velocity during imaging (29).

For most commercially available AFM systems, the positioning of the sample or measurement tip is provided by a piezoelectric tube scanner (8), which consist of a tube of piezoelectric material with segmented electrodes on the side, as shown in Figure 5.1. By applying an antiparallel voltage over these electrodes the piezoelectric tube bends, which generates the lateral scanning motion at the free end of the tube. Although these piezoelectric tube scanners provide high resolution, and are very cost-efficient, a major drawback are the weakly damped resonant-modes of the piezoelectric tube occurring at relatively low-frequencies (10; 34). When scanning at high speed these resonant modes cause scanner oscillations, which are a major source of image distortion. Recently, several improved scanner designs are reported in literature which are optimized for high-speed scanning, and consist of piezoelectric stack-actuators (4; 31).

To improve the scanner accuracy several control methods are investigated to compensate for the scanner non-linearities due to hysteresis, and creep, and to control the scanner oscillations stemming from the scanner's resonant modes. Using dedicated position sensors measuring the scanner displacement, these effects can be compensated for by applying direct feedback control (40; 28; 36). Especially for high accuracy metrological applications this method is preferred as the position of the sample with respect to the measurement probe can be directly traced back using the (interferometric) position sensors (23), given that the system is designed following the Abbé-principles (38). For qualitative imaging applications a high positioning precision is of major importance to achieve the desired image resolution. However, the accuracy requirements in qualitative imaging application are most often not as high as in metrological AFM. Therefore, for qualitative imaging applications the use of dedicated position sensors is not always very cost-efficient. Moreover, for high-speed AFM applications the noise levels of the position sensors are typically too high to be used for direct feedback control, as at the required feedback bandwidths the measurement noise would cause too strong a degradation of the instruments precision. As an alternative, feedforward compensation methods have been proposed to compensate for the non-linearities and scanner dynamics (10; 11; 34). As feedforward compensation methods work in an open-loop manner, the precision of the instrument is typically not compromised by measurement noise. A disadvantage of feedforward compensation methods, however, is that they require accurate knowledge of the dynamical behavior of the system, and are therefore not robust for system variations, occurring for instance when changing the sample and/or measurement tip.

In this chapter, two different control methods are discussed to cope with the lack of low-noise, and cost-efficient displacement sensors in high-speed AFM. Section 5.2.1 discusses a control method to improve the tracking of the scanner stage with an accuracy beyond the sensors' noise-floor by compensating for the scanner dynamics and non-linearities in an iterative learning control approach over several scan lines. Section 5.2.2 discusses a method to actively control the scanner dynamics without requiring dedicated position sensors by using the self-sensing capabilities of the piezoelectric actuators.

5.2.1 Iterative Learning Control

For AFM systems equipped with position sensors in the lateral scanning directions, the tracking errors stemming from scanner non-linearities and system dynamics can be compensated for by direct feedback control (40; 28; 36). The main advantage of feedback control methods is that they do not require detailed knowledge of the systems non-linear effects, such as hysteresis and creep, and provide certain robustness against variations of the system dynamics, occurring for instance when changing the sample and/or measurement probe. A disadvantage of feedback control, however, is that the feedback bandwidth is limited by the phase lag of the system stemming from for instance delays. Moreover, in feedback control the measurement noise of the position sensors is directly fed-back into the system, which can seriously compromise the instruments precision, especially at higher control bandwidths.

As an alternative, one can exploit the fact that the lateral scanning motion is a repetitive motion over several scan-lines, and therefore the tracking errors stemming from the non-linearities and system dynamics are also repetitive over those scan-lines. If sensors are available with sufficient resolution, these repetitive tracking errors can be measured during each scanning trail, and based on that, compensating control actions can be computed for future scanning trails (scan-lines). This is the principle of Iterative Learning Control (ILC) (5), which recently has been demonstrated in AFM (22; 41). The ILC algorithm can be seen as a 'quasi' closed-loop control method, as the measured tracking errors are not fed-back directly but over each scanning trail, without the limitations of direct feedback control. In this section it will be shows that with a small modification to the standard ILC algorithm, the accuracy of the instrument can be improved even beyond the noise floor of the position sensors.

Figure 5.2 shows the block-diagram of the Iterative Learning Control method, with the discretized open-loop system $P(q)$, learning-filter $L(q)$, and robustness-filter $Q(q)$. The signal sequence $u^j(k)$ is the input for the system, $x^j(k)$ is the output sequence of the system, $n^j(k)$ is the measurement noise, and $r(k)$ is the reference

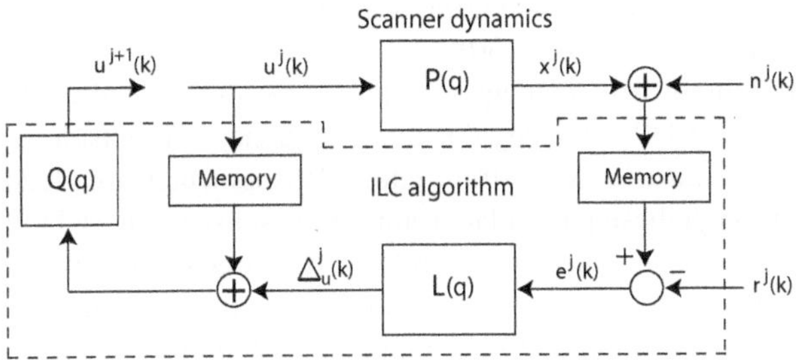

Fig. 5.2 Block diagram of Iterative Learning Control algorithm, with the open-loop system $P(q)$, learning filter $L(q)$, and robustness filter $Q(q)$. Signal $u^j(k)$ is the input sequence to the open-loop system, $x^j(k)$ is the output sequence of the system, $n^j(k)$ is the measurement noise, and $r(k)$ is the reference signal. Discrete time-stamp k is defined within the iteration-domain and superscript j denotes the iteration-number.

signal. Discrete time-stamp k is defined within the iteration-domain and superscript j denotes the iteration-number. The goal of ILC algorithm is to decrease the tracking error $e^j(k) = x^j(k) - r(k)$ asymptotically by iteratively 'learning' the optimal input sequence $\bar{u}(k)$. Critical in the design of the ILC algorithm is to determine the dynamical learning filter $L(q)$ (cf. Fig. 5.2) which converts the measured tracking error sequence from one trail ($e^j_{meas}(k) = x^j(k) + n^j(k) - r(k)$, cf. Fig. 5.2), into a compensating control action for the subsequent trails $\Delta^j_u(k) = L(q) \star e^j_{meas}(k)$. Most often this learning filter is chosen as a sub-optimal inverse of the identified linear dynamics of the open-loop system: $L(q) \approx \hat{P}^{-1}(q)$. The input sequence of the subsequent trail is now computed as $u^{j+1}(k) = Q(q) \star \left(u^j(k) + \Delta^j_u(k)\right)$, with robustness filter $Q(q)$ limiting the control action bandwidth in order to prevent instability of the algorithm due to un-modeled or uncertain dynamics at higher frequencies. Often the robustness filter $Q(q)$ is chosen as a non-causal, zero-phase low-pass filter, whose implementation is made possible due to the off-line computation of the updated input signal.

The ILC algorithm can be seen as a 'quasi' closed-loop control method, where the measured system output is not directly fed-back but by updating the open-loop input sequence over each trail period. Via this non-direct feedback scheme the ILC algorithm is capable of compensating for repetitive tracking errors due, for instance, to system non-linearities and dynamics. Note, however, that the measurement noise $n(t)$ (cf. Fig. 5.2) is still influencing the measurement of the tracking error. If not compensated for, the achievable accuracy in ILC is limited by the noise floor of the sensors (22). However, as the updated input signal $u^{j+1}(k)$ is computed off-line, a small modification to the standard ILC algorithm allows one to average out the sensor noise over several scan-lines before computing the corrective actions (30): $\Delta^j_u(k) = L(q) \star \left[\frac{1}{N}\sum_{i=1}^{N}(x^j_i(k) + n^j_i(k) - r(k))\right]$. Given that the sensor noise $n^j(k)$ is uncorrelated with the scanning signals, averaging the measurement data over several scan-lines can significantly reduce the influence of the measurement noise on the achievable scan accuracy with ILC, allowing the accuracy of the system to be improved beyond the noise floor of the sensor (30).

Figure 5.3 shows the measured displacement of a prototype high-speed AFM scanner (31), as a response to a 1030 Hz triangular input sequence. The displacement of this prototype scanner is measured via strain-gauge sensors which are glued onto the piezo elements. Although these particular strain-gauge sensors have a high noise level of 35 nm_{pp}, they provide high accuracy at relatively low cost. The goal of the ILC algorithm is to find the optimal input sequence in order to track a triangular reference signal of 1030 lines per second at a scan-range of 100 nm (dashed lines in Fig. 5.3), with an accuracy of ≤ 1 nm_{pp}. The ILC algorithm is programmed such that for each iteration the tracking error is determined by averaging the measurement data from the strain-gauge sensors over 512 scan-lines. Figure 5.4 shows the

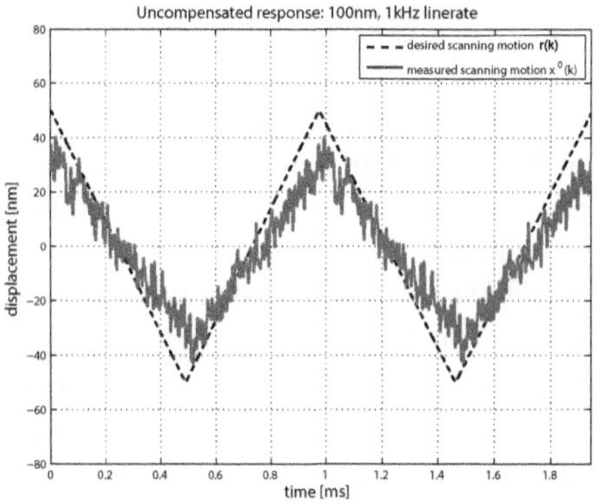

Fig. 5.3 Triangular 100 nm_{pp} 1030 Hz reference signal (dashed), and measured open-loop response of the high-speed AFM scanner using the strain-gauge sensor (grey) (30)

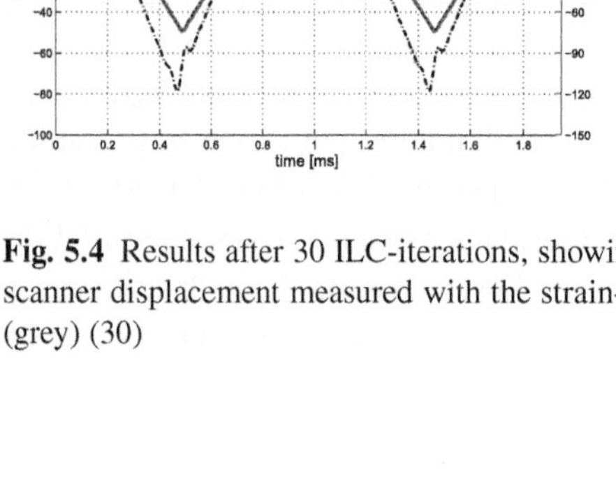

Fig. 5.4 Results after 30 ILC-iterations, showing resulting input signal (dashed-dotted) and scanner displacement measured with the strain-gauge sensors, averaged over 512 scan-lines (grey) (30)

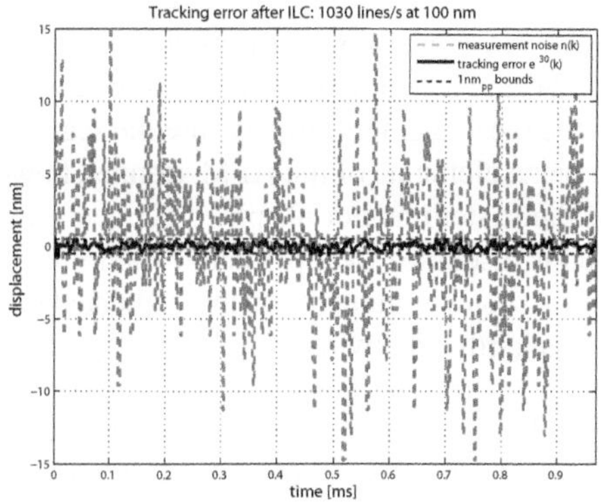

Fig. 5.5 Residual tracking error of the scanning motion shown in Figure 5.4, measured directly with the strain-gauge sensors (grey) and after averaging over 512 scan-lines (solid), demonstrating that the accuracy of the piezo-based positioning stage is improved beyond the noise-floor of the sensor (30)

resulting input sequence and averaged measurement of the scanner displacement after 30 iterations of the ILC algorithm. The oscillations in the input sequence before each turn-around point (cf. Fig. 5.4) demonstrate the compensation of the scanner dynamics of the AFM scanner. Figure 5.5 shows the measured tracking error after 30 iterations of the ILC algorithm averaged over 512 scan-lines, and a realization of the unaveraged measurement noise, showing that a tracking accuracy is achieved of less that 1 nm_{pp}, which is significantly less than the sensor noise.

Application of this control technique enables the use of more cost-efficient position sensors, and allows one to control the scanning motion in AFM at higher bandwidth without compromising the instrument precision by direct feedback of sensor noise, as compared with standard feedback control.

5.2.2 *Self-sensing Piezo Actuation*

The weakly damped resonances in piezo-based positioning stages can cause scanner oscillations when excited, which are a major source of image distortion in AFM (10; 34). To prevent excitation of the weakly damped resonances by the lateral scanning motion, feedforward control methods can be applied (10; 34), or Iterative Learning Control as discussed in Section 5.2. However, with these open-loop control methods the weakly damped resonances can still be excited by environmental noise. Moreover, in AFM the vertical feedback loop controlling the tip-sample interaction force can also excite the weakly damped resonances in the lateral plane via the strong dynamical coupling between the positioning axes, which occurs at the resonance frequencies (26).

The adverse effects of scanner oscillations can be significantly reduced by applying active damping to the resonant modes of the scanner, using position sensors to measure the scanner displacement, and applying active feedback control (40; 6). The use of dedicated position sensors, however, is not very cost-efficient, and for some AFM systems difficult to realize. In (24) the use of dedicated position sensors is avoided by using separate electrodes on the piezo elements to measure the scanner displacement. These modifications on the piezo elements of the AFM scanner, however, may compromise the scan-range of the system as not all piezo elements are used for actuation. In this section a method is discussed to apply active damping of both scanning axes of a piezo-based scanning stage, without requiring dedicated position sensors or modification of the systems hardware, by using self-sensing piezo actuation.

A piezoelectric element is a transducer, transforming energy from the electrical domain to the mechanical domain and vice versa. By applying a voltage over the electrodes of the piezo element, a force is generated within the piezo structure, which in the un-clamped case results in an elongation of the piezo element. Likewise, when applying an external force to the piezo structure, a voltage potential is generated over its electrodes, allowing the piezo element to be used as a force sensor. The idea of self-sensing piezo actuation is to use the piezo element both as an actuator and sensor simultaneously, as first described by (12), and recently applied in AFM (14; 21). To allow self-sensing actuation the piezo element can be connected in a capacitive bridge circuit, as shown in Figure 5.6, where the capacitor ratios in both branches are chosen equal. The voltage generated by the external forces acting on the piezo element can be directly measured across both branches of the bridge circuit (V_m, cf. Fig. 5.6), while in the same time the piezo element can be used as an actuator by applying a voltage over the bridge circuit (V_i, cf. Fig. 5.6).

Figure 5.7a (black lines) shows the Bode magnitude response of the displacement in both scanning axes of a piezoelectric tube scanner (E-scanner, Veeco, Santa Barbara, USA), measured with a vibrometer (Polytec, Waldbronn, Germany), clearly revealing the resonance peaks at 3 kHz in both axes as well the strong cross-couplings at those frequencies.

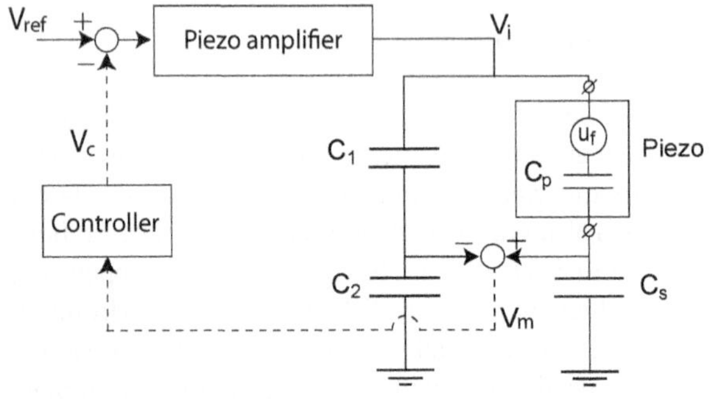

Fig. 5.6 Schematic description of self-sensing piezo actuation, showing the bridge circuitry, piezo amplifier, and feedback controller

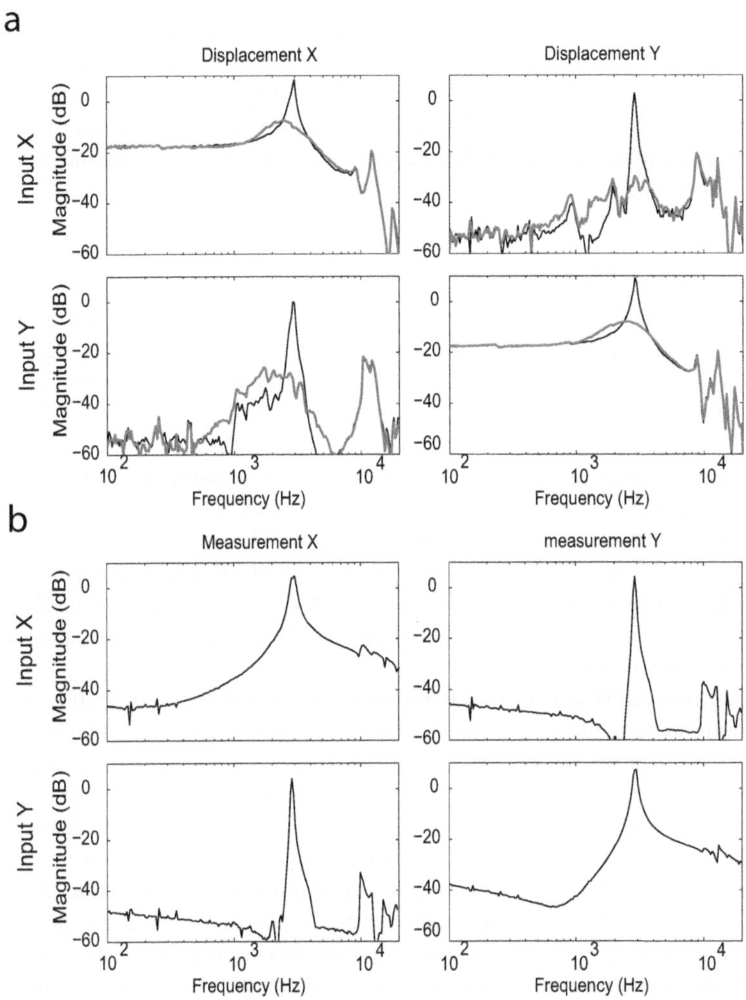

Fig. 5.7 (a) Bode magnitude response of the measured scanner displacement of a piezoelectric tube scanner measured with a vibrometer, uncontrolled (black) and controlled via self-sensing piezo actuation (grey). (b) Bode magnitude response of the measurement signals stemming from the self-sensing bridge circuitry in which the piezoelectric tube scanner is connected (21).

Figure 5.7b shows the Bode magnitude response of the measurements signals obtained from the self-sensing bridge circuitry in which the piezoelectric tube scanner is connected. The resonance frequencies are clearly resolved in the measurement signals, as well as the mechanical coupling at those frequencies. The main advantage of self-sensing piezo actuation is that it provides this direct information of the dynamical behavior of the scanner without requiring dedicated position sensors, which would increase the cost of the AFM system. The measurement signals in Figure 5.7b show a quasi second order derivative action with respect to the scanner displacement (Fig. 5.7a). This is explained by the fact that the measurement signals from the self-sensing bridge circuitry are generated by the external forces on the piezo-element, which are proportional to the accelerations of the scanner mass (20). As the transfer function towards the scanner acceleration shows a second order derivative action with respect to the scanner displacement, the signal to noise ratio of the measurement signal in self-sensing piezo actuation gets better for increasing frequency in comparison with normal position sensors. The obtained measurement signals from the self-sensing bridge circuitry are therefore particulary useful for selective damping of the high frequency resonant modes via active feedback control (cf. Fig. 5.6, dashed lines). To guarantee the integrity of the measurement signals obtained from the self-sensing bridge circuitry under influence of component mismatch and scanner hysteresis, an adaptive circuit balancing technique is utilized (9; 21).

Figure 5.7a (grey lines) shows the results of adding active damping to the scanner's resonance modes, by using the measurement signals from the self-sensing bridge circuitry for direct feedback control, showing a reduction of the resonance peaks by 18 dB and a reduction of the cross-couplings of 30 dB. In order not to compromise the accuracy and precision of the scanner at lower frequencies, a selective feedback controller is used, having a high feedback gain only in the frequency region of the resonance frequencies (21).

Figure 5.8 shows AFM images of a 0.476 μm line pitch calibration grating obtained with both the uncontrolled and the actively damped system. The images are obtained at a line scan rate of 122 Hz and at an image rotation of 45°, meaning that both scanning axes of the piezoelectric tube scanner are equally contributing to the fast scanning motion. In the uncontrolled case, the scanner oscillations stemming from the weakly damped resonances can be recognized by the non-straitness of the grid lines observed in both the topography and in the deflection image. Also the vertical lines in the friction image can clearly be observed which are due to the non-constant imaging speed as a result of the scanner oscillations. The images obtained with the actively damped system show significantly less image distortion, only showing some non-straitness of the grid lines at the left part of the image, which is the beginning of each scan-line where the scanning motion is reversed. Also the vertical bars in the friction image are not present in the actively damped case, meaning that the imaging speed is much more constant than in the uncontrolled case.

In summary, self-sensing piezo actuation allows active damping of both scanning axes of a piezoelectric tube scanner, without requiring dedicated position sensors or other hardware modifications. As implementation of this control method only requires a modification of the systems electronics, and leaves all other hardware intact, it is a very cost-efficient method to improve the performance of the AFM system.

In combination with the Iterative Learning Control discussed in Section 5.2.1, the active damping of the resonances might lead to better robustness and convergence

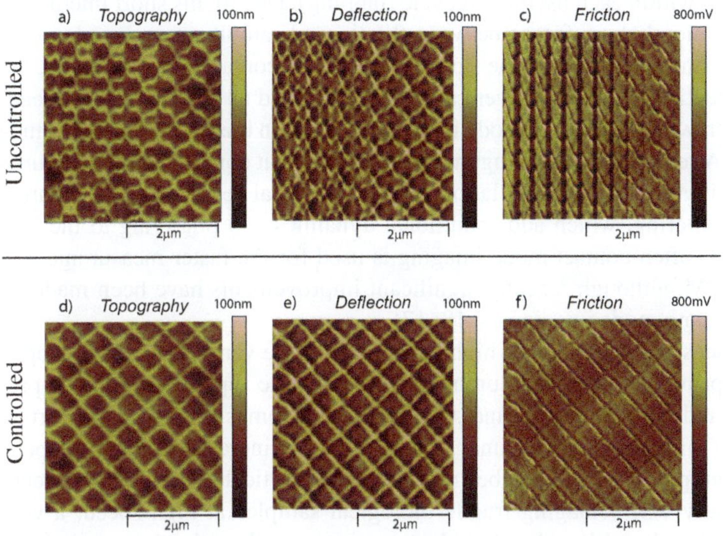

Fig. 5.8 AFM images of a calibration grid with line pitch of 0.476 μm, scanned with at an image rotation 45° and at a line-scan speed of 122 Hz. The resolution of the images is 256 × 256 pixels. The upper images are captured with the uncontrolled system and the bottom images are captured using the proposed control method for active damping of the tube scanner (21).

of the iterative algorithm, and also better rigidity of the system against stochastic disturbances. Moreover, self-sensing piezo actuation might be used to apply active damping of the resonances in the vertical positioning axis, which is used to control the tip-sample interaction force. This added damping in the vertical positioning axis may add robustness to the vertical feedback loop, and allow a higher control bandwidth as the gain margin of the control loop would be less restricted by the height of the resonance peaks.

5.3 Controlling the Tip-Sample Interaction Forces

In AFM, the tip-sample interaction forces are controlled in a feedback loop, detecting the deflection of the cantilever beam that holds the probing tip, and manipulating the vertical position of the tip and/or the sample (cf. Fig. 5.1). The feedback signal of the positioning stage during imaging is used as a measure of the actual sample topography. Only in few applications is the feedback control action omitted and the sample topography recorded in a *constant height* mode (29).

Measuring the tip-sample interaction force can be done in both contact and tapping mode (29). In contact mode imaging, the probing tip is permanently interacting with the sample surface while imaging. In contrast, in tapping mode AFM, the tip is oscillated above the sample surface, and interacts with the sample surface only for a short moment within each oscillation cycle (tapping). Due to this short interaction moments, the lateral shear forces between the tip and sample are reduced, which leads to less damage or wear of the tip and sample. In contact mode imaging, the measured deflection of the cantilever can be directly used as an input for the feedback controller, while in tapping mode imaging most often the oscillation amplitude is controlled. A restriction of tapping mode imaging is that converting the measured cantilever oscillation into an oscillation amplitude signal requires de-modulation and low-pass filtering, which adds additional dynamics and phase-lag to the system. Therefore, often contact mode imaging is used for the faster measurement in high-speed AFM, although recently significant improvements have been made on high-speed tapping mode imaging (3; 18; 17).

For both measurement modes a high bandwidth of the vertical feedback loop is important to provide a good measurement of the sample topography, and to prevent damage and wear of the tip and sample. This becomes even more important in the case of high-speed AFM imaging, as with the increased scanning speed also the topography variations to be tracked by the vertical feedback loop occur faster. Especially when imaging fragile biological samples at high speed, a very high ($\geq$ 100 kHz) closed-loop bandwidth is required to reduce the tip-sample force variations which otherwise might cause damage to the sample (15).

As compared with the classical proportional-integral (PI) controllers found in most commercially available AFM systems, a significant performance increase of the vertical feedback loop can be obtained by applying modern model-based control techniques (32). With model-based feedback control, a model of the scanner dynamics is directly taken into account in the design of the feedback controller (37). At higher frequencies, however, the dynamical behavior of most AFM systems is difficult to capture and also very sensitive to system variations. This dynamical uncertainty at higher frequencies poses a direct limitation on the achievable closed-loop bandwidth of the vertical feedback loop. Recently several high-speed prototype AFM systems have been developed with improved mechanical designs, optimized to push the resonant modes to as high frequencies as possible (4; 31), resulting in a tracking bandwidth of more that 100 kHz. Increasing the resonance frequencies of piezo-based positioning stages, however, most often comes at the cost of a reduction in positioning range. In the next section, a dual-actuated system design and control method is discussed which allows high-bandwidth control of the tip-sample interaction force, without sacrificing effective positioning range.

5.3.1 Dual Actuation

Figure 5.9 illustrates a dual-actuated AFM where the tip-sample force is controlled using a long-range, low-bandwidth actuator to provide the vertical positioning of the sample, and a short-range, high-bandwidth actuator providing the vertical positioning of the measurement tip. The principle of dual actuation has been successfully applied in the field of hard-disk drives (16), and has also been demonstrated on prototype AFM systems (39; 33; 13; 19).

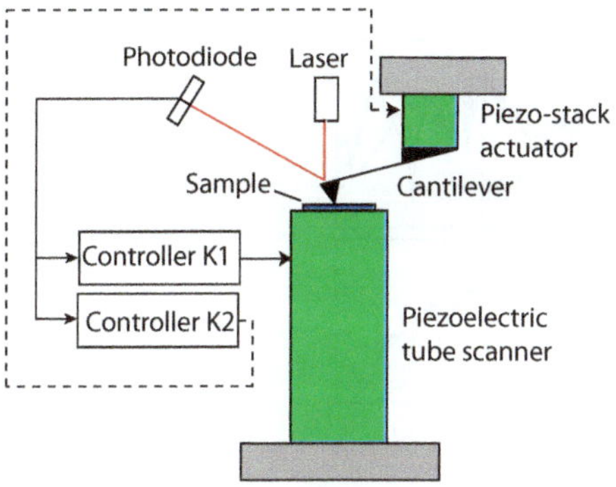

Fig. 5.9 Schematic description of a dual-actuated atomic force microscope

As most samples in AFM imaging are rather flat, the larger amplitude topography variations while scanning are most often stemming from a tilt of the sample and coupling of the scanning motion. The signal to be tracked by the vertical feedback loop in AFM therefore most often has a predominantly triangular shape with a base-frequency equal to the line-scan rate, combined with smaller amplitude high-frequency components stemming from the actual sample topography. The idea of dual actuation in AFM is that the larger amplitude, low-frequency topography variations can be tracked by the long-range, low-bandwidth actuator, and that the additional high-frequency topography variations can be tracked by a short-range, high-bandwidth actuator.

Figure 5.10 shows the measured and modelled frequency responses of a piezoelectric tube scanner (J-scanner, Veeco, Santa Barbara, USA) used to provide the vertical positioning the sample with a maximum range of 5 μm (Fig. 5.10a), and of a short-range piezoelectric plate actuator (CMAP12, Noliac, Kvistgaard, Denmark) which provides the vertical positioning of the measurement tip with a maximum range of 0.5 μm (Fig. 5.10b). In order to identify the dynamical behavior of the actuators under all working conditions, the measurements are repeated with different sample weights and measurement probes, from which two typical responses of both actuators are shown by the solid lines in Fig. 5.10. For the long-range piezoelectric tube scanner, the first resonance can be found at 8 kHz, whereas for the short-range piezoelectric plate actuator the first resonance can be found at 150 kHz. Based on the frequency response data from the various measurement trials, two 7th order models are fitted to capture the nominal dynamical behavior of the actuators (dashed lines, Fig. 5.10). The deviations of the frequency response measurements with respect to the nominal model allows one to determine the maximum allowable actuation bandwidth for both actuators at which closed-loop stability and performance can still be guaranteed under all working conditions. For the piezoelectric tube scanner this bandwidth is determined to be 6 kHz, and for the piezoelectric plate actuator this bandwidth is determined to be 40 kHz (19).

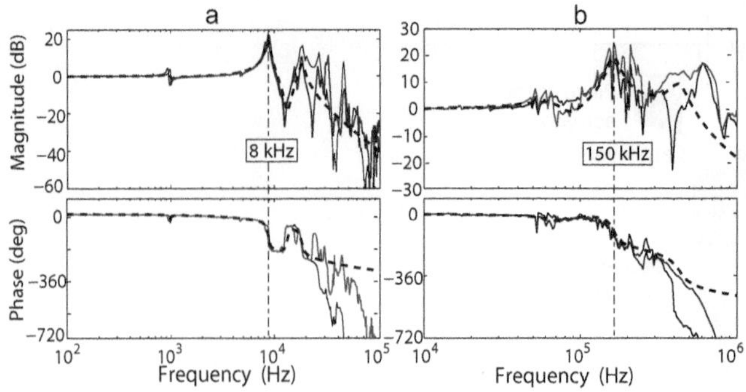

Fig. 5.10 (a) Bode plot of the piezoelectric tube scanner, measurements (solid, black and grey), and modeled (dashed, black). (b) Bode plot of the piezoelectric plate actuator, measurements (solid, red and blue) and modeled (dashed, black).

Based on the nominal model and the measured dynamical uncertainty of the system, a model-based feedback controller is designed to control the tip-sample force using the dual actuators. Critical in the design of a feedback controller for dual-actuated systems is the frequency region where the short-range actuator takes over from the long-range actuator (35). While one naturally would want to have the cross-over between the two actuators at as high a frequency as possible, pushing this cross-over frequency too close to the required roll-off frequency of the long-range actuator might lead a design in which both actuators start to strongly work against each other (19).

Figure 5.11a shows the modelled complementary sensitivity function of the designed closed-loop system, showing a tracking bandwidth of 40 kHz. The dashed lines in Fig. 5.11a give the contributions of both actuators, showing that the short-range actuator takes over from the long-range actuator at a cross-over frequency of 1.5 kHz.

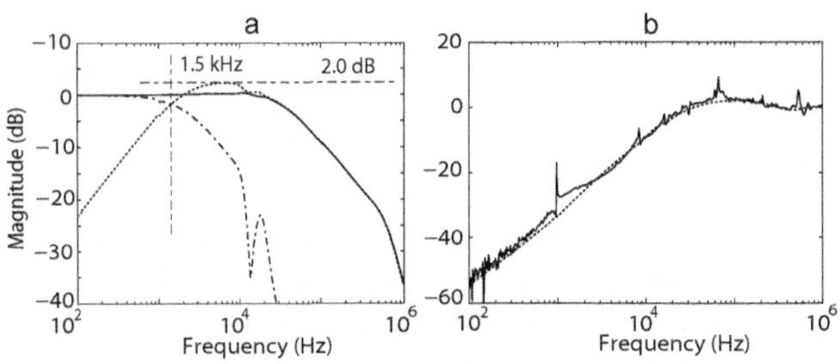

Fig. 5.11 (a) Simulated magnitude response of the contribution of the piezoelectric tube scanner (dashed-dotted) and of the piezoelectric plate actuator (dashed), and the total complementary sensitivity function (solid). (b) Magnitude response of the simulated (dashed) and measured (solid) sensitivity function of the system, showing a disturbance rejection bandwidth of about 20 kHz.

5.3.2 Imaging Results

Figure 5.11b shows the measured sensitivity function of the realized dual actuated prototype system, demonstrating an achieved disturbance rejection bandwidth of 20 kHz, which is about 30 times than what is achieved with the conventional single actuated system. Figure 5.12 shows AFM images of a silicon calibration grating with 1-μm-pitch square holes, imaged at 5 lines per second with the conventional single-actuated system, and with the newly designed dual-actuated AFM. The cantilever deflection images and cross sections in Fig. 5.12 show a significant reduction of the control error (cantilever deflection) in the dual-actuated case.

The faster control of the tip-sample interaction force results in more information being brought from the deflection image towards the topography image, which provides a more accurate measurement of the sample topography (15). The reduced force variations between the tip and the sample results in a more gentle probing of the sample, which is especially important when imaging fragile biological samples in high-speed AFM.

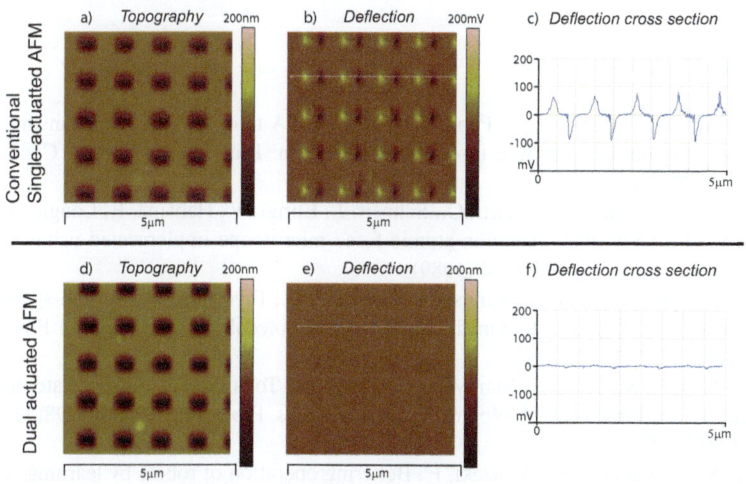

Fig. 5.12 AFM image of a 1-μm-pitch calibration grating, captured at a line scan rate of 5 Hz with the conventional single-actuated system (a, b, c) and with the newly designed dual-actuated system (d, e, f), showing a significant reduction of cantilever deflection in the dual actuated case

5.4 Conclusion

High-speed atomic force microscopy is a challenging control task as it requires positioning of the sample and measurement probe at high bandwidth and with (sub-) nanometer precision. Modern control methods can be utilized to make a significant improvement on the achievable imaging speed.

In controlling the lateral scanning motion, tracking errors stemming from scanner dynamics and non-linearities can be compensated for by iteratively learning the optimal input sequence over several scan lines, allowing a tracking accuracy beyond the noise-floor of the position sensors to be achieved. To compensate for the adverse effects of the weakly damped resonant modes of the positioning stage, self-sensing piezo actuation can be applied to actively dampen these modes without using dedicated position sensors, resulting in significantly less image artifacts when imaging at high speed.

To reduce the tip-sample force variations in high-speed AFM, dual-actuation is demonstrated to enable a significant improvement on the control bandwidth without compromising the effective positioning range. The improved feedback bandwidth results in significantly less variations of the tip-sample interaction force during imaging, and allows for faster AFM imaging without damaging the tip and/or sample.

Although the control methods in this article were demonstrated in atomic force microscopy, it is expected that similar performance improvements can be obtained when applying these control methods to other scanning probe techniques.

References

[1] Abramovitch, D., Andersson, S., Pao, L., Schitter, G.: A tutorial on the mechanisms, dynamics, and control of atomic force microscopes. In: Proc. Amer. Control Conf., pp. 3488–3502 (2007)

[2] Alexander, S., Hellemans, L., Marti, O., Schneir, J., Elings, V., Hansma, P., Longmire, M., Gurley, J.: An atomic-resolution atomic-force microscope implemented using an optical lever. J. Appl. Phys. 65, 164 (1989)

[3] Ando, T., Kodera, N., Naito, Y., Kinoshita, T., Furuta, K., Toyoshima, Y.: A high-speed atomic force microscope for studying biological macromolecules in action. Chem. Phys. Chem. 4(11), 1196–1202 (2003)

[4] Ando, T., Kodera, N., Takai, E., Maruyama, D., Saito, K., Toda, A.: A high-speed atomic force microscope for studying biological macromolecules. Proc. Nat. Acad. Sci. 98(22), 12468–12472 (2001)

[5] Arimoto, S., Kawamura, S., Miyazaki, F.: Bettering operation of robots by learning. J. Robotic Syst. 1(2), 123–140 (1984)

[6] Bhikkaji, B., Ratnam, M., Fleming, A., Moheimani, S.: High performance control of piezoelectric tube scanners. IEEE Trans. Control Syst. Technol. 15, 853–866 (2007)

[7] Binnig, G., Quate, C., Gerber, C.: Atomic force microscope. Phys. Rev. Lett. 56(9), 930–933 (1986)

[8] Binnig, G., Smith, D.: Single-tube three-dimensional scanner for scanning tunneling microscopy. Rev. Sci. Instrum. 57, 1688–1698 (1986)

[9] Cole, D., Clark, R.: Adaptive compensation of piezoelectric sensoriactuators. J. Intell. Mater. Syst. Struc. 5, 665–672 (1994)

[10] Croft, D., Devasia, S.: Vibration compensation for high speed scanning tunneling microscopy. Rev. Sci. Instrum. 70, 4600–4605 (1999)

[11] Croft, D., Shed, G., Devasia, S.: Creep, hysteresis, and vibration compensation for piezoactuators: Atomic force microscopy applications. AMSE J. Dyn. Syst. Meas. Control 123, 35–43 (2001)

[12] Dosch, J., Inman, D., Garcia, E.: A self-sensing piezoelectric actuator for collocated control. J. Intell. Mater. Syst. Struc. 3, 166–185 (1992)
[13] Fleming, A.: High-speed vertical positioning for contact-mode atomic force microscopy. In: Proc. IEEE/ASME Inter. Conf. Advan. Intell. Mechatron., Singapore, pp. 522–527 (2009)
[14] Fleming, A., Moheimani, S.: Sensorless vibration suppression and scan compensation for piezoelectric tube nanopositioners. IEEE Trans. Control Syst. Technol. 14, 33–44 (2006)
[15] Hansma, P., Schitter, G., Fantner, G., Prater, C.: High speed atomic force microscopy. Science 314, 601–602 (2006)
[16] Horowitz, R., Li, Y., Oldham, K., Kon, S., Huang, X.: Dual-stage servo systems and vibration compensation in computer hard disk drives. Control Engin. Prac. 15(3), 291–305 (2007)
[17] Jeong, Y., Jayanth, G., Jhiang, S., Menq, C.: Direct tip-sample interaction force control for the dynamic mode atomic force microscopy. Appl. Phys. Lett. 88, 204102 (2006)
[18] Kodera, N., Sakashita, M., Ando, T.: Dynamic proportional-integral-differential controller for high-speed atomic force microscopy. Rev. Sci. Instrum. 77, 083704 (2006)
[19] Kuiper, S., Fleming, A., Schitter, G.: Dual actuation for high speed atomic force microscopy. In: Proc. IFAC Mechatronics Conf., pp. 441–446 (2010)
[20] Kuiper, S., Schitter, G.: Self-sensing actuation and damping of a piezoelectric tube scanner for atomic force microscopy. In: Proc. Europ. Control. Conf. 2009 (2009)
[21] Kuiper, S., Schitter, G.: Active damping of a piezoelectric tube scanner using self-sensing piezo actuation. Mechatron. 20, 656–665 (2010)
[22] Leang, K., Devasia, S.: Design of hysteresis-compensating iterative learning control for piezo-positioners: Applications to atomic force microscope. Mechantron. 16, 141–158 (2006)
[23] Merry, R., Uyanik, M., van de Molengraft, R., Koops, R., van Veghel, M., Steinbuch, M.: Identification, control and hysteresis compensation of a 3 DOF metrological AFM. Asian J. Control 11, 130–143 (2009)
[24] Moheimani, S., Yong, Y.: Simultaneous sensing and actuation with a piezoelectric tube scanner. Rev. Sci. Instrum. 79, 073702 (2008)
[25] Picco, L., Bozec, L., Ulcinas, A., Engledew, D., Antognozzi, M., Horton, M., Miles, M.: Breaking the speed limit with atomic force microscopy. Nanotechnol. 18, 044, 030 (2007)
[26] Rifai, O., Youcef-Toumi, K.: Coupling in piezoelectric tube scanners used in scanning probe microscope. In: Proc. Amer. Control Conf. (2001)
[27] Rost, M., Crama, L., Schakel, P., Van Tol, E., van Velzen-Williams, G., Overgauw, C., Ter Horst, H., Dekker, H., Okhuijsen, B., Seynen, M., et al.: Scanning probe microscopes go video rate and beyond. Rev. Sci. Instrum. 76, 053710 (2005)
[28] Salapaka, S., Sebastian, A., Cleveland, J., Salapaka, M.: High bandwidth nanopositioner: A robust control approach. Rev. Sci. Instrum. 73, 3232 (2002)
[29] Sarid, D.: Scanning Force Microscopy: With Applications to Electric, Magnetic, and Atomic Forces. Oxford University Press, USA (1994)
[30] Schitter, G.: Improving the speed of AFM by mechatronic design and modern control methods. Technisches Messen. 76(5), 266–273 (2009)
[31] Schitter, G., Astrom, K., DeMartini, B., Thurner, P., Turner, K., Hansma, P.: Design and modeling of a high-speed AFM-scanner. IEEE Trans. Control Syst. Tech. 15(5) (2007)
[32] Schitter, G., Menold, P., Knapp, H., Allgower, F., Stemmer, A.: High performance feedback for fast scanning atomic force microscopes. Rev. Sci. Instrum. 72, 3320 (2001)

[33] Schitter, G., Rijkee, W., Phan, N.: Dual actuation for high-bandwidth nanopositioning. In: Proc. 47th IEEE Conf. Decis. Control 2008, pp. 5176–5181 (2008)
[34] Schitter, G., Stemmer, A.: Identification and open-loop tracking control of a piezoelectric tube scanner for high-speed scanning-probe microscopy. IEEE Trans. Control Syst. Technol. 12, 449–454 (2004)
[35] Schroeck, S., Messner, W.: On controller design for linear time-invariant dual-input single-output systems. In: Proc. Amer. Control Conf. 1999, vol. 6 (1999)
[36] Sebastian, A., Salapaka, S.: Design methodologies for robust nano-positioning. IEEE Trans. Control Syst. Technol. 13(6), 868–876 (2005)
[37] Skogestad, S., Postlethwaite, I.: Multivariable Feedback Control: Analysis and Design. John Wiley & Sons, Chichester (2005)
[38] Slocum, A.: Precision Machine Design. Society of Manufacturing (1992)
[39] Sulchek, T., Minne, S., Adams, J., Fletcher, D., Atalar, A., Quate, C., Adderton, D.: Dual integrated actuators for extended range high speed atomic force microscopy. Appl. Phys. Lett. 75, 1637–1639 (1999)
[40] Tamer, N., Dahleh, M.: Feedback control of piezoelectric tube scanners. In: Proc. 33rd IEEE Conf. Decis. Control 1994, vol. 2 (1994)
[41] Wu, Y., Zou, Q.: An iterative based feedforward-feedback control approach to high-speed AFM imaging. In: Proc. Amer. Control Conf. 2009, pp. 1658–1663. IEEE Press, Los Alamitos (2009)

Chapter 6
Non-raster Scanning in Atomic Force Microscopy for High-Speed Imaging of Biopolymers

Peter I. Chang and Sean B. Andersson

Abstract. Recent advances in actuator design, controller architecture, and system optimization have made possible video-rate atomic force microscopy. Despite this grand achievement, there are systems with dynamics that occur much faster even than video rate. Novel methods are needed to push imaging rates even higher. Here we present a scheme which is complementary to other high-speed approaches to atomic force: non-raster scanning. Using in real-time the data measured by the tip of the microscope coupled with basic information about the sample, we describe an algorithm that steers the tip of the microscope to remain in the regions of interest. The algorithm, designed for imaging biopolymers and other string-like samples, reduces overall imaging time not by increasing the speed of scanning but by reducing the total sampling area.

6.1 Introduction

Since the invention of atomic force microscopy (3), the technology has been used to increase our understanding across a wide range of fields, including molecular biology, medicine, materials science, and nanotechnology. It has been particularly useful in the field of biology due to its ability to operate in liquid, enabling researchers to study systems in their native environments. Despite its wide applicability, the usefulness of AFM in studying dynamics in systems with nanometer-scale features is severely hampered by the time scale of the imaging process. Current commercial AFMs typically require times on the order of minutes to produce a single high quality image. The need for higher imaging rates has led to a great deal of work in high speed AFM.

Approaches to high speed AFM can be broadly categorized into two main approaches. The first involves improving the physical hardware, such as using a balanced scanner to suppress mechanical resonances (2), using small cantilevers (19),

Peter I. Chang · Sean B. Andersson
Boston University, Boston, MA 02215, USA
e-mail: {itchang, sanderss}@bu.edu

E. Eleftheriou & S.O.R. Moheimani (Eds.): Cntrl. Tech. for Emerging Micro/Nanoscale Sys. LNCIS 413, pp. 101–117.
springerlink.com

and designing new high speed actuators (8; 18). The second involves the use of advanced control algorithms, including inversion-based feedforward control (20), combined feedback/feedforward design with adaptive control (10), and transient-based signal detection (14) (see also the review in (16)). Combinations of these schemes have yielded a few research systems that achieve video-rate AFM (see, e.g. the review article (17)).

While achieving video-rate has greatly enhanced the utility of the instrument and broadened the class of dynamics that can be studied using AFM, there are still dynamic events that exceed even this temporal resolution. For example, a class of motor proteins, the dynien motors, travels inside living cells at a measured rate of 1.5 μm/sec (11). For an AFM at current video-rate (25 frames/sec), in a single frame the dynein would have moved 60 nm. There is therefore a need to bring even better temporal resolution to the state of the art AFM systems.

All of the current approaches to high speed AFM view the imaging process as essentially open-loop and seek to move the tip more quickly through the raster-scan pattern. In our *local raster-scan* scheme, we seek to utilize the measurements acquired by the AFM tip to adjust the measurement process in real time and reduce the amount of sampling that needs to be done. The algorithm is designed for biopolymers and other similar samples which can be modeled as planar curves. The essential idea is to utilize the measurements to estimate the spatial evolution of the sample and then steer the tip to sample only this region. Overall imaging time is reduced not by increasing the tip speed but rather by decreasing the overall imaging area.

6.2 Overview of Local Raster Scanning

As illustrated in the block diagram in Fig. 6.1, the local raster scan algorithm is designed as a high-level feedback loop around the AFM system. The algorithm utilizes the data acquired by the AFM in real time to drive a model of the spatial evolution of the underlying sample. This model is then used to determine a tip trajectory such that the tip is scanned back and forth while moving along the path defined by the sample. As a result, data is acquired only local to the sample and the total imaging time is reduced purely by eliminating unnecessary measurements. As a guide to the algorithm, in this section we take around the loop and briefly describe each of the elements in Fig. 6.1.

6.2.1 AFM System Block

The AFM system block represents a physical AFM. It is abstractly described as consisting of two components: one containing the actuators, sensors, and controllers for performing lateral motion of the sample with respect to the tip and one containing the cantilever, tip, actuators, sensors, and controllers for performing motion in the vertical direction. From the point of view of the local raster scan algorithm, the AFM system is a black box which is assumed to faithfully follow the desired (x,y)-trajectory while producing (x,y)-measurements and z-measurements containing

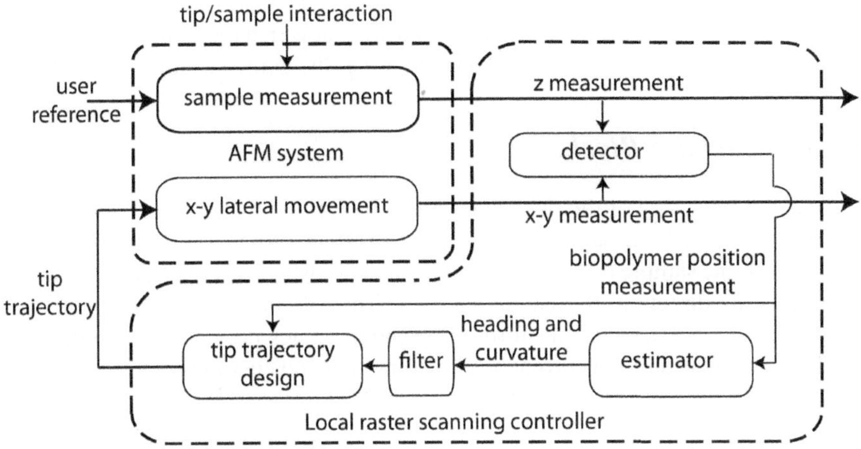

Fig. 6.1 Block diagram of the local raster scan control loop. Driven by the data acquired by the AFM, the *detector* block determines the current position of the sample in the scan. These positions are used by the *estimator* block to determine the geometric parameters driving the spatial evolution of the path of the sample. After *filtering*, these values are fed to the *tip trajectory* block which estimates the evolution of the sample and, from that, the desired trajectory of the tip.

information about the sample. The details of this block, such as whether the scanner is a frame-in-frame stage or a tube actuator, whether the AFM is operated in constant force, AC, or alternative sensing mode, and details on the cantilever are unimportant to the basic design of the local raster scan algorithm. (For an overview of AFM from a controls perspective, see (1).)

Such details do, of course, influence the performance of the local raster scan algorithm. For example, novel actuator designs and control architectures would allow the AFM system to follow the desired (x,y)-trajectory with high fidelity at a high tip velocity, given a z-measurement scheme that reflect the sample at those speeds. In addition, details about the content of the z-measurement directly influence the design on detector block since the information about the sample may be encoded in the signal with different ways. Given a detector block, however, the remainder of the algorithm is independent of such information.

6.2.2 *Detector Block*

The local raster scan algorithm is designed to scan the tip along a biopolymer or similar sample, moving repeatedly from the substrate up onto the sample and back again. The role of the detector is to determine the point of transition between substrate to sample. There are a variety of ways to do this, depending on the data available. As described in detail in Sec. 6.3.1, in this work we assume the z-measurement provides the height of the sample. Combining this data with the

(x,y)-measurements, the detector essentially builds the local topology of the sample and then uses that topology to estimate the position at which the transition occurred. This position is passed to the next block and drives the entire scanning algorithm.

6.2.3 *Estimator Block*

The desired trajectory of the tip is determined by estimating the future spatial evolution of the underlying sample based on the positions provided by the detector. To achieve this, the sample is modeled as a planar curve whose evolution is determined by its heading direction, θ, and curvature, κ. The role of the estimator block, described in Sec. 6.3.2, is to transform the position values provided by the detector block into estimates of the current heading direction and curvature.

6.2.4 *Filter Block*

Noise in the original AFM measurements propagates into the position values estimated by the detector. The estimator block then calculates numerical derivatives of that data, thereby amplifying the noise. To compensate for this effect, the raw estimates are run through the filter block. Using a simple dynamic model for the evolution of the curvature and heading direction, described in Sec. 6.3.3, this block uses a Kalman filter to produce the optimal estimates of these curve parameters.

6.2.5 *Tip Trajectory Design Block*

The final block in the feedback loop transforms the heading and curvature estimates into the desired tip trajectory. The block first uses the estimates to propagate the model of the planar curve forward in space from the last detected position of the sample. The tip trajectory is then specified as a sinusoidal variation along this planar curve in which the amplitude, A, and spatial frequency, ω, are specified by the user. The result is a smooth curve that on average tracks the underlying sample, leading to an image of width A and resolution $(1/\omega)$. The details on trajectory determination are provided in Sec. 6.3.4.

6.3 Controller Details

The details of each of the blocks in Fig. 6.1 are given in this section. The entire scheme begins with the detection of the sample.

6.3.1 *Sample Detection*

Consider the motion of the tip across a cylindrical sample of constant diameter d, depicted in Fig. 6.2. The method of detecting the transition from substrate to sample

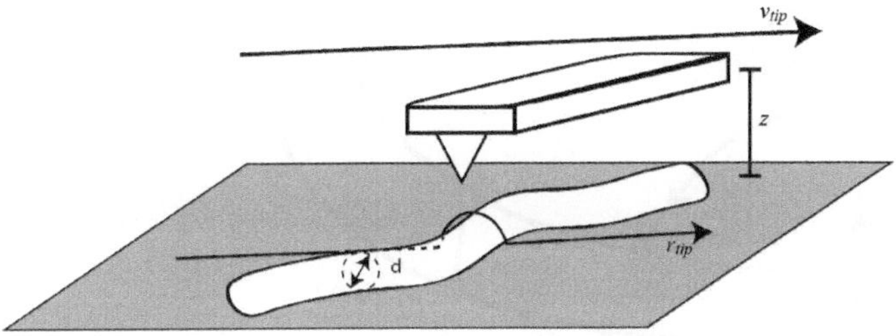

Fig. 6.2 Illustration of an AFM tip crossing a cylindrical sample with diameter d. Under the assumption that the tip speed, v_{tip}, is slow enough, the height data directly indicates whether the tip is on or off the sample.

or sample to substrate in the data is dependent upon the specific application and upon the data available to the detector. Under the assumption that the AFM is imaging a sample in intermittent contact (or tapping) mode, detectors can be designed based on direct measurements of the cantilever position or on information generated by the z-controller, including the height, amplitude and phase signals. For example, using the transient force detection scheme in (15), transitions could be detected within a few oscillations of the cantilever. Such an approach would allow for detection to occur at tip speeds far higher than the limits imposed by the need for accurate imaging.

In practice, accessing the cantilever data can be challenging. In this work we assume that the goal is to produce a high-quality image. Under this assumption, the tip speed should be kept slow enough such that the height data acquired by the AFM will accurately reflect the local topology. Because the local raster scan algorithm keeps the tip in the vicinity of the sample, it can be assumed that the substrate around the sample is relatively flat with the sample itself rising off the base. As a consequence of these two assumptions, a simple detector can be designed simply by establishing a threshold value, τ_h, relative to the local measurements. Heights below the threshold are assumed to be on the substrate while those above the threshold are on the sample. In practice, the choice of τ_h would be informed by *a priori* knowledge of the type of sample (e.g., the height of a DNA strand when imaging in liquid is approximately 1.5 nm (12)) as well as by user experience with typical noise levels in the imaging process.

Given a means of detecting transitions, one must then decide what part of the sample is to be represented by the planar curve. Consider the schematic of a sample and tip path shown in Fig. 6.3. One possibility is to use the centerline of the sample as the planar curve to be tracked by the local raster scan algorithm and a scheme for doing this has been previously presented in (5). For samples with varying width, however, trying to estimate the centerline can lead to widely varying tip trajectories that are difficult for the actuators to follow. As an alternative, one can choose to track a single edge of the polymer. Consider the motion of the tip from the point $p[k]$ in Fig. 6.3 to point $p[k+1]$. Under the assumption that the tip trajectory fully covers the sample, there is a sequence of $down - up - down$ transitions between

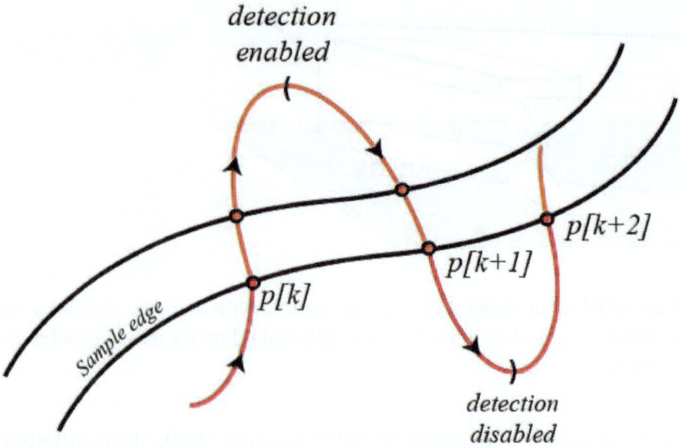

Fig. 6.3 Based on the signals available, a detector can be designed to indicate when the tip moves from the substrate to sample or from the sample to substrate. Using high-level logic, one can then choose to select only transitions on one side of the sample. Shown here are three detected crossing events, moving from $p[k]$ to $p[k+2]$. In order to prevent chattering between on and off due to measurement noise at the edge of the sample, detection is disabled for a portion of the trajectory.

sample and substrate. Using a state machine to count these transitions, the crossings on one side of the sample can be isolated.

Noise in the measurements can trigger false transitions in the state machine and ultimately false detections. Consider, for example, a simple threshold detector. The noise can cause chattering in the detector as the tip crosses between sample and substrate. To avoid this, after a transition event the detector is disabled for a portion of the tip trajectory. Typically detection would be re-enabled again at the maximum of the sinusoidal component of the tip trajectory. In the example shown, the state detector would then only count a *down-up* transition followed by a *up-down* transition to track the right-hand edge of the sample.

When the detector determines that a transition event along the desired edge has occurred, the position is recorded in $p[k]$ and the value passed along to the estimation block.

6.3.2 Estimation of Curvature and Heading Direction

As described in detail in Sec. 6.3.4, the path of the underlying sample from $p[k]$ can be estimated based on the heading direction, $\theta[k]$, and curvature $\kappa[k]$ at $p[k]$. The role of the estimator is to determine the values of $\theta[k]$, $\kappa[k]$ from the history of crossings.

Consider a sequence of three crossings, $p[k-2]$, $p[k-1]$, and $p[k]$ as illustrated in Fig. 6.4. The heading direction at crossing k is by definition the angle of the

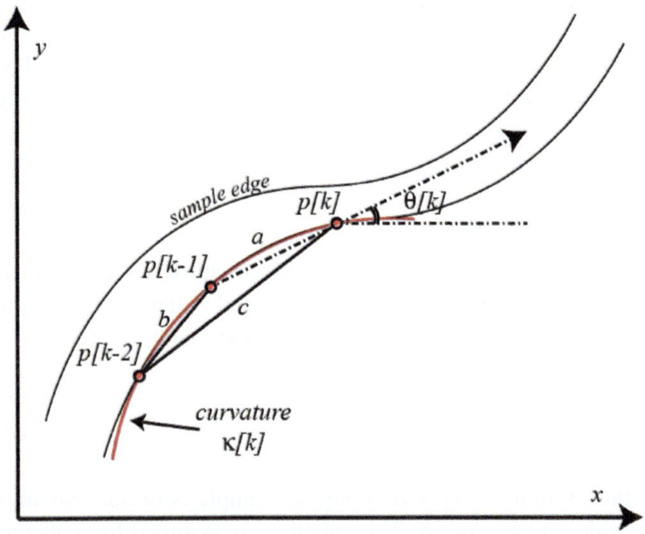

Fig. 6.4 Estimating $\theta[k]$ and $\kappa[k]$. The heading direction is estimated by calculating the angle of the vector connecting the previous transition point, $p[k-1]$ to the current point $p[k]$ while the curvature is estimated from the radius of the circle passing through the points $p[k-2]$, $p[k-1]$, and $p[k]$.

tangent vector to the sample path at that point with respect to a fixed frame. It can be estimated geometrically as follows. Define

$$\delta p[k] = p[k] - p[k-1] = \begin{pmatrix} \delta x[k] \\ \delta y[k] \end{pmatrix}. \tag{6.1}$$

The heading direction at $p[k]$ is then estimated to be

$$\theta[k] = \arctan 2(\delta y[k], \delta x[k]). \tag{6.2}$$

The curvature at $p[k]$ is estimated from the previous three transitions, $p[k]$, $p[k-1]$ and $p[k-2]$ using Heron's formula (4). That is,

$$\kappa[k] = \pm 4\sqrt{\frac{(l-a)(l-b)(l-c)}{abc}}. \tag{6.3}$$

Here a, b and c are the Euclidian lengths of the three sides of the triangle defined by $p[k-2], \ldots, p[k]$, and $l = (a+b+c)/2$ as the semiperimeter of that triangle. The sign of $\kappa[k]$ is defined to be positive when the normal vector to the curve points inside the circle passing through the three transition points.

Since the estimates of both $\theta[k]$ and $\kappa[k]$ are essentially numerical derivatives of the sample path with respect to arclength, any noise in the measurements of the transition positions (originally arising from noise in the AFM measurements) will be amplified. To mitigate the effect of this noise, the estimates are sent to the filtering block.

6.3.3 *Filtering*

To filter the measurements of the heading direction and curvature, we model their spatial evolution using a discrete time stochastic differential equation. Under the assumption of constant curvature between update steps, the heading direction is driven directly by the curvature according to

$$\theta[k+1] = \theta[k] + \Delta s \kappa[k] + w_\theta[k] \tag{6.4}$$

where $\Delta s = 2\pi/\omega$ is the (nominal) step size along the sample between crossings and w_θ is a Gaussian white noise process. The curvature is assumed to evolve as a random walk, driven by a Gaussian white noise process w_κ. Defining a new system state as $X = [\theta \ \kappa]^T$ and adding in a measurement model yields the stochastic system

$$X[k+1] = FX[k] + W[k], \tag{6.5a}$$
$$Y[k] = HX[k] + V[k]. \tag{6.5b}$$

where

$$F = \begin{pmatrix} 1 & \Delta s \\ 0 & 1 \end{pmatrix}, \quad H = \begin{pmatrix} 1 & 0 \\ 0 & 1 \end{pmatrix}, \quad W[k] = \begin{pmatrix} w_\theta[k] \\ w_\kappa[k] \end{pmatrix}, \quad V = \begin{pmatrix} v_\theta[k] \\ v_\kappa[k] \end{pmatrix}. \tag{6.6}$$

Here V is Gaussian white noise process capturing the noise in the estimator. A Kalman filter is applied to this system, yielding the filter equations

$$\hat{X}[k+1|k] = F\hat{X}[k|k] + E[W], \tag{6.7a}$$
$$\Sigma[k+1|k] = F\Sigma[k|k]F^T + \Sigma_W, \tag{6.7b}$$
$$K[k+1] = \Sigma[k+1|k]H^T(H\Sigma[k+1|k]H^T + \Sigma_V)^{-1}, \tag{6.7c}$$
$$\hat{X}[k+1|k+1] = \hat{X}[k+1|k] + K[k+1](Y[k+1] - H\hat{X}[k+1|k]), \tag{6.7d}$$
$$\Sigma[k+1|k+1] = (I - K[k+1]H)\Sigma[k+1|k], \tag{6.7e}$$

where $\hat{X}[k|k]$ is the *a posteriori* state estimate and $\Sigma[k|k]$ is the *a posteriori* error covariance.

6.3.3.1 Comments on the Input and Measurement Noise

The noise process W represents random fluctuations in the heading direction and curvature of the underlying sample and arises from the physical characteristics of that sample. To date, we have utilized an approach in which the mean and variance of W are taken to be constant and selected in an *ad hoc* manner. Better performance is to be expected, however, if the choice of parameters is informed by *a priori* knowledge of the underlying sample. Biopolymers, for example, are known to twist and turn and a variety of structural models have been developed to describe this bending behavior, such as the freely-jointed chain and the worm-like chain (9). These behaviors are captured by the choice of input noise parameters.

The noise process V captures noise propagated from the AFM measurements into the detection of the crossing positions and from there into the estimation of the curvature and heading direction. While we currently use an *ad hoc* selection of measurement noise parameters, the choice of mean and variance based on the initial measurement noise is a topic of ongoing research.

6.3.4 Tip Trajectory

6.3.4.1 Frenet–Serret Frame

As discussed in Sec. 6.2, at the core of the local raster scan algorithm is a model of the sample as a planar curve. The evolution in two dimensions of such a curve can be described using a Frenet–Serret frame approach. This method uses a system of ordinary differential equations, driven by the curvature, that describes the evolution of the path in space (13). Denoting the curve as $r(\cdot)$, the equations are

$$\frac{dr}{ds}(s) = q_1(s), \tag{6.8a}$$

$$\frac{dq_1}{ds}(s) = \kappa(s)q_2(s), \tag{6.8b}$$

$$\frac{dq_2}{ds}(s) = -\kappa(s)q_1(s), \tag{6.8c}$$

where s is arclength along the curve, $q_1(s)$ is the tangent vector to the curve at s, $q_2(s)$ is the normal vector to the curve at s, and $\kappa(s)$ is the curvature at s (see Fig. 6.5). The heading direction at s, denoted $\theta(s)$, is defined to be the direction of the tangent vector at that point.

Alternatively, given the heading direction, the vectors $q_1(s)$ and $q_2(s)$ can be defined by

$$q_1(s) = \begin{pmatrix} \cos\theta(s) \\ \sin\theta(s) \end{pmatrix}, \quad q_2(s) = \begin{pmatrix} -\sin\theta(s) \\ \cos\theta(s) \end{pmatrix}. \tag{6.9}$$

With a constant curvature, the path defined by (6.8) is a circle of radius $\frac{1}{\kappa}$. Note that our scheme assumes the curvature remains constant between crossings of the tip over the sample and thus the sample is modeled as a sequence of arcs of circles.

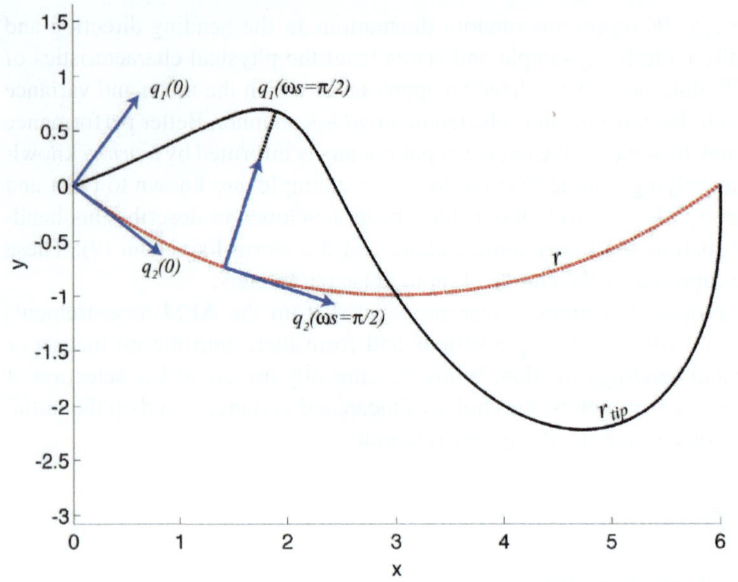

Fig. 6.5 Illustration of the local raster scan AFM tip trajectory. Both the path of the tip (solid black) and sample (dashed red) begin at the origin. The sample curve evolves with a constant curvature of $\kappa = 0.2$ from an initial heading direction of 53°. The tip trajectory is shown over one period and with an amplitude of $A = 1.2$ units and a spatial frequency of $\omega = 0.49$ radius/unit. The Frenet–Serret frame of the sample curve is shown at $s = 0$ and at $s = 2\pi/\omega$.

6.3.4.2 Defining the Tip Trajectory

Given the sample path, the tip trajectory, denoted $r_{tip}(s)$, is designed by adding a sinusoidal variation to $r(\cdot)$,

$$r_{tip}(s) = r(s) \pm A\sin(\omega s)q_2(s), \tag{6.10}$$

where A, in units of length, and ω, in units of 1/length, are user parameters that define the amplitude and spatial frequency respectively, of the tip scanning motion. The choice of positive or negative sign in (6.10) is alternated at each detection of the transition between sample and substrate (c.f. Sec. 6.3.5).

An example of the pattern produced by our scheme is shown in Fig. 6.5. The underlying curve (dashed red) begins at the origin with an initial heading direction of $\theta = 53°$ and a constant curvature of $\kappa = 0.2$. The tip trajectory (solid black) begins at the origin as well and is illustrated over one cycle for a choice of $A = 1.2$ units and $\omega = 0.49$ radius/unit. Also illustrated is the initial Frenet–Serret frame and the frame at $s = \pi/(2\omega)$.

The tip trajectory in (6.10) is specified with respect to the arclength of the (estimated) path of the sample. In practice, the trajectory must be specified with respect to time. The relationship between arclength and time is given by

$$v_{tip}t = \int_0^s \sqrt{(1 - A\kappa\sin(\omega\sigma))^2 + A^2\omega^2\cos^2(\omega\sigma)}\,d\sigma, \tag{6.11}$$

where v_{tip}, the speed of the tip, is assumed to be constant. Given the current value of t, (6.11) must be solved for s before the tip trajectory can be determined. Details can be found in (6).

6.3.4.3 User-Defined Parameters

The two user parameters A and ω influence the shape of r_{tip}. A is the amplitude of the sinusoidal scanning and is analogous to the scan range in the standard raster imaging procedure. In the local raster scan, however, this range is defined relative to the path $r(\cdot)$, ensuring the tip sweeps out an image of width A around the sample. In general, the size of A should be chosen to ensure that the tip trajectory fully covers the width of the sample to be imaged. In practice, one typically knows the type of sample to be imaged. Thus, the choice of A can be informed by *a priori* knowledge of the theoretical width of the sample. We note, however, that the algorithm can be modified to replace a fixed value of A with a detection-based scheme in which its value is determined based on an evolving estimate of the sample width.

The parameter ω defines the spatial frequency of the tip trajectory and is analogous to the resolution of the image. Increasing the value of ω decreased the distance between crossings of the tip trajectory and the sample path, leading to an increase in resolution in the sample data.

It should be noted that error in the measured transition locations, in the heading direction, or in the curvature, or a breakdown in the assumptions such as that of constant curvature between transition locations can lead to loss of tracking. The choice of A and ω influence the robustness of the algorithm to such errors. Guidelines on choosing A and ω to ensure tracking in the face of error in the curvature estimate can be found in (7) and a more general approach is the subject of ongoing research.

6.3.5 Summary

Given an initial position of the sample path, $r(0)$, an initial heading direction, $\theta(0)$, and a value of the curvature κ, the predicted path of the sample is calculated from (6.8) under the assumption of constant curvature. Based on this prediction, the tip trajectory is determined from (6.10) (with a choice of positive or negative sign in the equation). This trajectory is sent as a reference trajectory to the scanner controller to drive the tip along this path.

When a detection of a transition is made, new values of θ and κ are generated and filtered. These values, as well as the new measurement of the position of the sample, are passed to the trajectory determination block. These measurements are used to correct the current estimate of the predicted sample path, leading to an instant jump both in the value of $r(s)$ (to the new transition position, $p[k]$) and to the value of $\theta(s)$ (to the new estimate $\hat{\theta}[k]$). From (6.10), these discontinuities lead to a discontinuity in the designed tip trajectory. To minimize the size of this discontinuity, upon detection of a transition the arclength parameter is reset to 0 and the sign in (6.10) is switched. This is summarized in the following algorithm.

Algorithm 1. [Local raster-scan]

0. *Initialize* $k = 0$.
1. *Set* $s = 0$.
2. *Set the initial conditions* $r(0) = p[k]$, $\theta(0) = \hat{\theta}[k]$, $\kappa = \hat{\kappa}[k]$.
3. *Predict the sample path* $r(\cdot)$ *and the desired tip trajectory* $r_{tip}(\cdot)$ *from* (6.8) *and* (6.10), *respectively.*
4. *Steer the tip according to* $r_{tip}(\cdot)$.
5. *Monitor the measured* z *data until a transition is detected.*
6. *Set* $p[k+1]$ *to the detected transition position,*
7. *Estimate* $\theta[k+1]$ *and* $\kappa[k+1]$ *from* (6.2) *and* (6.3), *respectively. Filter using the Kalman filter in* (6.7) *to produce* $\hat{\theta}[k+1]$ *and* $\hat{\kappa}[k+1]$.
8. $k \leftarrow k+1$. *Go to 1.*

We note that in practice the scheme would be initialized either based on a preliminary, low-resolution raster-scan, or in an automatic fashion. Auto-initialization could proceed by the following approach. A raster pattern is followed until a transition is detected. The tip is then scanned along a small circle to find a second transition. This pair of transition locations defines both the initial position $p[0]$ and the initial heading direction $\theta[0]$. The initial curvature is set to 0.

6.4 Simulation Experiments

To illustrate the algorithm we performed simulation experiments using data acquired from a standard raster-scan image of DNA. A solution of $\lambda-$DNA was purchased from New England BioLabs (Ipswich, MA, USA) and diluted to a concentration of 1.25 μg/mL in 5 mM $NICl_2$ and 50 mM HEPES buffer, adjusted to a pH of 6.8. A small droplet was placed on freshly cleaved mica and incubated for five minutes. The mica surface was then gently rinsed with deinoized water to wash away unbounded DNA and, finally, dried in air.

Imaging was performed in air using an Asylum Research MFP 3-D with an Olympus AC-240 cantilever operating in intermittent contact mode. The line scan rate was set to 1 Hz, the image resolution to 256 by 256 pixels, and the image area to a square with 400 nm sides. The resulting height image is shown in Fig. 6.6. Note that for the given imaging parameters, the tip speed is approximately 800 nm/sec and that the image took 4 minutes and 12 seconds to acquire. The measured height of the DNA of approximately 0.8 nm is within the expected range (see, e.g. (12)). The image also contains residual salt islands with heights of about 0.1 nm.

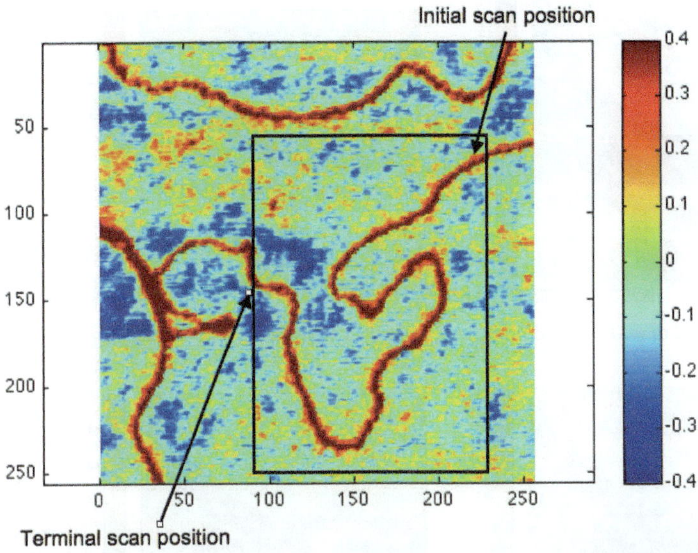

Fig. 6.6 Height retrace image of λ – DNA sample acquired in air using raster scanning. Image size is 400 by 400 nm and has a 256 by 256 resolution. Total scan time is 4 minutes and 12 seconds. Measured DNA height is 0.8 nm.

The local raster scan algorithm was implemented in Matlab and executed using the measurements from the raster scan image in Fig. 6.6 as the z-measurement output from the AFM block (Fig. 6.1). Overlaid on Fig. 6.6 is a box illustrating the region over which the local raster scan algorithm was applied. The scan was begun from the upper right corner and proceeded along the DNA until the sharp turn at the middle of the left edge of the box. Note that due to the limited resolution in the original raster data, the local raster algorithm was unable to accommodate the large curvature at the sharp turn. In practice, the spatial frequency ω could be set sufficiently large for this turn to be tracked.

In the simulations, a simple threshold detector at a height of 0.36 nm was used to detect sample-substrate transitions. The tip velocity was set to 12.5 μm/sec. Through trial and error, the input and measurement noise parameters in the model (6.5) were set to be zero mean with covariances of

$$\Sigma_w = \begin{pmatrix} 0.0685 & 0 \\ 0 & 0.0009 \end{pmatrix}, \quad \Sigma_v = \begin{pmatrix} 1.10 & 0 \\ 0 & 0.250 \end{pmatrix}. \tag{6.12}$$

Note that the discrete nature of the raster scan image led to increased measurement noise in the detection of the crossing positions and therefore in the measurement of the curvature and heading direction.

To ensure that a region surrounding the DNA was sampled by the tip and to illustrate the effect of the parameter A, the amplitude was set to either $A = 12$ nm or $A = 15$ nm, both substantially larger than the DNA width in the image. To illustrate

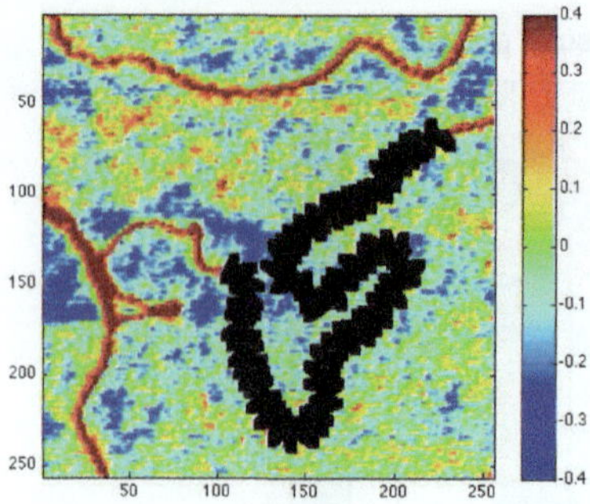

(a) $A = 12$ nm, $\omega = 0.625$ rad/nm

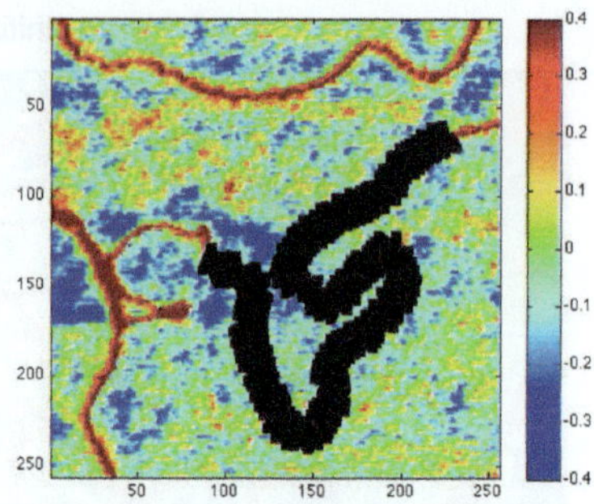

(b) $A = 12$ nm, $\omega = 1.256$ radn/nm

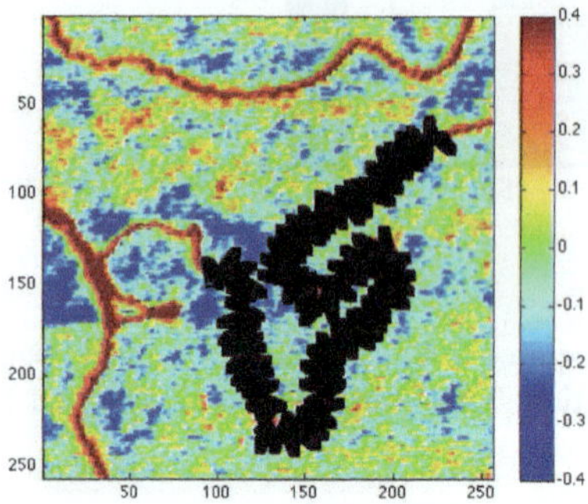

(c) $A = 15$, nm $\omega = 0.625$ rad/nm

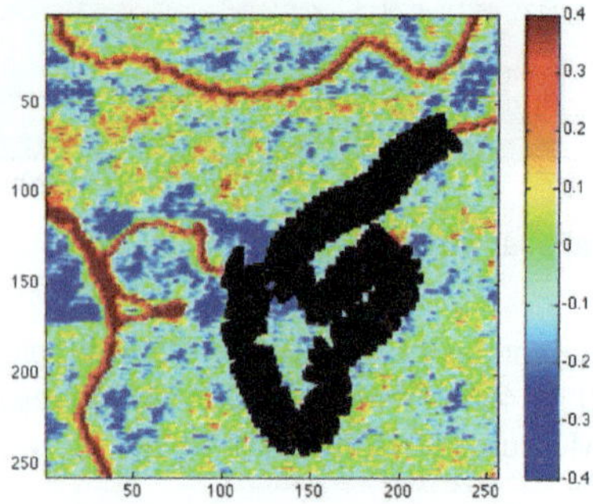

(d) $A = 15$ nm, $\omega = 1.256$ rad/nm

Fig. 6.7 Simulation results: local raster scan on a DNA sample. All units are in nm. The top images used an amplitude of $A = 12$ nm; bottom images used $A = 15$ nm. Left images used $\omega = 0.625$ rad/nm; right images used $\omega = 1.26$ rad/nm. The resulting trajectory of the tip is superimposed on the original raster scan image for comparison.

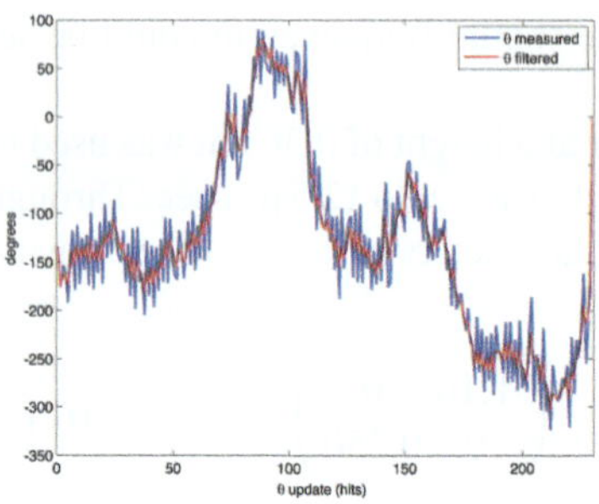

(a) θ filter result

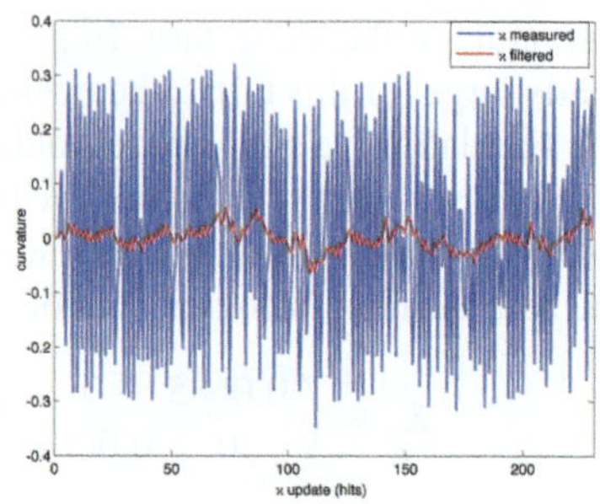

(b) κ filter result

Fig. 6.8 Results of estimation and filtering of the heading direction and curvature measurements along the trajectory of the simulated run shown in Fig. 6.7(d). The large noise on κ relative to that of θ is indicative both of the numerical derivative nature of the estimator as well as of the discrete nature of the underlying data set used in the simulations.

the effect of changing the spatial frequency, scans were performed with $\omega = 0.628$ rad/nm and $\omega = 1.26$ rad/nm, corresponding to nominal step sizes along the DNA of 5 nm and 2.5 nm, respectively. The results of four scans with these different choices of A and ω are shown in Fig. 6.7 by superimposing the resulting tip trajectory on the original raster image. Note that the algorithm did not have access to the full image but was simply sampling from the raster-scan data set to drive the local raster-scan trajectory.

In Fig. 6.8(a,b) we show the raw and filtered estimates of the heading direction and curvature along the scan trajectory of Fig. 6.7(d). The noise on the measurement of κ is clearly much larger than in θ, due in part to the higher-order numerical derivative employed in transforming the crossing positions into curvature estimates as compared to heading direction estimates. The discrete nature of the underlying data set increased the noise somewhat with respect to what would be expected in a true continuous sample.

The simulation lacked explicit models of the dynamics, capturing them roughly through the use of an actual raster-image. As a result, a direct comparison of the time to capture the data between the raster scan image and the non-raster approach is not useful. One can assume, however, that since the same machine would be used for both scans, the tip speed would be the same. Comparing the total path length traveled by the tip, then, is a good estimate of the relative performance. Consider the raster scan image, limited to the boxed area in Fig. 6.6. The reduced scan area is 233 nm $\times$ 307 nm with a resolution of 143 $\times$ 197 pixels. Since the tip must perform a trace and retrace in each line, the total distance traveled is 87.9 μm.

One way to show the efficiency of local raster scanning is to directly compare the total scan time for local raster scanning to raster scanning. However, this comparison would not have ruled out the difference in tip traveling velocity, thus we compare the total scan length for both algorithms instead. Among the four examples shown for the non-raster, the choice with the largest values of A and ω, shown in Fig. 6.7(d), led to the maximum distance traveled, with a total value of only 7.17 μm. The result is an order of magnitude reduction in total path length and therefore an (expected) order of magnitude reduction in imaging time. It should be noted that this comparison holds true regardless of the hardware; should a high-speed AFM be used that allows for much higher tip speeds while still producing high-quality images, the same order of magnitude reduction would be achieved by the non-raster approach.

6.5 Conclusion and Future Work

We have presented a non-raster scanning algorithm for AFM systems designed to image biopolymers and other string-like samples. The algorithm reduces the overall scanned area by automatically tracking the sample and reduces the total scan time by scanning the area of interest only. The essential concept is to close the high-level loop in the AFM and, as a result, the method is complementary to current high-speed

AFM schemes. The scheme was illustrated through a simulation involving true data as captured by a raster-scan image.

With the foundation in place, we are pursuing several lines of inquiry in addition to physical implementation. Among the most important to the application of the method are the development of effective techniques for choosing the parameters in the Kalman filter and methods for choosing the values of A and ω so as to ensure the entire sample is imaged (under appropriate assumptions).

Acknowledgements. This work was supported in part by the National Science Foundation under grant no. CMMI-0845742.

References

[1] Abramovitch, D.Y., Andersson, S.B., Pao, L.Y., Schitter, G.: A tutorial on the mechanisms, dynamics and control of atomic force microscopes. In: Proceedings of the American Control Conference, pp. 3488–3502 (2007)

[2] Ando, T., Kodera, N., Takai, E., Maruyama, D., Saito, K., Toda, A.: A high-speed atomic force microscope for studying biological macromolecules. In: Proceedings of the National Academy of Sciences of the United States of America, vol. 98(22), pp. 12468–12472 (2001)

[3] Binnig, G., Quate, C.F., Gerber, C.: Atomic force microscope. Physical Review Letters 56(9), 930–933 (1986)

[4] Calabi, E., Olver, P.J., Shakiban, C., Tannenbaum, A., Haker, S.: Differential and numerically invariant signature curves applied to object recognition. International Journal of Computer Vision 26(2), 107–135 (1998)

[5] Chang, P.I., Andersson, S.B.: A maximum-likelihood detection scheme for rapid imaging of string-like samples in atomic force microscopy. In: Proceedings of the 48th IEEE Conference on Decision and Control, held jointly with the 28th Chinese Control Conference (CDC/CCC 2009). pp. 8290–8295 (2009), http://10.1109/CDC.2009.5400136

[6] Chang, P.I., Andersson, S.B.: Smooth trajectories for imaging string-like samples in AFM: A preliminary study. In: American Control Conference, pp. 3207–3212 (2008)

[7] Chang, P.I., Andersson, S.B.: Theoretical bounds on a non-raster-scan method for tracking string-like samples. In: Proceedings of the American Control Conference, pp. 2266–2271 (2009)

[8] Fleming, A.J.: Dual-stage vertical feedback for high-speed scanning probe microscopy. IEEE Transactions on Control Systems Technology 99, 1–10 (2010), http://10.1109/TCST.2010.2040282

[9] Howard, J.: Mechanics of Motor Proteins and the Cytoskeleton. Sinauer Associates, Inc. (2001)

[10] Butterworth, J A., Pao, L. Y., Abramovitch, D.Y.: Adaptive-delay combined feedforward/feedback control for raster tracking with applications to AFMs. In: Proceedings of the American Control Conference, pp. 5738–5744 (2010)

[11] Kural, C., Kim, H., Syed, S., Goshima, G., Gelfand, V.I., Selvin, P.R.: Kinesin and dynein move a peroxisome in vivo: A tug-of-war or coordinated movement? Science 308(5727), 1469–1472 (2005)

[12] Moreno-Herrero, F., Colchero, J., Baró, A.M.: DNA height in scanning force microscopy. Ultramicroscopy 96(2), 167–174 (2003)

[13] Pressley, A.: Elementary Differential Geometry. Springer, Heidelberg (2005)
[14] Sahoo, D.R., Sebastian, A., Salapaka, M.V.: Transient-signal-based sample-detection in atomic force microscopy. Applied Physics Letters 83(26), 5521–5523 (2003)
[15] Sahoo, D.R., Sebastian, A., Salapaka, M.V.: Harnessing the transient signals in atomic force microscopy. International Journal of Robust and Nonlinear Control 15, 805–820 (2005)
[16] Salapaka, S.M., Salapaka, M.V.: Scanning probe microscopy. Control Systems Magazine 28(2), 65–83 (2008)
[17] Schitter, G., Rost, M.J.: Scanning probe microscopy at video-rate. Materials Today 11(1-2), 40–48 (2008)
[18] Schitter, G., Astrom, K., DeMartini, B., Thurner, P., Turner, K., Hansma, P.: Design and modeling of a high-speed AFM-scanner. IEEE Transactions on Control Systems Technology 15(5), 906–915 (2007)
[19] Viani, M.B., Schäffer, T.E., Paloczi, G.T., Pietrasanta, L.I., Smith, B.L., Thompson, J.B., Richter, M., Rief, M., Gaub, H.E., Plaxco, K.W., Cleland, A.N., Hansma, H.G., Hansma, P.K.: Fast imaging and fast force spectroscopy of single biopolymers with a new atomic force microscope designed for small cantilevers. Review of Scientific Instruments 70(11), 4300–4303 (1999)
[20] Yan, Y., Zou, Q., Lin, Z.: A control approach to high-speed probe-based nanofabrication. Nanotechnology 20(17), 175301 (2009)

Chapter 7
High-Bandwidth Intermittent-Contact Mode Scanning Probe Microscopy Using Electrostatically-Actuated Microcantilevers

Deepak R. Sahoo, Walter Häberle, Abu Sebastian,
Haralampos Pozidis, and Evangelos Eleftheriou

Abstract. A critical issue in scanning probe microscopy (SPM) in the intermittent-contact (IC) mode is the achievable bandwidth, which is limited because of the high quality factor of the cantilevers. Cantilevers for IC-mode SPM must have high stiffness for stable operation, which necessitates high quality factors for high force sensitivity, and thus results in a slow response time. Here we present an IC SPM method that achieves high bandwidth by using electrostatically-actuated cantilevers with low stiffness and low quality factor. Reliable IC operation is achieved by shaping the input signal applied to the cantilever for electrostatic actuation. By keeping the oscillation amplitude small, high-frequency operation is possible and the tip-sample interaction force is reduced, which in turn prolongs the lifetime of tip and sample. For high-bandwidth imaging, the cantilever deflection signal is sampled directly at each oscillation cycle using input-based triggering. Our experimental results demonstrate the efficacy of the proposed scheme. In particular, in long-term scanning experiments, the tip diameter was maintained over a remarkable 140 m of tip travel. Moreover, as no demodulation electronics are needed, compact SPM devices using this method could be developed, including devices that employ large arrays of cantilevers in parallel operation for high throughput.[1]

Deepak R. Sahoo
University of Bristol,
H. H. Wills Physics Laboratory,
Tyndall Avenue, BS8 1TL,
Bristol, United Kingdom
e-mail: Deepak.Sahoo@bristol.ac.uk

Walter Häberle · Abu Sebastian · Haralampos Pozidis · Evangelos Eleftheriou
IBM Research - Zurich, Säumerstrasse 4,
8803 Rüschlikon, Switzerland
e-mail: {wha,ase,hap,ele}@zurich.ibm.com

[1] This chapter is an expanded version of our paper "High-throughput intermittent-contact scanning probe microscopy," which was published in Nanotechnology 21, 075701 (2010; doi: 10.1088/0957-4484/21/7/075701), used by permission of IOP Publishing Ltd.

E. Eleftheriou & S.O.R. Moheimani (Eds.): Cntrl. Tech. for Emerging Micro/Nanoscale Sys. LNCIS 413, pp. 119–135.
springerlink.com

7.1 Introduction

Scanning probe microscopy (SPM) methods are used in many areas, ranging from maskless lithography to semiconductor metrology and ultrahigh-density data storage (1; 2; 3; 4; 5). However, for such applications to become widespread, two main issues have to be overcome, namely, low throughput and poor durability (i.e. tip and sample wear). Of course, to improve the throughput and the durability of SPM-based devices, one could build devices in which a large number of cantilevers are operated in parallel. However, even in such implementations, the throughput and the durability required for each cantilever remain a challenge.

For practical SPM-based devices, a throughput per cantilever in excess of 100 kpixel/s is desired, which is approximately two orders of magnitude higher than that in current commercial SPM setups. Likewise, an acceptable lifetime of such devices means that the cantilevers should be able to endure more than 100 m of continuous scanning, which is also about two orders of magnitude higher than the durability of commercial silicon-tip cantilevers. Clearly, a high-bandwidth and noninvasive imaging technique is required. Note that high-speed operation in current SPM setups is typically demonstrated at low lateral resolution or in small scan areas. Moreover, in general, tip and sample wear due to long-term scanning are not addressed in such setups.

High-speed SPM setups are typically based on contact-mode (CM) operation because of the associated fast response of the cantilever. The cantilever tip remains in constant contact with the sample, behaving like a beam that is clamped at both ends, which results in a high-bandwidth response on the order of hundreds of kHz. High-speed CM-SPM images are obtained by directly sampling the cantilever deflection signal (6; 7). This sampling technique is also attractive for use in parallel SPM devices because there only limited electronic resources are available per cantilever.

The stiffness k of CM cantilevers is usually on the order of 0.1 N/m to reduce the tip-sample loading force during imaging. However, the constant contact of the tip with the sample results in high lateral forces, giving rise to severe tip and sample wear. Several schemes have been proposed to reduce CM wear. Terris et al., for example, achieved low CM tip wear by building a feedback scheme in which a piezoresistive actuator is integrated on the cantilever to achieve low loading forces (8). However, in their application, the tip diameter was greater than 100 nm, whereas current applications require durably sharp tips with diameters of less than 10 nm. Recently, an approach was proposed in which the tip-sample force was modulated at high frequency during CM operation (9), which appears promising, but it is not entirely clear how modulating the normal forces will affect the sample properties.

Intermittent-contact (IC) mode operation is known to be less harsh on both the tip and the sample (10; 11; 12). Tapping mode (TM) is the most common form of IC mode operation (13). The cantilevers for TM are typically designed with a stiffness k on the order of 10 N/m and a quality factor Q on the order of 100. Reliable IC operation requires a high cantilever stiffness because in each oscillation cycle the tip has to overcome the tip-sample adhesion to move away from the sample surface. However, high stiffness is concomitant with low force sensitivity. For TM operation, the force sensitivity can be increased by increasing the Q factor of the cantilever, but at the price of a deterioration in the response time.

Using small cantilevers with a high resonant frequency in IC-mode operation and a peak-to-peak detector on the deflection signal, Ando et al. demonstrated high-speed (video-rate) imaging (12). However, the scan range was small, less than $1\,\mu m^2$, and their SPM setup and optical detection system were not designed for parallel SPM. Hong et al. (14) and Onaran et al. (15) used electrostatically-actuated cantilevers in TM and sensed the cantilever deflection by means of optical beam deflection and interferometry, respectively. Park et al. employed cantilevers with thermoelectric sensors in TM (16), inducing cantilever oscillation with a dither piezo. Krauss et al. (10) and Su et al. (11) reported reduced tip and sample wear using TM, at a low imaging bandwidth in the range of 1 kHz. In addition to its low operating speed, TM also requires demodulation electronics to extract the amplitude, phase and frequency information of each cantilever, rendering it unsuitable for parallel SPM devices for complexity reasons. Therefore, we investigated the possibility of achieving high-bandwidth IC-mode SPM by using cantilevers with low stiffness for high sensitivity and a low Q factor for high bandwidth that, moreover, are amenable for parallel operation

For parallel SPM, a large number of cantilevers is mass-fabricated with integrated sensors and actuators. Cantilevers with integrated piezoelectric or bimorph actuation with piezoresistive sensing as well as with electrostatic actuation using capacitive sensing have been proposed for parallel SPM (17; 18; 19). To demonstrate the proposed IC method, we considered cantilevers with integrated electrostatic actuation and thermoelectric sensing (20).

Realizing an effective noninvasive and high-bandwidth IC mode for parallel SPM applications poses several challenges. To be able to use low normal forces and thus reduce wear, small-amplitude oscillation and a low cantilever stiffness are essential. Small-amplitude oscillation is also beneficial for achieving high speed and hence high bandwidth. Therefore the problem of maintaining small-amplitude oscillation with low-stiffness cantilevers needs to be addressed. Moreover, for electrostatic actuation with low voltages and for high-resolution thermoelectric sensing, the tip height has to be less than approx. 1 μm, which implies a lower quality factor owing to squeeze-film damping. This, in turn, will significantly impact both the cantilever oscillation and the readout mechanism. In the following, we describe how these challenges can be resolved by using a cantilever that is electrostatically actuated and has integrated thermoelectric sensors.

7.2 SPM Setup for IC-Mode Operation

The IC method is demonstrated on a SPM setup that uses electrostatically-actuated cantilevers with integrated thermoelectric sensors.

A scanning electron micrograph of the cantilever is shown in Fig. 7.1. The cantilevers are fabricated with integrated electrostatic actuation, a concept that is widely used in MEMS, and thermoelectric sensors (20). The cantilevers are approx. 100 μm

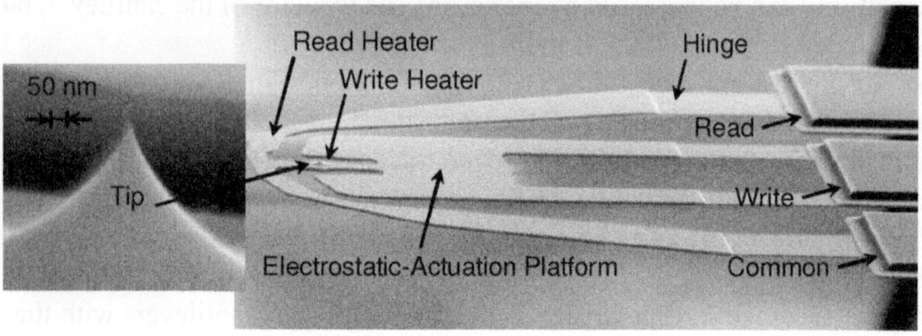

Fig. 7.1 Scanning electron micrograph of the cantilever with integrated thermoelectric sensors and electrostatic actuation platform. Reprinted from (21).

long, 60 μm wide and 1.5 μm thick. At the hinge, the cantilever is only 0.5 μm thick to impart low stiffness. Its nominal stiffness is approx. 0.15 N/m. A platform fabricated near the tip enables electrostatic actuation. The tip is approx. 600 nm high, i.e., rather low compared with commercial tips (which typically are $> 10\,\mu\text{m}$ high), to enhance electrostatic actuation and thermoelectric sensing. It is ultra-sharp, with a diameter of approx. 10 nm at the apex.

The deflection of the cantilever can be sensed by using either of the two integrated thermoelectric sensors, denoted "write heater" and "read heater", which are located at the base of the tip and on one of the cantilever legs, respectively. The thermoelectric sensors are regions of lower doping than the rest of the cantilever to make them more resistive. The read and write heaters are typically biased by applying a DC voltage between the "common" and the "read" or the "write" leg of the cantilever, respectively. During scanning, the sample is being moved below the cantilever, and the air gap between the sensors and the sample surface is modulated because of the tip traversing the sample topography, which changes the temperature and thus the resistance of the sensors.

To measure the cantilever position with respect to the sample surface, the current flowing through the common leg is monitored. In addition, it is possible to heat the sample surface and modify it thermomechanically using the write heater and the cantilever tip. These features were used at a later stage in the experiments to create and sense nano-indentations in a thin polymer film.

The SPM setup is shown schematically in Fig. 7.2. The cantilever is fixed, and the sample movable by means of a three-axis scanner. A conductive substrate under the sample enables electrostatic actuation. To actuate the cantilever, a voltage signal is applied between the write leg of the cantilever and the substrate. In typical SPM setups, the sample can be moved towards the cantilever using the z-scanner until contact between the tip and the sample is established, such that the cantilever is bent upwards during contact.

In contrast, in our setup, the sample is placed below the cantilever with a desired tip-sample separation, and the cantilever is actively bent towards the sample to make tip-sample contact by means of an electrostatic force. CM imaging is performed in

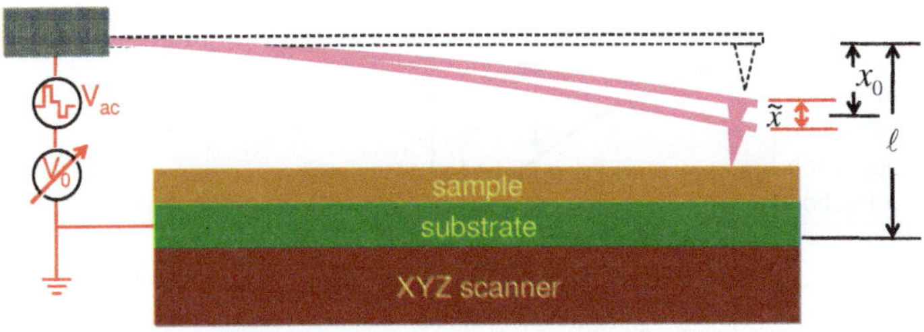

Fig. 7.2 Schematic of SPM setup and the intermittent-contact method. Reprinted from (21).

constant-height mode, in which a DC voltage is applied between cantilever and substrate to establish tip-sample contact, and the sample is raster scanned using the x- and y-scanners. The read heater is biased with a DC voltage V_R, and the cantilever deflection signal from the read sensor is sampled directly to form the imaging signal. Alternatively, CM images can be obtained using the write sensor. The working principle of the SPM setup is described in detail in Section 7.4.

7.3 Methodology

The proposed IC-mode SPM method is described next with the help of Fig. 7.2. The cantilever is actively bent towards the sample by applying a DC voltage V_0 between the write leg (ground) and the substrate underneath the sample to induce a mean deflection x_0. An alternating voltage signal $V_{ac}(t)$ is added to V_0 to achieve a small oscillation of amplitude $\tilde{x}$. $V_{ac}(t)$ is chosen such that the net voltage $V(t) = V_0 + V_{ac}(t)$ does not change sign at any time t.

The dynamics of the cantilever used in the simulation model is given by

$$m\ddot{x}(t) + c\dot{x}(t) + kx(t) = \frac{\varepsilon_0 A}{2(\ell - x(t))^2} V(t)^2 + \phi(x(t)), \qquad (7.1)$$

where m is the effective mass, c the damping coefficient, k the spring constant and $x(t)$ the cantilever deflection. In addition, ε_0 is the electric permittivity of air, $V(t)$ is the net voltage applied to the cantilever for electrostatic actuation, ℓ is the effective separation and A the area between cantilever and substrate underneath the sample. Finally, $\phi(\cdot)$ denotes the nonlinear tip-sample interaction force. For brevity, we will sometimes drop the time dependence of the functions $x(t)$ and $V(t)$. The electrostatic force is modeled by assuming that the cantilever and the substrate behave as two oppositely charged parallel plates. Equation (7.1) shows that the mechanical restoring force on the cantilever increases linearly with x, whereas the electrostatic pull-in force increases as a nonlinear function of x. Accordingly, the cantilever will crash into the sample if the deflection $x > \ell/3$ (22). Therefore, we ensure that the cantilever is operated in the safe region, i.e., that $x(t) < \ell/3$ at all times t to guarantee smooth IC operation.

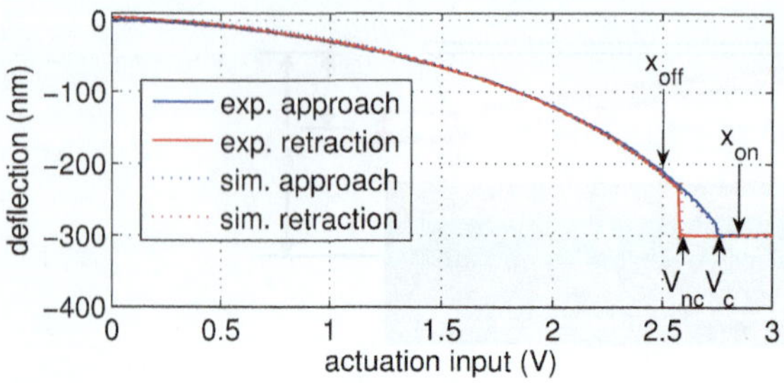

Fig. 7.3 Comparison of simulated and experimental approach-retract curves obtained by electrostatic actuation of the cantilever. The experimental deflection signal is obtained using the read sensor. Reprinted from (21).

A frequency sweep and an approach-retract curve using electrostatic actuation were used to obtain the parameters of the cantilever and the electrostatic force parameters in Eq. (7.1) experimentally, see Fig. 7.3 (23). This yielded a resonant frequency $\omega_0 = \sqrt{k/m}$ of 50.7 kHz, a quality factor $Q = \omega_0 m/c$ of 0.73 and an electrostatic force constant $K_{\text{esf}} = \varepsilon_0 A/2m$ of the cantilever near the sample surface of 5.28×10^{-22} Nm^2/V^2. For the simulations, a piecewise linear model for the tip-sample interaction force $\phi(\cdot)$ was used with repulsive and attractive (adhesive) spring constants of 10.5 and 1.2 N/m, respectively (24).

For IC operation, a cantilever with the approach curve shown in Fig. 7.3 is typically actuated by applying a periodic square waveform whose amplitude varies from the retract voltage V_r (e.g., 2.5 V) to the approach voltage V_a (e.g., 2.9 V). This is usually implemented by adding an offset voltage $V_0 = 2.7$ V to a pulsed signal with 0.2 V amplitude. The highly damped cantilever ($Q = 0.73$) will oscillate between the off-sample position x_{off} and the on-sample position x_{on}, as indicated in Fig. 7.3. Because the approach and retract forces are small, such an actuation signal will result in slow IC operation. The cantilever is also prone to "stick" to the soft sample surfaces because its stiffness is very low (< 0.15 N/m, owing to a weakening of the electrostatic field gradient). Note that the stiffness of commercially available IC-mode cantilevers typically is on the order of 10 N/m to achieve reliable IC operation.

7.3.1 Input Shaping

In our IC-mode method, we address the problem of reliability in small-amplitude operation and the slow operating speed by shaping the input actuation signal, as will be described next.

Suppose that the tip of the cantilever is in contact with the sample, such that its deflection is given by x_{on}. For IC operation, the cantilever is retracted from the sample surface by applying a voltage V_r between cantilever and substrate. The corresponding restoring (retraction) force is given by $F_r = kx_{\text{on}} - \varepsilon_0 A V_r^2/2(\ell - x_{\text{on}})^2$ and can be increased by reducing V_r. The maximum instantaneous restoring force given

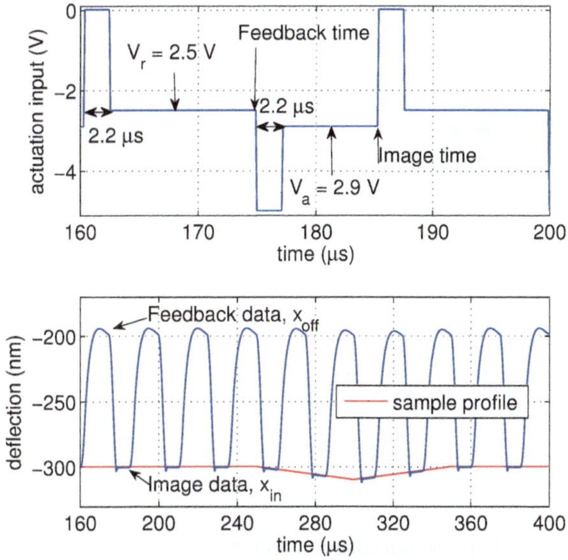

Fig. 7.4 Simulation results: The shaped actuation input (top) and the corresponding deflection of the cantilever (bottom) in the proposed IC-mode scheme. Reprinted from (21).

by $F_r\,|_{V_r=0} = kx_{on}$ typically is sufficient to overcome the adhesion force for different sample surfaces. Therefore, to improve the speed of IC operation, it is possible to reduce the retraction time needed to change the cantilever deflection from x_{on} to x_{off} significantly by applying $V_r = 0$ V. To achieve small-amplitude oscillation between x_{on} and x_{off}, $V_r = 0$ V is applied for a short period of time (2.2 μs in the simulation example), followed by the steady-state retraction voltage $V_r = 2.5$ V corresponding to the off-sample position x_{off}. The results simulating the retraction input shaping are shown in Fig. 7.4.

Likewise, the input signal can be shaped for approach during an IC oscillation cycle. The input actuation force during approach for a cantilever deflection of x is given by $F_a = \varepsilon_0 A V_a^2 / 2(\ell - x)^2 - kx$ and can be increased by increasing V_a. The approach time can be reduced significantly by applying the maximum allowed input voltage ($V_{max} = 5$ V in the example of Fig. 7.4) to provide an initial, large acceleration to the cantilever. Then, before the cantilever contacts the surface, the magnitude of the approach voltage can be reduced to the steady-state approach voltage $V_a = 2.9$ V corresponding to the on-sample position x_{on} in Fig. 7.3. For small tip-sample interaction forces, this has the added benefit of reducing the cantilever velocity to zero at the time of contact.

The cantilever model in Eq. (7.1) can be used to calculate the time interval for applying the retract and the approach voltage pulses to achieve effective IC operation. However, it is advantageous to fine-tune the shaped retract and approach voltage pulses and their corresponding switching times in test experiments to achieve the desired IC performance. The IC operation of the cantilever shown in Fig. 7.4

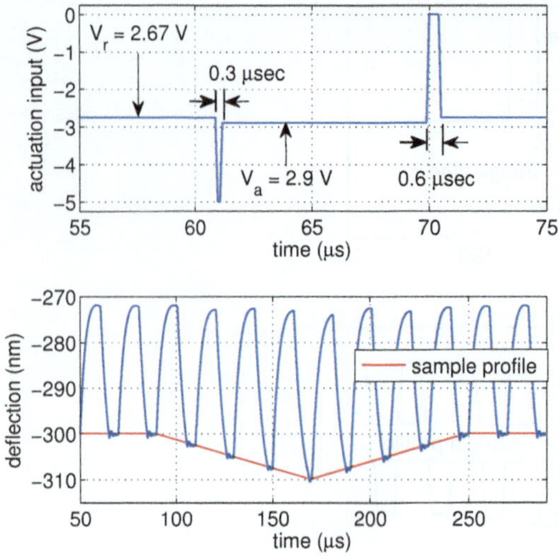

Fig. 7.5 Simulation: IC operation at 50 kHz within adhesion region

corresponds to an operating frequency of 40 kHz. Clearly, the operating speed can be further increased by reducing the tip-sample contact time. In the simulation results shown in Fig. 7.5, the operating frequency is increased to 50 kHz by reducing the peak-to-peak oscillation amplitude to 25 nm. Figure 7.5 also shows that the retract voltage $V_r = 2.67$ V falls into the adhesion region in Fig. 7.3. Therefore, input shaping can be exploited to overcome the effect of tip-sample adhesion. Note also that the cantilever could not be operated beyond 25 kHz without employing input shaping because of the tip-sample adhesion.

An additional advantage of the proposed method, revealed by the simulation results in Fig. 7.4, is that it is not necessary to use complex demodulation electronics to obtain the imaging data. The topography signal can be captured by directly sampling the cantilever deflection signal when the tip is in contact with the sample because the deflection signal follows the topography precisely. We also performed simulations to verify that the low quality factor and the nonlinear electrostatic actuation force indeed contribute to this behavior. Thus, the topography signal is obtained by sampling the deflection signal just before the actuation signal V_r is set to 0 V for retraction. This implementation simplifies the electronic circuitry considerably and is thus particularly well suited for SPM using large arrays of cantilevers in parallel.

7.3.2 Parallel IC Operation

Key to operating a large array of cantilevers reliably in IC mode is that the same operating voltages and switching times are used for all cantilevers.

The approach voltage V_a for IC operation is typically set to a value greater than the minimum actuation voltage V_c required to bring the cantilever tip in contact with the sample surface (see Fig. 7.3). V_c is independent of the variations in the surface forces and can be calculated from Eq. (7.1) with the tip-sample interaction force $\phi = 0$. It depends on the device parameters k, c, K_{esf} and ℓ. Thus, V_a can be set to a constant value independently of the sample type if k, c, K_{esf} and ℓ remain constant. In the case of an implementation with parallel cantilevers, fabrication tolerances may result in variations in the individual k, c, K_{esf} and ℓ values of the cantilevers. However, as in general batch fabrication is used, these variations are expected to be small. Accordingly, approach experiments on each cantilever are performed to obtain the individual V_c values of the cantilevers prior to imaging, and a common approach voltage V_a is set to a value greater than the maximum V_c value obtained. Likewise, the variation in the switching times of the approach short pulse (i.e., the 5 V pulse in Fig. 7.4) among the cantilevers is expected to be minimal because of batch fabrication. To ensure reliable small-amplitude parallel IC operation, a common switching time corresponding to the maximum time duration of the approach short pulse can be selected.

The retract voltage V_r is typically set to a value less than the maximum actuation voltage V_{nc} required to pull the cantilever tip out of contact from the sample surface (see Fig. 7.3). Unlike V_c, V_{nc} strongly depends on the variations in the tip-sample interaction force ϕ. For example, a larger adhesion force may cause the cantilever tip to remain in contact with the sample surface for a longer period of time during retraction, see Fig. 7.3. The surface forces can vary significantly over the scan area of the sample. However, the variation in V_{nc} does not affect small-amplitude IC operation if the actuation signal is shaped as shown in Fig. 7.4. To achieve a desired minimum oscillation amplitude, common values are chosen for V_r and for the switching time for the retract short pulse (i.e., the 0 V pulse in Fig. 7.4). These values may either be chosen with the help of Eq. (7.1), assuming $\phi = 0$, or tuned online during parallel IC operation.

7.3.3 Feedback for Reliable IC Operation

Small-amplitude oscillation is desired in the proposed IC scheme to improve the operating speed and reduce the tip-sample loading force. In practice, however, small-amplitude oscillation is difficult to maintain because of tip-sample adhesion, micro-scale irregularities and contamination of the sample. As we have seen, shaping of the input voltage signal can effectively addresses these issues. In addition, a feedback scheme can also be employed to maintain reliable small-amplitude IC operation.

Note that in Fig. 7.4 the deflection signal corresponding to the cantilever position x_{off} is largely independent of the topography of the sample. Instead, it is a good indication of the off-contact position. The off-contact signal x_{off} can be sampled just prior to the application of $V_{\max}$ and can be used to obtain a fixed cantilever-sample separation. This is achieved by using a feedback regulator that modulates the offset voltage V_0 and/or the retraction voltage V_r applied between cantilever and substrate.

The bandwidth of the feedback scheme can be estimated by linearizing the non-linear cantilever response given by Eq. (7.1) near the sample surface. For feedback operation, the measured and controlled variable is $x_{\text{off}}(nT)$ and the input variable is $V_0(nT)$, where T is the duration of IC operation and n the index of the oscillation cycle. In each oscillation cycle n, $V_0(nT)$ is computed and applied to the cantilever. The low quality factor of the cantilever, which is close to the value corresponding to the critical damping factor of a second-order system, together with a high resonant frequency and input shaping enable the cantilever to reach the desired position within one natural oscillation cycle. The bandwidth of the feedback scheme may therefore be as large as or higher than the bandwidth of the cantilever, which is already high. To compute $V_0(nT)$, a proportional and integral controller was used in the experiments.

We tested the feedback scheme experimentally. For this experiment, the sample was fixed underneath the cantilever, with a tip-sample separation of 300 nm. The voltages for IC operation, $V_r = 2.5$ V and $V_a = 2.9$ V, were obtained from Fig. 7.3. In the experiment, a $V_0 = 2.7$ V from a computer-controlled digital-to-analog converter (DAC) circuit and a square waveform of 25 kHz frequency and 0.2 V amplitude from a waveform generator were added and applied between cantilever and substrate. The deflection signal of the cantilever was measured by the thermoelectric read sensor with a bias voltage of $V_r = 2.25$ V. Note that during the experiment the sample was not scanned and no input shaping was employed. IC operation without and with V_0-based voltage control is shown in the upper and the lower panel of Fig. 7.6, respectively. Specifically, the results without feedback control (upper panel) reveal a time window during which the tip could not come out of contact with the sample (polymer) surface during IC operation. If V_0 control is implemented, the

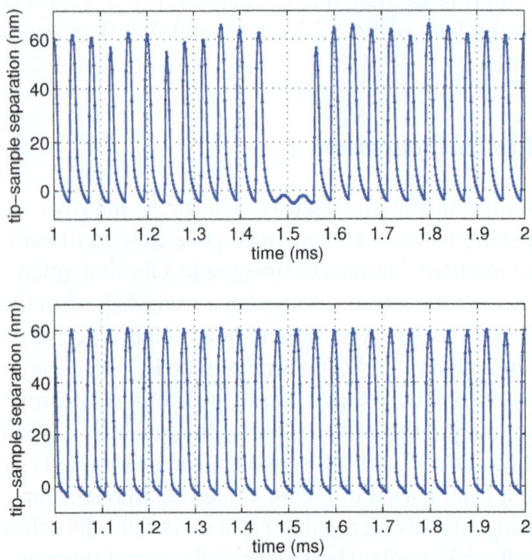

Fig. 7.6 Deflection signals from an IC experiment without (top) and with feedback (bottom). Without the feedback loop, the cantilever spontaneously adheres to the polymer surface.

deflection signal and thus the trajectory of the tip motion are periodic and homogeneous because of the feedback (lower panel).

The off-contact signal $x_{\rm off}$ from the thermoelectric sensor also captures the drifts in the SPM setup that affect the separation between cantilever and sample as well as any slope of the sample surface during scanning. Such low-frequency effects may be corrected by a feedback scheme that varies the input reference signal applied to the z-scanner controller. In this way, a uniform oscillation of the cantilever over time and over the scan area of the sample surface can be maintained.

7.4 Experimental Results

Experiments were carried out on the SPM setup described in Section 7.2 to demonstrate the high-speed capability and the noninvasive nature of the proposed IC-mode imaging method. To demonstrate high-speed imaging, a polymer surface imprinted with nano-indentations at ultra-high density was used. The wear effects on the tip and the sample were assessed by performing continuous scanning over extended periods of time to demonstrate the noninvasive nature of the method.

In the SPM setup used, the cantilever is fixed and the sample can be moved by means of the x-, y- and z-scanners. First, an approach-retract experiment was performed to fix the sample position in the vertical (z) direction to obtain a desired tip-sample separation for smooth IC operation. A triangular reference voltage signal with a frequency of 1 Hz was applied to the z-scanner controller (from Physik Instrumente (PI)) to move the sample towards the cantilever until tip-sample contact occurred and then to retract it to its original position. The absolute position of the z-scanner was captured from the embedded (PI) capacitive sensors. The cantilever position with respect to the sample was measured using the integrated thermoelectric sensor, i.e., the "read heater" shown in Fig. 7.1. A DC voltage of 2.25 V was applied between the "read" and the "common" (virtual ground) legs of the cantilever to bias the sensor, and the corresponding current was captured as the "thermal signal".

The relative position of the z-scanner and the corresponding reference voltage signal applied to the z-scanner controller are plotted against the thermal signal in Fig. 7.7. The tip-sample contact position can be detected accurately from the thermal signal. The horizontal axis in Fig. 7.7 is offset so that its origin is at the position where contact between the tip and the sample occurs. Thus, the in-contact and the out-of-contact positions are denoted by positive and negative values, respectively. At this length scale, the out-of-contact thermal signal in Fig. 7.7 varies linearly with the tip-sample separation because of the associated change of the air gap between the read heater and the sample. Note that the resistance of a thermoelectric sensor varies linearly with the air gap because of sample-dependent cooling. The slope of the in-contact thermal signal is significantly smaller in Fig. 7.7 because the separation between cantilever and sample did not vary as the tip was in contact with the sample. However, the air gap changes slightly because the bending of the cantilever gives rise to a rotation of the read heater, which results in a slow change in the thermal signal. At the end of the approach-retract experiment, a precalculated DC reference

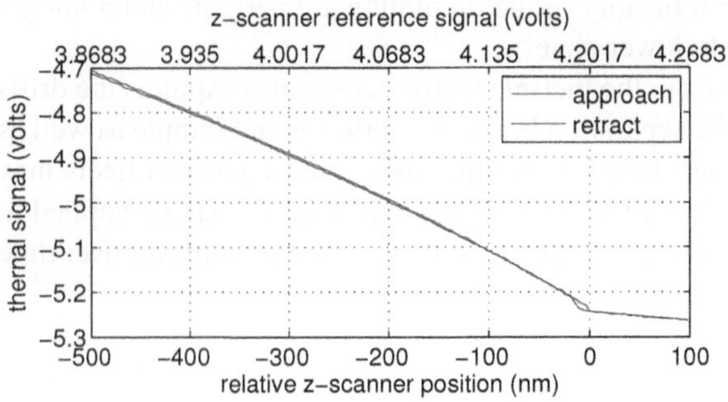

Fig. 7.7 An approach-retract curve obtained by moving the *z*-scanner

voltage (i.e., 4.0017 V, from Fig. 7.7) was applied to the *z*-scanner controller to obtain a fixed initial tip-sample separation of 300 nm for the experiments described next.

The next experiment performed was an approach-retract experiment using electrostatic forcing. To bias the read heater, a DC voltage of 2.25 V was applied between the "read" and the "common" (virtual ground) leg of the cantilever. A positive triangular voltage signal of 1 Hz frequency was applied between the "write" (ground) leg (see Fig. 7.1) and the substrate underneath the sample such that the cantilever was bent towards the sample until tip-sample contact occurred; then it was retracted to its original position. A plot of the thermal signal captured during the experiment versus the input voltage signal is shown in Fig. 7.3, where the nonlinear effect of the input voltage signal on the cantilever deflection described by Eq. (7.1) can be observed. Unlike the approach-retract experiment using the *z*-scanner, the thermal signal remained flat when the cantilever was in contact with the sample. This implies that when electrostatic forcing is used, the thermal signal will be free of artifacts due to the bending of the cantilever. The tip-sample contact position can be detected accurately from the thermal signal in Fig. 7.3. The voltages required to bring the cantilever into and out of contact with the sample were $V_c = 2.76$ V and $V_{nc} = 2.58$ V, respectively. The operating voltages and the timing for IC operation described in Section 7.3 were then chosen as shown in Fig. 7.4.

Next, we identified the cantilever dynamics when operating near the sample surface. The cantilever was first bent towards the surface by applying a DC voltage of $V_0 = 2.6$ V between the write (ground) leg of the cantilever and the substrate underneath the sample. This corresponds to a tip-sample separation of approx. 60 nm (see Fig. 7.3). Then, a 50-mV amplitude chirp signal (V_{ac}) from an Agilent 33250A network analyzer was added to V_0 as shown in Fig. 7.2. The frequency response data computed by the analyzer and a third-order transfer function fit are shown in Fig. 7.8. The first-order component of the transfer function fit was considered as the model of the thermoelectric sensor. The -3 dB bandwidth of the sensor was approx. 8.4 kHz. The second-order component of the transfer function fit was considered as the linear model of the cantilever near the surface as shown in Eq. (7.1). The

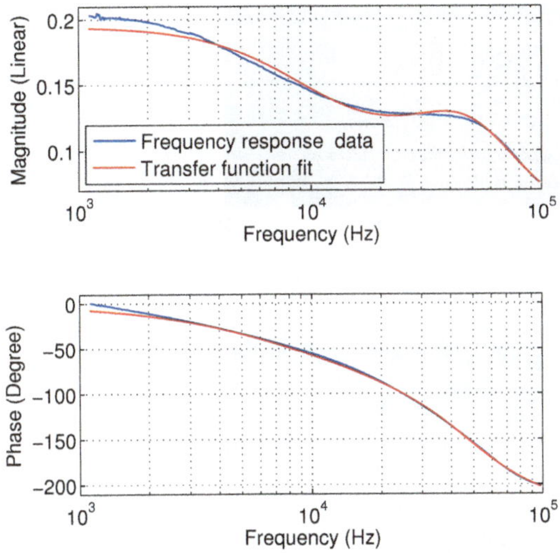

Fig. 7.8 Experimental frequency response data and third-order transfer function fit

equivalent resonant frequency and the quality factor of the cantilever model were 50.7 kHz and 0.73, respectively. These parameters were used in the simulations described in Section 7.3.

After the initial calibration experiments, imaging and wear experiments in IC mode were performed. For imaging, a sample was chosen that could show the resolution achievable by the cantilever tip during continuous scanning: For this purpose, nano-indentations were formed in a thin polymer film (polyaryletherketone) spuncast onto a silicon wafer as described in (5). From a data-storage point of view, the presence and the absence of indentations correspond to the logical symbols "1" and "0", respectively. The minimum distance between any two indentations was 29.33 nm (both symbol and line pitch), which corresponds to 1 Tb/in^2 of storage density after encoding the stored data with a suitable (d,k) modulation code (20).

To image this modified surface, IC-mode imaging with input shaping, described in Section 7.3.1 and illustrated in Fig. 7.4, at an operating frequency of 40 kHz was used. The values chosen for the proportional and the integral gain for V_0-based feedback control, see in Section 7.3.3, were $K_p = -5 \times 10^{-2}$ and $K_i = -1 \times 10^{-6}$, respectively. In each oscillation cycle of the cantilever, the thermoelectric sensor signal was sampled once for the imaging signal. The scanner speed was 146.7 μm/s, which translates into a spatial resolution of 3.67 nm per pixel of the image. The distance between consecutive logical symbols was 14.67 nm, resulting in a readback rate of 10 kSymbol/s at the scanner speed chosen. The image of the polymer surface obtained using IC-mode SPM is shown in the upper panel of Fig. 7.9 and the corresponding topography signal from one line (section view) in the lower panel.

To determine the invasiveness of our imaging scheme, the imaging process as described above was continued for several days to collect long-term tip and sample wear data. The relative humidity was maintained below 5%. To compensate for

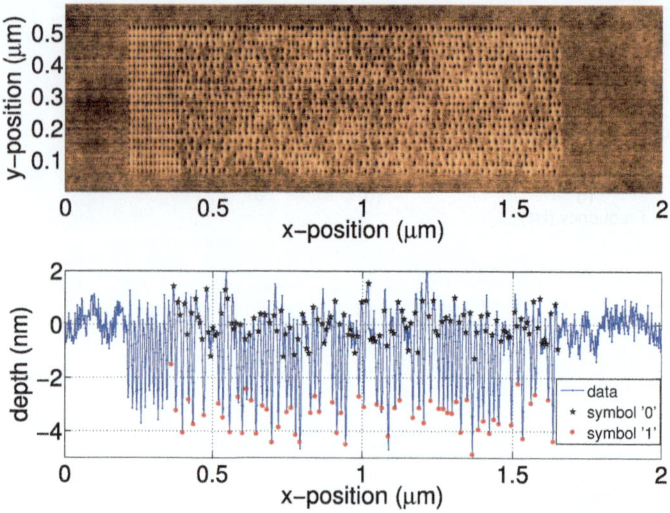

Fig. 7.9 SPM image obtained using the IC-mode method (top) and the corresponding topography signal (bottom). Reprinted from (21).

the long-term drift in the SPM setup caused by temperature fluctuations, the feedback scheme based on the off-contact signal x_{off}, see Section 7.3.3, was applied to the z-scanner with integral gain $K_i = -1 \times 10^{-7}$. In accordance with standard read-endurance performance specifications in nonvolatile devices, the long-term experiment was terminated after approx. 10^{10} logical bits had been acquired. The symbols "0" and "1" were detected from the imaging (topography) signal, and the difference between the corresponding signal values yielded the indentation "amplitude" signal.

Figure 7.10 is a plot of the normalized average indentation amplitude signal over the entire scan area and for the entire duration of the long-term IC imaging experiment. By the end of the experiment, the tip had traveled a distance of approx. 140 m over the polymer surface in IC mode, and the scanned area of the polymer surface had been imaged a total of 6632 times. Figure 7.10 shows that, after an initial

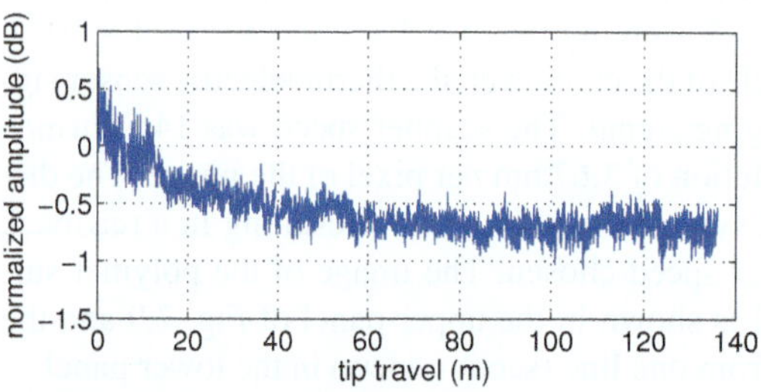

Fig. 7.10 Normalized average indentation amplitude in dB during a long-term scanning experiment. Reprinted from (21).

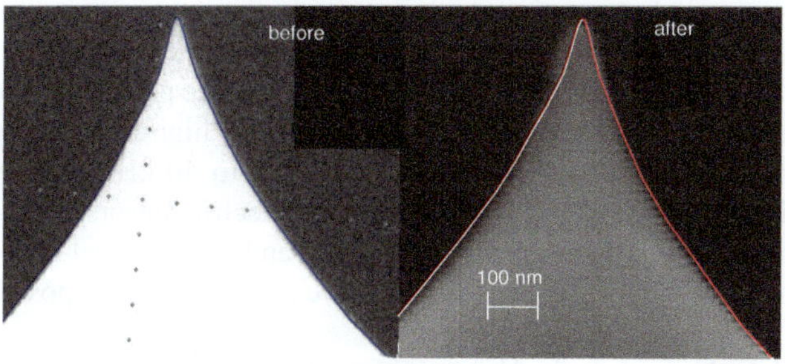

Fig. 7.11 Scanning electron micrograph of the tip before (left) and after (right) a long-term scanning experiment. Reprinted from (21).

transient period, the indentation amplitude signa degraded very slowly. This signal is a measure of the quality of the imaging signal and hence of both the sharpness of the tip and the retained quality of the polymer surface. The estimated amplitude loss rate is 0.014 dB/decade (with decade here referring to the number of logical bits imaged), which is a substantial improvement compared with the 1.8 dB/decade typically observed in contact-mode operation. This result is truly remarkable, especially in view of the fact that a part of the signal decay is caused by the inherent relaxation of indentations in a polymer medium (25). To illustrate the noninvasive nature of the proposed IC-mode SPM, Fig. 7.11 shows scanning electron micrographs of the tip before and after the IC experiment: Apart from a slight contamination, the tip remained almost unchanged during the experiment. In summary, although a standard silicon tip has been used to image a surface at high resolution for approx. 140 m, only minimal tip and sample wear resulted thanks to using the new IC mode of operation.

7.5 Conclusion

Electrostatic actuation can be exploited for high-bandwidth intermittent-contact (IC) imaging because useful imaging data can be captured at the oscillation frequency of the cantilever using input-based triggering. For reliable small-amplitude operation, input shaping and off-contact-signal-based feedback can be applied to eliminate the effect of tip-sample adhesion, which increases the speed of operation and reduces the tip-sample interaction force. In experiments, high-bandwidth imaging at a rate of 40 kpixel/s has been demonstrated using a cantilever that oscillates at 40 kHz. An improvement in bandwidth of more than one order of magnitude has been obtained over imaging in amplitude-modulated tapping mode using cantilevers of comparable resonant frequency, which typically offers an imaging rate of 1 kpixel/s (corresponding to an amplitude demodulation bandwidth of 1 kHz). Minimal tip-sample interaction has been demonstrated in long-term scanning experiments, in which 140 m of equivalent tip travel were achieved with only minimal tip and sample wear.

Thermoelectric sensors and electrostatic actuators provide a convenient way to integrate sensors and actuators on a cantilever microstructure, which is key for building compact and parallel SPM devices. Moreover, the IC scheme presented can be used in massively parallel architectures such as are needed for high-throughput SPM. The imaging bandwidth in our experiments was limited to 40 kHz mainly because of the low sensing bandwidth of the integrated thermoelectric sensors. However, thermoelectric sensors with higher sensing bandwidths and system-level techniques to improve the sensing bandwidth even further have already been proposed (26; 27).

The imaging rate of the scheme described here scales with the resonant frequency of the cantilever. We expect to achieve an imaging rate in excess of 100 kpixel/s with the cantilever described in (26), so that parallel images with an aggregate rate of 10 Mpixel/s can be obtained using an array of one hundred such cantilevers operating in parallel.

The proposed technique is particularly suitable for industrial SPM applications because of its combination of imaging bandwidth improvement and minimal invasiveness, and the fact that it is tailored for cantilever arrays. We believe that this constitutes a significant step towards promoting a widespread use of SPM devices in areas such as semiconductor metrology, lithography and data storage.

Acknowledgements. We thank the nanofabrication group of IBM Research - Zurich, in particular Michel Despont and Ute Drechsler, for the design and fabrication of the cantilevers used in this work. We also thank Peter Bächtold and Charlotte Bolliger for their help with the electronic design and in preparing the manuscript, respectively. This work was partly supported by the European Research Council under the project ProTeM.

References

[1] Pires, D., et al.: Science 328(5979), 732 (2010)
[2] Knoll, A.W., et al.: Advanced Materials 22(31), 3361 (2010)
[3] Wei, Z., et al.: Science 328(5984), 1373 (2010)
[4] Oliver, R.A.: Reports on Progress in Physics 71, 076501 (2008)
[5] Pantazi, A., et al.: IBM Journal of Research and Development 52(4/5), 493 (2008)
[6] Picco, L.M., et al.: Nanotechnology 18, 044030 (2007)
[7] Schitter, G., Rost, M.J.: Materials Today 11, 40 (2008)
[8] Terris, B.D., Rishton, S.A., Mamin, H.J., Ried, R.P., Rugar, D.: Applied Physics A. Material Science & Processing 66(2), S809 (1998)
[9] Lantz, M.A., Wiesmann, D., Gotsmann, B.: Nature Nanotechnology 4, 586 (2009)
[10] Kraussa, P.R., Choub, S.Y.: Applied Physics Letters 71, 3174 (1997)
[11] Su, C., Huang, L., Kjoller, K., Babcock, K.: Ultramicroscopy 97, 135 (2003)
[12] Ando, T., Uchihashi, T., Fukuma, T.: Progress in Surface Science 83, 337 (2008)
[13] Bhushan, B. (ed.): Springer Handbook of Nanotechnology. Springer, Heidelberg (2006)
[14] Hong, J.W., Khim, Z.G., Hou, A.S., Park, S.: Applied Physics Letters 69(19), 2831 (1996)
[15] Onaran, A.G., et al.: Review of Scientific Instruments. 77, 023501 (2006)

[16] Park, K., Lee, J., Zhang, Z.M., King, W.P.: Review of Scientific Instruments. 78, 043709 (2007)

[17] Minne, S.C., Manalis, S.R., Quate, C.F.: Applied Physics Letters 67, 3918 (1995)

[18] Rangelow, I.W., et al.: Microelectronic Engineering 84, 1260 (2007)

[19] Miller, S.A., Turner, K.L., MacDonald, N.C.: Review of Scientific Instruments 68, 4155 (1997)

[20] Pozidis, H., Häberle, W., Wiesmann, D., Drechsler, U., Despont, M., Albrecht, T., Eleftheriou, E.: IEEE Transactions on Magnetics 40(4), 2531 (2004)

[21] Sahoo, D.R., Häberle, W., Sebastian, A., Pozidis, H., Eleftheriou, E.: Nanotechnology 21(7), 075701 (2010)

[22] Lu, M.S.C., Fedder, G.K.: Journal of Microelectromechanical Systems 13(5), 759 (2004)

[23] Agarwal, P., Sahoo, D., Sebastian, A., Pozidis, H., Salapaka, M.: IEEE Journal of Microelectromechanical Systems 19(1), 83 (2010)

[24] Sebastian, A., Salapaka, M.V., Chen, D., Cleveland, J.P.: Journal of Applied Physics 89(11), 6473 (2001)

[25] Knoll, A., Wiesmann, D., Gotsmann, B., Dürig, U.: Physical Review Letters 102, 117801 (2009)

[26] Rothuizen, H., Despont, M., Drechsler, U., Hagleitner, C., Sebastian, A., Wiesmann, D.: In: Proceedings IEEE 22nd Intl. Conf. on Micro Electro Mechanical Systems (MEMS 2009), pp. 603–606 (2009)

[27] Sebastian, A., Wiesmann, D., Baechtold, P., Rothuizen, H., Despont, M., Drechsler, U.: In: Proc. 2009 Intl. Solid-State Sensors, Actuators and Microsystems Conf. (TRANSDUCERS 2009), pp. 1963–1966 (2009)

Chapter 8
Systems and Control Approach to Electro-Thermal Sensing

Abu Sebastian

Abstract. Electro-thermal sensors serve as low-cost, highly integrable deflection sensors in scanning probe applications. They also find application as position sensors for nanopositioning applications. These sensors consist of microfabricated silicon structures with integrated heating elements. The dynamics of these sensors are well captured by a feedback model consisting of a linear thermal system and a non-linear memoryless operator. During sensing, the signal being measured perturbs the thermal system. This perturbation can be analyzed to derive the sensing transfer function. This systems approach is particularly well suited for the experimental identification of electrothermal sensors. Moreover, the sensitivity and bandwidth of these sensors can be further enhanced by external feedback, thus illustrating the applicability of feedback control to shape a sensors sensing dynamics.

8.1 Introduction

Heated probes find applications in areas ranging from microthermal analysis to the study of nanoscale heat transport (1; 2). They can be used to generate fast heat pulses and to locally heat surfaces with high spacial resolution. They can also serve as low-cost, highly integrable topography sensors in scanning-probe applications (3; 4; 5). A scanning electron micrograph of one such integrated topography sensor is shown in Fig. 8.1(a). A simplified schematic is shown in Fig. 8.1(b). Parts of the cantilever are designed to be micro-heaters having an area of approximately 4 μm × 4 μm. The heaters are produced by varying the doping levels and can be heated by applying an electrical current through the cantilever legs and the heater. Typically, constant voltage excitation is employed.

The heat transportation is primarily through the air gap between the heater and the sample and through the cantilever legs. The micro-heaters are heated to a certain

Abu Sebastian
IBM Research - Zurich, 8803 Rüschlikon, Switzerland
e-mail: ase@zurich.ibm.com

E. Eleftheriou & S.O.R. Moheimani (Eds.): Cntrl. Tech. for Emerging Micro/Nanoscale Sys. LNCIS 413, pp. 137–152.
springerlink.com

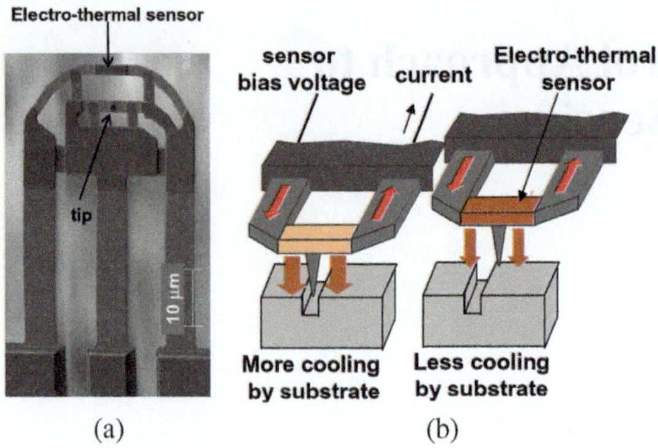

Fig. 8.1 (a) Scanning electron micrograph of a micro-fabricated silicon cantilever with integrated electrothermal sensor. (b) Schematic of the sensor illustrating the mode of operation.

temperature, and the subsequent topography sensing is based on the following two premises. (1) The heat conduction through air depends on the distance of the sensor from the substrate. (2) The resistance of the heater is a strong function of the temperature. As the cantilever tip follows the topography, the distance of the micro-heater from the substrate gets modulated by the topographical variations. This translates to a difference in cooling. The subsequent change in the electrical resistance is measured as a change in the current in the case of constant voltage excitation.

Owing to the seemingly complicated dynamics, the interplay between thermal and electrical responses and finite-size effects, numerical simulations are widely preferred over analytical approaches for investigating the operation of microheaters (6; 7). One approach to analytical modeling is to start from physical principles governing the heat transfer and the electrical characteristics of doped silicon and then arrive at dynamical relations between the signals of interest (see (8)). The closeness of the model to physically interesting variables make this approach particularly interesting. However, the intricate design of the microheaters, fabrication tolerances, and finite size effects make the direct application of this approach less practical. A systems approach is particularly well suited to model these sensors. Such a systems-inspired model of the micro-heater was presented in (9) and (10).

8.2 Systems Model

The main components of the systems model, see Fig. 8.2, are a linear operator relating temperature with input power and a nonlinear operator relating the electrical resistance with temperature. The linear operator is modeled by $\mathbf{T}_{TP_x}$ and it captures the dynamics of thermal conduction as a function of the power dissipated in the heating elements when the heater sample separation equals x. $\mathbf{T}_{TP_x}$ captures the thermal system for that particular x. The memoryless nonlinear relationship between temperature and the electrical resistance is modeled by $g(.)$ which is typically a bell-shaped

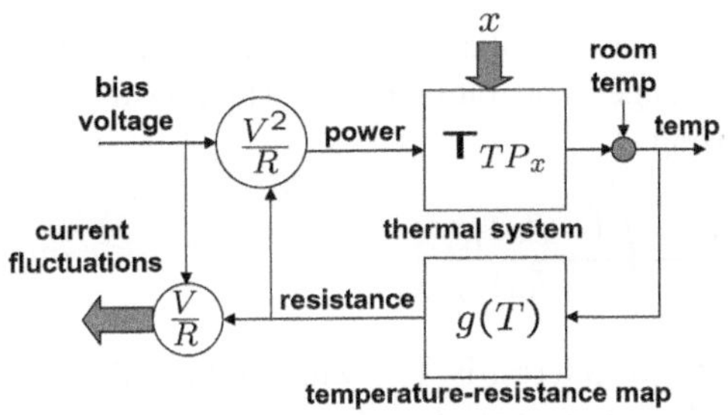

Fig. 8.2 The model of the electrothermal sensor derived from a systems perspective. The primary components are a linear thermal system and a nonlinear memoryless operator relating electrical resistance to temperature.

curve with the resistance value reaching a maximum at a certain temperature $T_{\max}$. Below $T_{\max}$, the resistance increases with temperature because of a corresponding decrease in the mobility of the majority carriers. However, above $T_{\max}$, the resistance becomes smaller with increasing temperature owing to the predominance of the thermally activated increase of the carrier density. The signal that is measured experimentally is the current, i.e., the input voltage divided by the resistance of the heater.

For small-signal analysis, the various nonlinear blocks are linearized about an operating point using the Taylor series expansion. The resulting linear model is shown in Fig. 8.3, where $V = V_0 + \tilde{V}$, $I = I_0 + \tilde{I}$, $R = R_0 + \tilde{R}$, $T = T_0 + \tilde{T}$, and $P = P_0 + \tilde{P}$. Such a linear model facilitates the calculation of transfer functions relating various signals of interest. One of them is the transfer function relating the current fluctuation with the input voltage fluctuation at a certain heater-sample separation x and a certain operating voltage, denoted by $\mathbf{T}_{\tilde{I}\tilde{V}_x}$. The importance of this transfer function stems from the fact that it is easier to measure experimentally and that it can be used to obtain $\mathbf{T}_{TP_x}$. From the linear model, $\mathbf{T}_{\tilde{I}\tilde{V}_x}$ is given by

$$\mathbf{T}_{\tilde{I}\tilde{V}_x} = \frac{1}{R_0}\left(\frac{1 - I_0^2 g'(T_0)\mathbf{T}_{TP_x}}{1 + I_0^2 g'(T_0)\mathbf{T}_{TP_x}}\right), \tag{8.1}$$

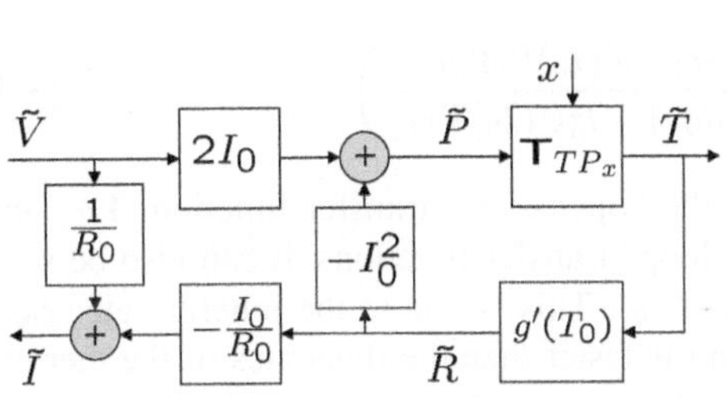

Fig. 8.3 Linearized version of the electrothermal sensor model

where $g'(T_0)$ denotes the slope of $g(.)$ evaluated at T_0. Note that the linear assumption for $\mathbf{T}_{\tilde{I}\tilde{V}_x}$ holds only for small fluctuations in voltage and current, whereas $\mathbf{T}_{TP_x}$ is assumed to be linear over the whole range of operating points. The transfer function, $\mathbf{T}_{TP_x}$ can be derived from $\mathbf{T}_{\tilde{I}\tilde{V}_x}$ using

$$\mathbf{T}_{TP_x} = \frac{1}{I_0^2 g'(T_0)} \left(\frac{1 - R_0 \mathbf{T}_{\tilde{I}\tilde{V}_x}}{1 + R_0 \mathbf{T}_{\tilde{I}\tilde{V}_x}} \right). \tag{8.2}$$

For electrothermal sensing, the operator of interest is $\mathbf{T}_{\tilde{I}\tilde{x}_x}$ which relates the fluctuation in current to the fluctuation in heater-sample separation about an operating point of the heater-sample separation, x_0. The heater-sample separation, $\tilde{x}$, modifies the operator $\mathbf{T}_{TP_x}$. This perturbation of the thermal system is detected by monitoring the current fluctuations. Electro-thermal sensing can be viewed in the context of the general class of problems in which the perturbation of a linear time-invariant system is detected by monitoring signals within a feedback configuration it is part of. The efficiency (sensitivity) and speed (bandwidth) of detection of this change can be studied using linear transfer functions. For simplicity, one can assume that $\tilde{x}$ perturbs the gain of the operator $\mathbf{T}_{TP_x}$ alone. This is clearly an oversimplification and later on it will be shown that even the dynamic component of $\mathbf{T}_{TP_x}$ changes significantly with $\tilde{x}$. However, it can be shown that for the specific microheaters we are investigating, the gain perturbation is primarily responsible for sensing. So we persist with the assumption and let the dynamics of $\mathbf{T}_{TP_x}$ be described by

$$\mathbf{T}_{TP_x}(s) = K(x)\frac{b(s)}{a(s)}, \tag{8.3}$$

where $K(x)$ is the gain of $\mathbf{T}_{TP_x}$ as a function of the heater-sample separation, x, and $b(s)$ and $a(s)$ are polynomials of s. Then linearizing $\mathbf{T}_{TP_x}$ about x_0 gives

$$\mathbf{T}_{TP_x}(s) = K(x_0)\frac{b(s)}{a(s)} + K'(x_0)\frac{b(s)}{a(s)}\tilde{x}. \tag{8.4}$$

Hence under the assumption that only the gain of $\mathbf{T}_{TP_x}$ is modulated by $\tilde{x}$, $\tilde{x}$ enters the feedback loop as another input. The resulting system is depicted in Fig. 8.4. The sensing transfer function is given by

$$\mathbf{T}_{\tilde{I}\tilde{x}_x} = \frac{K'(x)}{K(x)} \left(\frac{-I_0}{R_0} \frac{g'(T_0) P_0 \mathbf{T}_{TP_x}}{1 + I_0^2 g'(T_0) \mathbf{T}_{TP_x}} \right). \tag{8.5}$$

Note that $\mathbf{T}_{TP_x}$ can be thought of as the "open-loop" transfer function. $\mathbf{T}_{\tilde{I}\tilde{V}_x}$ and $\mathbf{T}_{\tilde{I}\tilde{x}_x}$ can be thought of as the "closed-loop" transfer functions. It can also be seen that $\mathbf{T}_{\tilde{I}\tilde{x}_x}$ has a higher bandwidth than $\mathbf{T}_{TP_x}$. This is due to the inherent electrical feedback. Hence electrothermal sensing is faster than the dynamics of the thermal system. This inherent electrical feedback serves as the motivation for the external feedback technique presented in Section 8.5.

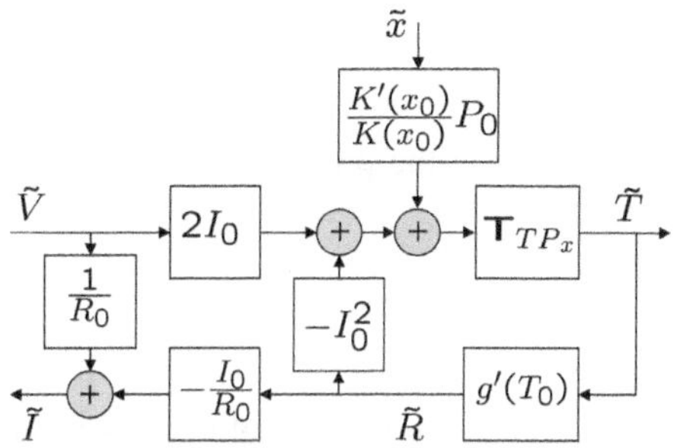

Fig. 8.4 The model for the electrothermal sensor in which the perturbation of $\mathbf{T}_{TP_x}$ is cast as an input to the feedback system

The systems model is particularly amenable to experimental identification of the electrothermal sensor. The first step in the identification process is the identification of $g(.)$. In a typical experiment, a ramp voltage signal is applied to the heater and the resulting current is measured. From the voltage and current signals, the power vs. resistance map is obtained. From our modeling assumption, the relationship between power and temperature is linear. Moreover, $T_{\max}$ is known from the doping levels of the micro-heater. Using these two facts, the temperature vs. resistance map can be obtained.

The next step in the identification process is that of the linear time invariant operator $\mathbf{T}_{TP_x}$. A DC voltage with a small-amplitude noise signal added on top is input to the heater. The corresponding current fluctuation is measured. The resistance fluctuation can be obtained from the voltage and current measurements. Using the identified relationship between temperature and resistance, $g(.)$, the temperature fluctuation can be obtained as $\tilde{T} = g^{-1}(R) - g^{-1}(R_0)$. The power fluctuation is measured as $\tilde{P} = VI - V_0 I_0$. From $\tilde{P}$ and $\tilde{T}$, the frequency response of $\mathbf{T}_{TP_x}$ can be obtained. A proper stable transfer function can be used to fit the experimentally measured frequency response. We call this method the direct method. An alternative approach to the identification of $\mathbf{T}_{TP_x}$ is to fit a stable transfer function to the frequency response obtained for $\mathbf{T}_{\tilde{I}\tilde{V}_x}$ and then to use the relationship given by Eq. (8.2) to obtain $\mathbf{T}_{TP_x}$. We call this approach the indirect approach. Note that for this approach, the noise signal has to be small enough for $\mathbf{T}_{\tilde{I}\tilde{V}_x}$ to be sufficiently linear, whereas for the former approach the smallness of the noise signal is not very critical.

The systems approach is also suitable for the analysis of the resolution of these sensors. The predominant noise sources in electrothermal sensing are the thermal noise of the silicon resistor and $1/f$ noise (11). The $1/f$ noise contribution is more difficult to predict. This conductance fluctuation noise is typically described by a model developed by Hooge that relates the $1/f$ noise to the number of carriers in the bulk of the resistor and assumes that it is proportional to the dissipated power. Let n_{R_j} represent the thermal noise and let n_{R_f} denote the $1/f$ noise component. Both random processes are thought of as equivalent resistance fluctuations. Let $S_{n_{R_j}}$ be the power spectral function corresponding to n_{R_j} given by

$$S_{nR_j}(f) = \frac{4k_B T_0 R_0^2}{P_0}, \tag{8.6}$$

where k_B is the Boltzmann constant. Let S_{nR_f} denote the power spectral function corresponding to rn_f given by

$$S_{nR_f}(f) = \frac{\alpha}{fN_{carr}}, \tag{8.7}$$

where α denotes the Hooge factor and N_{carr} denotes the number of carriers in the bulk of the resistor. The net resistance noise is the sum of these two noise sources given by $n_R = n_{R_j} + n_{R_f}$.

From Fig. 8.4, the resulting current fluctuation corresponding to an input resistance fluctuation is given by

$$\begin{aligned}\tilde{I} &= \frac{-I_0/R_0}{1 + I_0^2 g'(T_0)\mathbf{T}_{TP_x}} n_R \\ &= \mathbf{T}_{\tilde{I}n_{R_x}} n_R.\end{aligned}$$

Note that the measured resistance fluctuation is slightly less than the "input" resistance fluctuation given by

$$\tilde{R} = \frac{1}{1 + I_0^2 g'(T_0)\mathbf{T}_{TP_x}} n_R. \tag{8.8}$$

This effective "cooling" is another effect of the inherent electrical feedback. The signal-to-noise ratio is given by

$$SNR = 20\log\left(\frac{\sqrt{\int_{BW} |\mathbf{T}_{\tilde{I}\tilde{x}_x}(f)|^2 df}}{\sqrt{\int_{BW} |\mathbf{T}_{\tilde{I}n_{R_x}}(f)|^2 S_{n_R}(f) df}}\right). \tag{8.9}$$

8.3 Electro-Thermal Topography Sensor

In this section we present the identification of an electrothermal topography sensor. This sensor is part of a thermo-mechanical probe used for probe-based data storage (10; 12). The identification procedure described in Section 8.2 is employed. Hence we need to identify $\mathbf{T}_{TP_x}$ as a function of the heater-sample separation. From these measurements, the sensing transfer function $\mathbf{T}_{\tilde{I}\tilde{x}_x}$ can be evaluated.

The static relationship between the temperature and resistance was obtained as described in Section 8.2 (see Fig. 8.5(a)). Next $\mathbf{T}_{TP_x}$ is identified at various values of tip-sample separation. The thermal system identified when the tip is in contact with the sample (which means a heater-sample separation of ≈ 500 nm) is given by

$$\mathbf{T}_{TP_x}(s) = \frac{378.7}{3 \times 10^{-6}s + 1}.$$

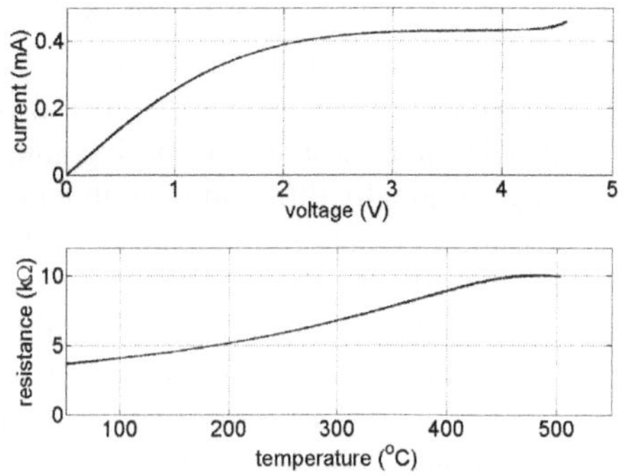

(a)

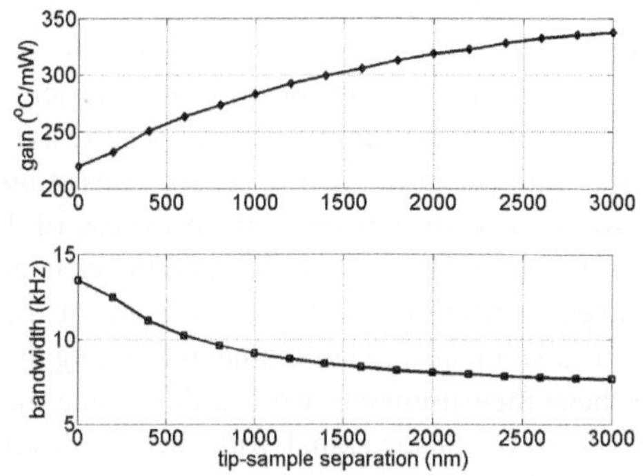

(b)

Fig. 8.5 (a) The static voltage vs. current relationship and resistance vs. temperature relationship ($g(.)$) corresponding to the topography sensor. (b) Gain and bandwidth variation of the thermal system as a function of the sensor-sample separation.

As the heater-sample separation is increased, intuitively, the linear thermal system should become slower because the conduction path through the air gap between the heater and the substrate is getting longer. Moreover, as the heater is getting thermally more isolated, the heater should heat up to a higher temperature for a given power input. This is precisely what is observed experimentally. The change in the gain and bandwidth is shown in Fig. 8.5(b).

To obtain the sensing transfer function the relationship given by Eq. (8.5) is used to derive $\mathbf{T}_{\tilde{I}\tilde{x}_x}$. From the gain variation of the thermal system as a function of heater-sample separation (Fig. 8.5(b)), we obtain $\alpha = K'(x)/K(x)$. Using (8.5), $\mathbf{T}_{\tilde{I}\tilde{x}_x}$ is evaluated at various heater-sample separations. In particular $\mathbf{T}_{\tilde{I}\tilde{x}_x}$ while the tip is in contact with the sample is given by

$$\mathbf{T}_{\tilde{I}\tilde{x}_x}(s) = \frac{-9.844 \times 10^{-6}}{2.281 \times 10^{-6}s + 1}.$$

Note that Eq. (8.5) is derived with the assumption that the change in the gain of the thermal system is primarily responsible for sensing. However, from Fig. 8.5(b), we see that there is also a significant change in the bandwidth of the thermal system. A simulation was performed to obtain the sensing transfer function that also accounts for the varying bandwidth of the thermal system besides the varying gain. The resulting sensing transfer function was only marginally different from that obtained using (8.5). This further confirms the assumption that the gain change of the thermal system is primarily responsible for sensing. The sensing transfer functions were evaluated at other tip-sample separations and the results are shown in Fig. 8.6. It can be seen that the sensitivity and bandwidth both deteriorate as the heater-sample separation increases. Relation (8.5) also provides a straightforward means of ascertaining the sensitivity and bandwidth variations as a function of bias voltage or power.

The sensing transfer function obtained using the analytical technique is compared with that obtained via direct measurement. To directly measure the sensing transfer function the following experiment was performed. The micro-cantilever was actuated electro-statically in very close proximity with the sample surface. The resulting cantilever motion was measured using an optical deflection sensor consisting of a laser and a position-sensitive photo-diode with a bandwidth in excess of 1 MHz. Given that the mechanical resonant frequency of the micro-cantilever is at 80 kHz we can assume that the optical sensor provides a faithful measure of the cantilever motion. The cantilever motion is also measured simultaneously using the integrated electrothermal sensor. From these measurements we can determine the sensing transfer function relating electrothermal sensor signal with the cantilever movement detected optically. A comparison between the directly measured sensing transfer function and that obtained analytically using electrical measurements is shown in Fig. 8.7. Also presented is the sensing transfer function obtained using elaborate finite element simulations (12). There is remarkable agreement between all three measurements.

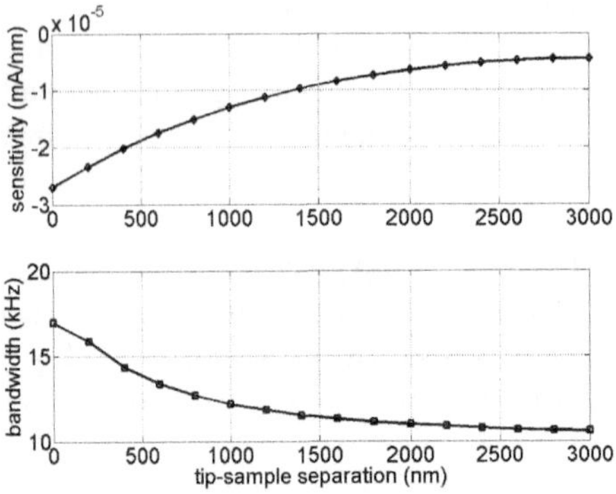

Fig. 8.6 Variation of the gain and bandwidth of $\mathbf{T}_{\tilde{I}\tilde{x}_x}$ as a function of sensor-sample separation

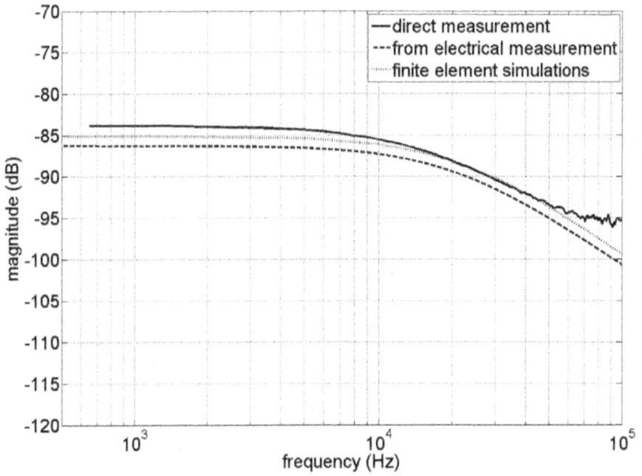

Fig. 8.7 Comparison between the sensing transfer functions obtained using analytical means, direct measurement and finite element simulations

8.4 Electro-Thermal Position Sensor

In (13), Lantz *et al.* introduced the application of electrothermal sensors as nanoscale position sensors. These position sensors were used to measure the motion of a micro-electro-mechanical micro-scanner (see (14)) along the X and Y scan directions. The sensing bandwidth was approximately 5 kHz and the resolution was less than a nanometer over this bandwidth. Closed-loop control of the micro-scanner based on the thermal position sensors is described in (15; 16).

As shown in Fig. 8.8, the position sensors are positioned directly above the scan table such that they partially overlap the scan table. They are relatively large heaters compared with the sensors described in the preceding sections. Displacement of the scan table results in a change of the overlapping area, which results in a change in the heat conduction path. As in the case of the electrothermal topography sensor, this perturbation to the thermal system can be sensed in an electrical fashion. Hence the analysis of the sensing dynamics of these position sensors is identical to that of the topography sensor. The only difference is that the position sensors are typically operated in a differential configuration. As the scan table moves, the overlapping area of one of the sensors decreases while that of the other sensor increases. The two sensors are denoted as positive and negative sensors. In addition to the obvious advantage of having an increased SNR, this also influences the sensing dynamics significantly. It was experimentally observed that the differential sensor is much more linear than an individual sensor. The analytical description of the differential sensing transfer function presented later clearly explains the linearizing effect of differential sensing.

If we assume that each sensor is an operator mapping the scan table motion, $\tilde{x}$, to the corresponding current fluctuation $\tilde{I}$, differential sensing can be viewed as shown

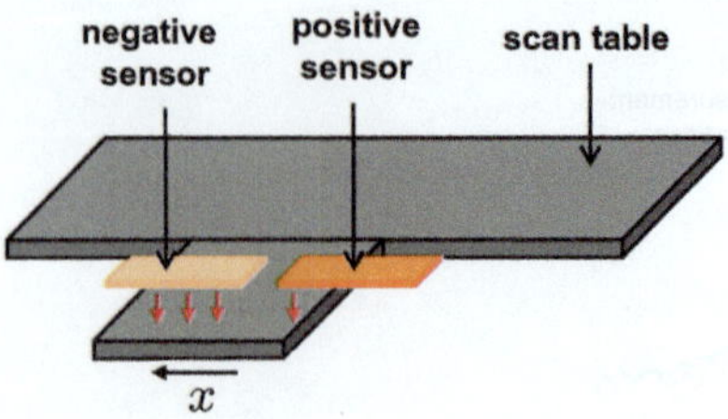

Fig. 8.8 Schematic of a pair of electrothermal position sensors used to measure the position of a nanopositioner. These sensors are positioned directly above the scan table whose lateral motion they are sensing. Displacement of the scan table results in a change in the overlapping area and is sensed in an electrothermal manner.

in Fig. 8.9. Let us assume that first-order transfer functions capture the two sensing transfer functions corresponding to the positive and negative sensors denoted by

$$\mathbf{T}^{P}_{\tilde{I}\tilde{x}_x} = \frac{-K_p(x)}{\tau_p(x)s+1} \tag{8.10}$$

and

$$\mathbf{T}^{N}_{\tilde{I}\tilde{x}_x} = \frac{K_n(x)}{\tau_n(x)s+1}. \tag{8.11}$$

Then the differential sensing transfer function, $\mathbf{T}^{PN}_{\tilde{I}\tilde{x}_x}$, is given by

$$\mathbf{T}^{PN}_{\tilde{I}\tilde{x}_x} = -\frac{[K_p(x)+K_m(x)]\left(\frac{\tau_p(x)K_m(x)+\tau_m(x)K_p(x)}{K_p(x)+K_m(x)}s+1\right)}{(\tau_p(x)s+1)(\tau_m(x)s+1)}. \tag{8.12}$$

The gain of $\mathbf{T}^{PN}_{\tilde{I}\tilde{x}_x}$ is the sum of the local gains of $\mathbf{T}^{P}_{\tilde{I}\tilde{x}_x}$ and $\mathbf{T}^{N}_{\tilde{I}\tilde{x}_x}$. Moreover the newly formed zero cancels out the difference in the pole positions as a function of the scanner position. Hence the differential sensor is much more linear along the travel range of the micro-scanner than an individual sensor.

To identify the sensitivity and sensing bandwidth of the thermal position sensors, we have to identify the sensing transfer functions of the individual positive and

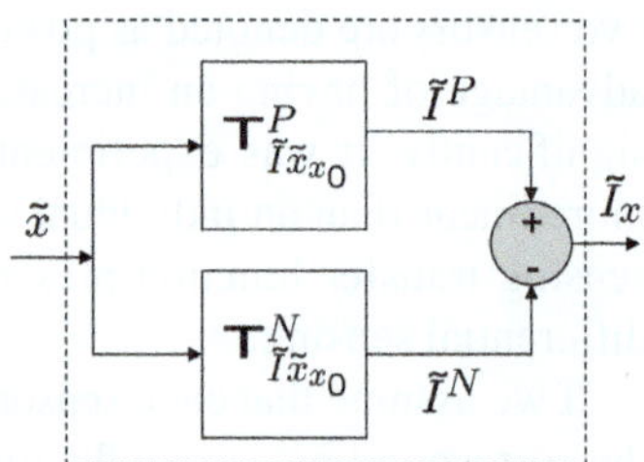

Fig. 8.9 Block diagram representation of the differential sensing configuration

negative sensors. As done in the previous section, the thermal systems and the non-linear temperature-dependent resistance relationships are identified.

The relationship between the current and voltage and the identified temperature-dependent resistance, $g(.)$, for the two thermal position sensors are shown in Fig. 8.10(a). It can be seen that the resistance values are relatively small compared with the much smaller electrothermal topography sensor presented earlier. Moreover, $g(.)$ is much more linear compared to that of the topography sensor. The thermal systems $\mathbf{T}_{TP_x}$ corresponding to both the thermal position sensors are identified at different micro-scanner positions, and are well captured by first-order transfer functions at all positions.

There is a large variation in the gain and bandwidth owing to the large motion range of the micro-scanner and the corresponding large change in the overlap. The resulting variation in gain and bandwidth as a function of the micro-scanner position is shown in Fig. 8.10(b). It can be seen that as the scanner moves from –50 μm to

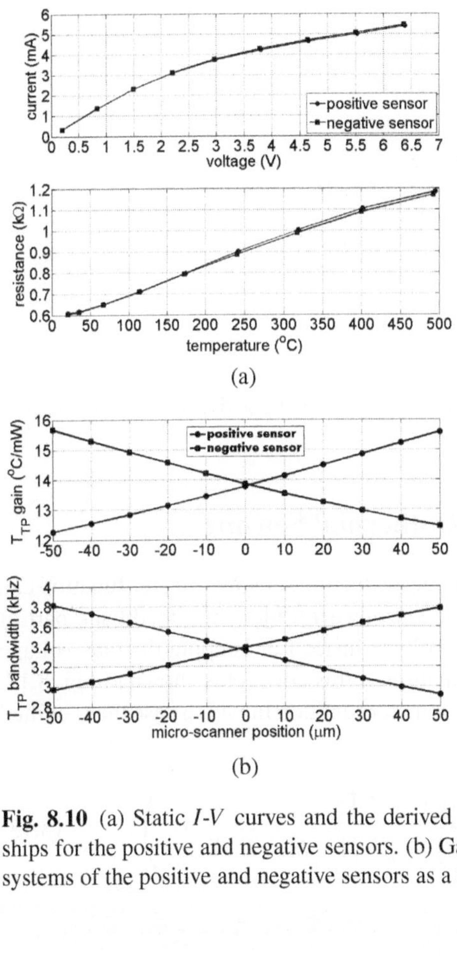

Fig. 8.10 (a) Static *I-V* curves and the derived temperature-dependent resistance relationships for the positive and negative sensors. (b) Gain and bandwidth variations of the thermal systems of the positive and negative sensors as a function of the micro-scanner position.

50 μm, the gain of $\mathbf{T}_{TP_x}$ increases for the positive sensor whereas it reduces for the negative sensor. This is intuitive because, in a differential configuration, as the overlap of one sensor with the scan table reduces, the overlap of the other sensor with the scan table increases.

Using the relationship given by (8.5) and the gain variations depicted in Fig. 8.10(b), the sensing transfer function $\mathbf{T}_{\tilde{I}\tilde{x}_x}$ is obtained for both the positive and negative sensors at different positions of the scanner. Ideally, for a linear sensor across the entire travel range of the scanner, $\mathbf{T}_{\tilde{I}\tilde{x}_x}$ need to be invariant across the entire range. However, for individual position sensors, $\mathbf{T}_{\tilde{I}\tilde{x}_x}$ varies significantly when evaluated at different scanner positions. The differential sensing transfer function is evaluated using the individual sensing transfer functions. The sensitivity and sensing bandwidth variations as a function of the micro-scanner position are shown in Fig. 8.11. Owing to the averaging effect of differential sensing, the differential sensing transfer function is remarkably linear with an average sensitivity of -2.5×10^{-3} mA/μm and bandwidth of 4 kHz.

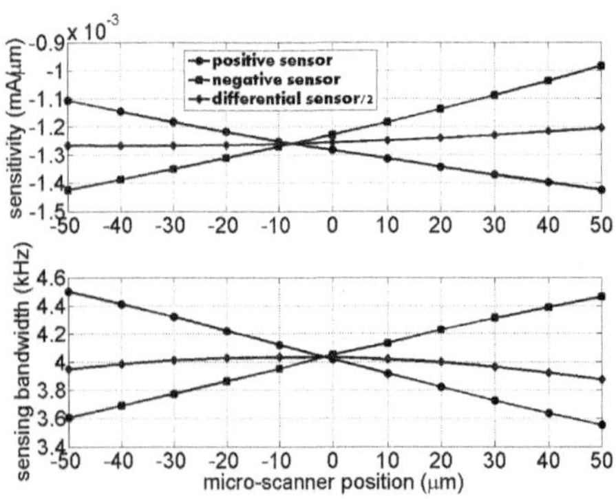

Fig. 8.11 Sensitivity and sensing bandwidth variations of the positive and negative sensors compared with those of the differential sensor

8.5 Feedback-Enhanced Electrothermal Sensing

The inherent electrothermal feedback suggests that the sensing transfer function can be "shaped" using feedback. One simple yet powerful idea is to feed back part of the sensed current to modulate the bias voltage instead of keeping the bias voltage constant (17). A block diagram representation of this concept is shown in Fig. 8.12. The feedback block, K_{FB} can be a linear transfer function in which case the modified sensing transfer function is given by

$$\mathbf{T}_{\tilde{I}\tilde{x}_x} = \frac{K'(x)}{K(x)} \left(\frac{\frac{-I_0}{R_0} g'(T_0) P_0 \mathbf{T}_{TP_x}}{\left(1 + I_0^2 g'(T_0) \mathbf{T}_{TP_x}\right) - \frac{K_{FB}}{R_0} \left(1 - I_0^2 g'(T_0) \mathbf{T}_{TP_x}\right)} \right). \tag{8.13}$$

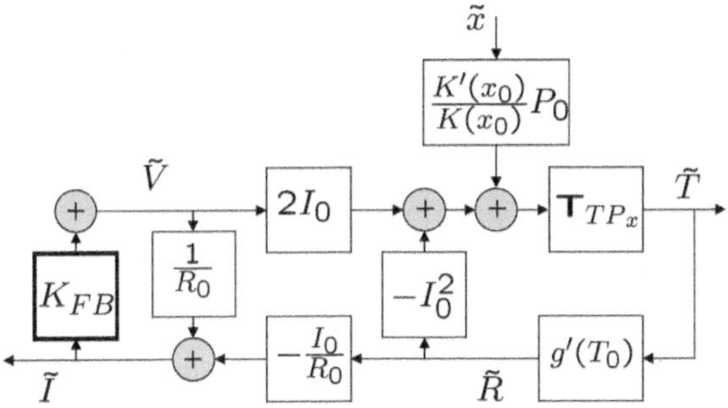

Fig. 8.12 Block diagram representation of feedback enhanced electrothermal sensing

This presents us with the opportunity to "shape" the sensing transfer function in a desirable manner. However, even a constant K_{FB} can significantly influence the sensing transfer function. Assuming that the thermal system is governed by first order dynamics, Eq. (8.13) shows that the sensing bandwidth increases monotonically towards infinity as we increase the feedback gain K_{FB} from zero to R_0 accompanied by an increase in the sensitivity. Furthermore, it can be shown that the feedback system is stable as long as K_{FB} is less than R_0. To assess whether this increase in bandwidth and sensitivity translates into an increased resolution, the impact of the feedback scheme on the electrical noise needs to be taken into account. The primary noise sources in electrothermal sensing are "resistance noise" and "current measurement noise". As mentioned earlier, the "resistance noise" primarily consists of $1/f$ noise and Johnson noise. The current measurement noise consists of the noise from the measurement circuit, noise resulting from the non-co-location of the sensor and the measurement circuitry, and the finite resolution of analog to digital converters. It can be shown that the external feedback scheme results in an increase in resolution in addition to bandwidth and sensitivity provided the current measurement noise is dominant.

Using the identified system in Section 8.3, the sensing transfer function was simulated with and without external feedback of varying magnitude. The resulting sensing transfer functions are depicted in Fig. 8.13. It can be seen that there is a steady increase in the bandwidth and sensitivity as the feedback gain is increased.

To experimentally verify this result, the sensing transfer function with and without feedback has to be measured. A similar experiment as that described in Section 8.3 was performed. Figure 8.14(a) shows the mechanical frequency response of the micro-cantilever accurately measured using the optical deflection sensor (dotted line) and as detected by the electrothermal sensor with and without feedback (dashed and solid line, respectively). Note that the apparent absence of a first resonance in the mechanical frequency response is due to the close proximity of the cantilever to the sample surface — roughly the tip height of 600 nm — which results in significant squeeze film damping. The feedback gain used was $0.82R_0$ where R_0 was 4.7 kΩ. It can be seen that there is a significant enhancement in bandwidth

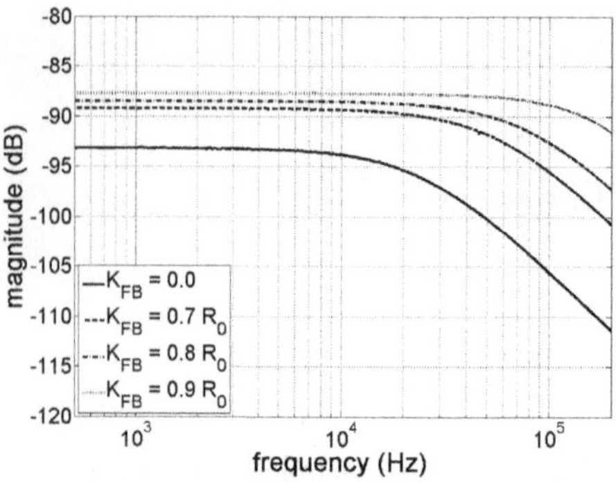

Fig. 8.13 Calculated sensing transfer functions for different values of feedback gain

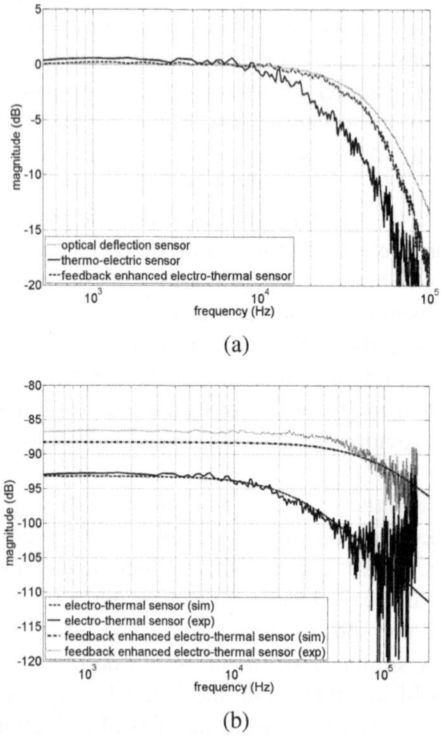

Fig. 8.14 (a) The mechanical frequency response of the micro-cantilever obtained using an optical deflection sensor and the integrated electrothermal sensor (with and without feedback). (b) Experimentally measured sensing transfer functions of an electrothermal sensor with and without external feedback.

as evidenced by the fact that the feedback enhanced electrothermal sensor can sense the cantilever deflection almost as well as the optical sensor in terms of bandwidth.

The resulting sensing transfer functions are shown in Fig. 8.14(b). Both the enhancement in sensitivity and bandwidth are clearly visible and there is remarkable agreement with the simulations. The sensitivity was improved by a factor of 2, and the –3 dB bandwidth was improved from 19 kHz to 73 kHz. This is an example of feedback control being employed to alter the sensing characteristics of a sensor.

8.6 Conclusion

Electro-thermal sensors serve as deflection sensors in scanning probe applications and as position sensors in nanopositioning applications. These sensors consist of silicon structures with integrated heating elements. The integrated heating elements are created by lightly doping certain regions of the heavily doped silicon structure. The electrothermal sensors are of very low cost and can easily be integrated into a micro-electro-mechanical structure. As topography sensors they are excellent for parallel operation of multiple cantilevers in an array configuration. A systems approach is particularly well suited for the modeling and experimental identification of these sensors. The dynamics of these sensors are well captured by a feedback model which consists of a linear thermal system and a non-linear memoryless operator. The non-linear operator captures the temperature dependent resistance relationship of doped silicon. During sensing, the signal being measured perturbs the linear thermal system. This perturbation can be analyzed to derive the sensing transfer function corresponding to these sensors. This tractable model can be used to experimentally identify the electrothermal topography and position sensors. The position sensors are typically operated in a differential configuration. The remarkable linearity of these differential sensors can be explained. Moreover, by feeding back a part of the sensed current, the sensitivity and bandwidth of electrothermal sensors can be enhanced. This also demonstrates the applicability of feedback control to shape a sensors sensing dynamics.

Acknowledgements. I gratefully acknowledge the contributions from D. Wiesmann to this work. I would like to thank M. Despont, U. Drechsler, M. Lantz and H. Rothuizen for the design and fabrication of the micro-cantilevers and position sensors used in this work. Special thanks go to P. Bächtold for his help with the electronic circuits. I would also like to thank A. Pantazi, Ch. Hagleitner, and U. Dürig for fruitful discussions, and H. Pozidis and E. Eleftheriou for their support of this work.

References

[1] Pollock, H.M., Hammiche, A.: J. Phys. D: Appl. Phys. 34, R23 (2001)
[2] Cahill, D.G., Ford, W.K., Goodson, K.E., Mahan, G.D., Majumdar, A., Maris, H.J., Merlin, R., Phillpot, S.R.: J. Appl. Phys. 93(2), 793 (2003)

[3] King, W.P., Kenny, T.W., Goodson, K.E., Cross, G., Despont, M., Dürig, U., Rothuizen, H., Binnig, G., Vettiger, P.: Appl. Phys. Lett. 78, 1300 (2001)
[4] Eleftheriou, E., Antonakopoulos, T., Binnig, G.K., Cherubini, G., Despont, M., Dholakia, A., Dürig, U., Lantz, M.A., Pozidis, H., Rothuizen, H.E., Vettiger, P.: IEEE Trans. Magnetics 39(2), 938 (2003)
[5] Pantazi, A., Sebastian, A., Antonakopoulos, T., Bächtold, P., Bonaccio, T., Bonan, J., Cherubini, G., Despont, M., DiPietro, R.A., Drechsler, U., Dürig, U., Gotsmann, B., Häberle, W., Hagleitner, C., Hedrick, J.L., Jubin, D., Knoll, A., Lantz, M.A., Pentarakis, J., Pozidis, H., Pratt, R.C., Rothuizen, H., Stutz, R., Varsamou, M., Wiesmann, D., Eleftheriou, E.: IBM J. Res. Develop. 52(4/5), 493 (2008)
[6] Lee, J., Beechem, T., Wright, T.L., Nelson, B.A., Graham, S., King, W.P.: J. Microelectromech. Syst. 15(6), 1644 (2006)
[7] Park, K., Lee, J., Zhang, Z.M., King, W.P.: J. Microelectromech. Syst. 16(2), 213 (2007)
[8] Dürig, U.: J. Appl. Phys. 98, 044906 (2005)
[9] Wiesmann, D., Sebastian, A.: In: Proceedings of the IEEE MEMS Conference, pp. 182–185 (2006)
[10] Sebastian, A., Wiesmann, D.: J. Microelectromech. Syst. 17(4), 911 (2008)
[11] Hagleitner, C., Bonaccio, T., Rothuizen, H., Lienemann, J., Wiesmann, D., Cherubini, G., Korvink, J., Eleftheriou, E.: IEEE J. Solid-State Circuits 42(8), 1779 (2007)
[12] Rothuizen, H., Despont, M., Drechsler, U., Hagleitner, C., Sebastian, A., Wiesmann, D.: In: Proceedings of the IEEE MEMS Conference, pp. 603–606 (2009)
[13] Lantz, M.A., Binnig, G.K., Despont, M., Drechsler, U.: Nanotechnology 16, 1089 (2005)
[14] Lantz, M.A., Rothuizen, H., Drechsler, U., Haeberle, W., Despont, M.: J. Microelectromech. Syst. 16(1), 130 (2007)
[15] Pantazi, A., Sebastian, A., Cherubini, G., Lantz, M., Pozidis, H., Rothuizen, H., Eleftheriou, E.: IEEE Trans. Control Syst. Technol. 15(5), 824 (2007)
[16] Sebastian, A., Pantazi, A., Pozidis, H., Eleftheriou, E.: IEEE Control Syst. Mag. 28(4), 26 (2008)
[17] Sebastian, A., Wiesmann, D., Bächtold, P., Rothuizen, H., Despont, M., Drechsler, U.: In: Proceedings of the International Conference on Solid-State Sensors, Actuators and Microsystems, pp. 1963–1966 (2009)

Chapter 9
Motion Controller for Atomic Force Microscopy Based Nanobiomanipulation

Ning Xi, Ruiguo Yang, King Wai Chiu Lai, Bo Song,
Bingtuan Gao, Jian Shi, and Chanmin Su

Abstract. Nanomanipulation with Atomic Force Microscopy (AFM) is one of the fundamental tools for nanomanufacturing. The motion control of the nanomanipulation system requires accurate feedback from the piezoelectric actuator and a high-frequency response from the control system. Since a normal AFM control system for scanning motion is not suitable for control of arbitrary motion, we therefore modified the hardware configuration to meet the demand of nanomanipulation control. By identifying the necessary parameters using system identification methods, we built up a new dynamic model for the modified configuration. Based on the new model and configuration, we designed and implemented a control scheme as motion controller for AFM nanomanipulation operation. The aims are to analyze various factors in the control of the AFM-based nanomanipulation system. By integrating the original AFM controller with the external Linux real-time controller, we achieved a stable system with high-frequency response. Several problems have been addressed based on the new control scheme, such as high frequency response, robust feedback control and non-linearity, etc. Finally this Multiple-Input Single-Output (MISO) system is validated by a real-time nanomanipulation task. It is proved to be an effective and efficient tool for the controlling of the nanobiomanipulation operation by cutting the intercellular junction of human keratinocytes.

9.1 Introduction

Atomic Force Microscopy (AFM) has often been used as high-resolution imaging tool to visualize the topography and characterize the surface properties of

Ning Xi · Ruiguo Yang · King Wai Chiu Lai · Bo Song · Bingtuan Gao
Department of Electrical and Computer Engineering,
Michigan State University, East Lansing, MI
e-mail: xin@egr.msu.edu

Jian Shi · Chanmin Su
Bruker-Nano, Inc, Santa Barbara, CA
e-mail: chanmin.su@bruker-nano.com

E. Eleftheriou & S.O.R. Moheimani (Eds.): Cntrl. Tech. for Emerging Micro/Nanoscale Sys. LNCIS 413, pp. 153–168.
springerlink.com

materials and matters at small scale. However, due to the nature of AFM imaging, by nanoscale force interaction, it was soon realized that it can also be used to control and change matter at the nanoscale. Nanomanipulation has been established ever since an IBM logo was formed by manipulating xenon atoms (1).

The AFM-based nanomanipulation technique has been a quite useful tool since then in nanomanufacturing as demonstrated by its ability to perform nanolithography (2), assembly (3; 4) and manipulating nano-particles (5). Numerous applications of nanomanipulation have been reported. The assembly of nano-structures has been achieved by Li and coworkers. In the nanoassembly process, not only nanoparticles have been pushed, but also more complex structures like nano-rods and nano-tubes have been manipulated by an AFM tip. This may eventually lead the way to the nanomanufacturing process. An AFM based nanomanipulation system is able to push the carbon nano-tubes (CNTs) onto electrodes and to make mechanical contact between the two. This makes the investigation of the characteristic properties of CNTs much more convenient (6). Besides, it holds potential in biological research, like drug delivery and cell mechanics (7). Single-strand intermediate filaments have been stretched by a controlled AFM tip to measure their tensile strength (8).

The AFM tip can be controlled to mechanically push, pull and cut nanoscale objects. In this sense, it works as a robotic arm in its task space by precise computer control. Therefore traditional concepts such as feedback, stability, and frequency response can all be integrated into these miniatured sensory robotic systems (9). The development of a robust AFM-based nanomanipulation system therefore increasingly relies on those concepts and related techniques. In recent years, extensive work has been done by researchers trying to integrate tele-operation (10) and haptic devices into the whole picture (11).

We have developed the augmented reality control system for nanomanipulation (12). A haptic device (Phantom, SensAble Technologies, Inc., Woburn, MA, USA) was employed as an interface media between the operator and the controller, through which position input can then be passed to the AFM cantilever. The coordinate of the joystick will be mapped onto the nano-scale space and the position of the

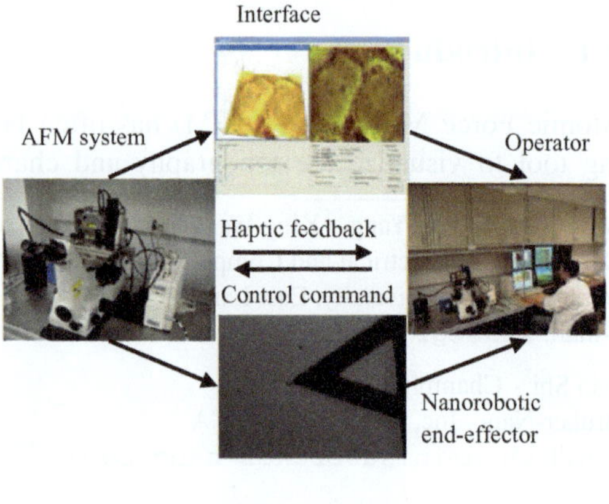

Fig. 9.1 AFM based nanomanipulation system configuration

joystick will be converted to control voltage and applied to the piezoelectric actuator. Therefore, the image obtained from the AFM scanning can be used to direct the manipulation operation, while the interaction of the tip and the nano-scale objects can be modeled to give the operator the force feedback (Fig. 9.1).

The main problem in implementing such a nanomanipulation system lies in the control of the piezoelectric actuator motion. The control problem can be formulated into two categories, the modeling and control of the piezoelectric tube and the scheme design and implementation for the whole system. The former has been addressed by many researchers (13). Nonlinearity issues, such as creep and hysteresis on most piezoelectric tubes that drive most AFMs have been compensated by control algorithms (14).

In terms of the control scheme, the latter has received little attention. In manipulating nanometer-scale objects, there are even larger spatial uncertainties than for atomic-scale matter (15). Besides, for biomolecular-related manipulations, the material properties such as viscoelasticity will pose special problems. Precise position control and a high-frequency response are crucial to the successful manipulation of those targets; they will be the focus of this paper. Previous work (16) on this issue uses a single-input control mechanism with custom-designed external controller. With that scheme, only sub-micron precision and few-Hz frequency can be achieved. Therefore it mainly works for non-biological samples such as nanoparticles and nanowires. The nature of the biological matters requires the development of a new control scheme.

This paper is organized in the following way. In Sect. 9.2, the system modifications in terms of hardware reconfiguration and system identification are discussed. Then Sect. 9.3 delineates the issues involved in stable control of a high-frequency response system with position feedback. By proposing and analyzing the control issues, Sect. 9.4 presents the system performance of a multi-input and single-output (MISO) control scheme. In Sect. 9.5, experimental verifications of the system in nanobiomanipulation tasks are presented for dissecting intercellular junctions between keratinocytes. Sect. 9.6 concludes the paper and lays out future directions.

9.2 System Modification and Identification

AFM operates by using a cantilever with a sharp tip to scan across sample surfaces to form a topography image. A piezoelectric tube actuates the cantilever and the tip both horizontally—the XY scanning direction—and vertically—the Z movement direction. The nanoscale interaction force between the scanning AFM tip and the sample surface will bend the cantilever. A laser which will be reflected from the back of the cantilever is able to record this deflection with a position-sensitive device (PSD). The Z-piezo actuator will move accordingly to keep the tip at a constant distance from or in contact with the sample depending on the mode used. The schematic drawing of the working principle of AFM is shown in Fig. 9.2.

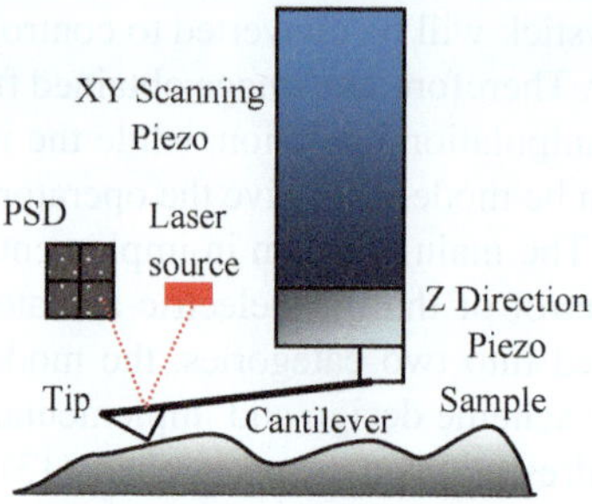

Fig. 9.2 Schematic of AFM working principle

9.2.1 AFM Imaging

The AFM system that we use is Bioscope and the controller is Nanoscope IV (Bruker-Nano, Santa Barbara, CA, USA). When doing scanning in either the X or the Y direction, the AFM control system just inputs a triangle wave voltage (shown in Fig. 9.3) ranging from −10 V to 10 V; after amplification, the voltage will be in the range from −110 V to 110 V. This voltage will be applied directly to the piezoelectric transducer which will do the scanning. The position of the scanning tip can be obtained and fed back to the controller for the closed-loop compensation. The output voltage from the AFM controller (−10 V to 10 V) will correspond to a scanning region of 100 μm. Shown in Fig. 9.4 is the control scheme of the AFM scanning and imaging system. The control scheme for doing nanomanipulation will be much more complex and flexible.

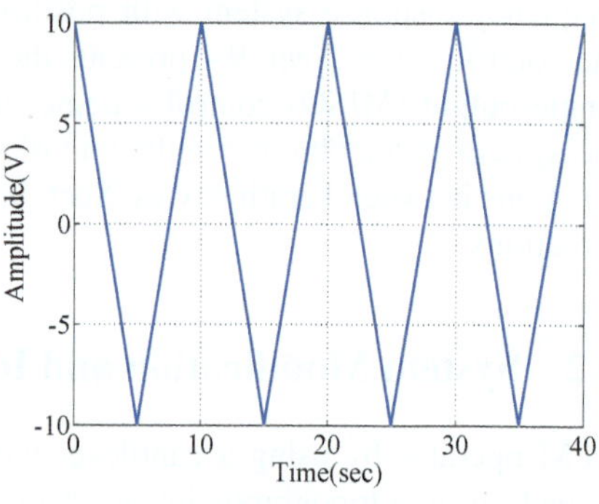

Fig. 9.3 AFM scanning wave

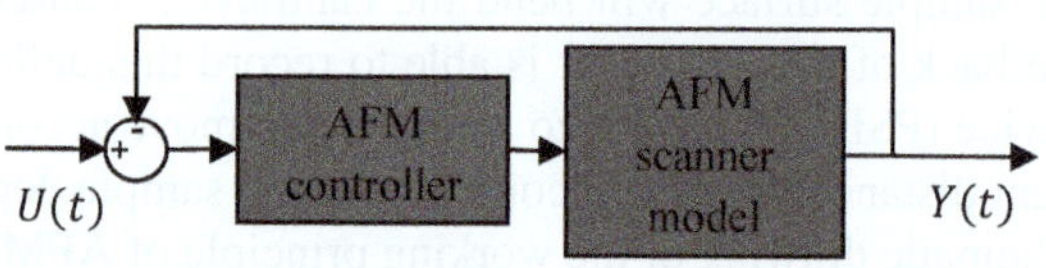

Fig. 9.4 AFM scanning control scheme

9.2.2 AFM Nanomanipulation

When performing nanomanipulation tasks, the AFM tip, which is mounted on the piezoelectric actuator, will be moved by the specially designed voltage stream rather than a zig-zag profile. The benefit of this is that we can pattern our own control voltage to achieve a particular task. But this patterned voltage stream cannot be merged to the controller as $U(t)$ in Fig. 9.4. The AFM control system for scanning motion is not suitable for manipulation purposes. Therefore the overall hardware configuration should be modified to apply the position information to the AFM tip effectively and efficiently.

9.2.3 Hardware Modification

There are two ways to add position input to the scanner. The software provides a macro mechanism which allows developers to compile a dynamic link library (.dll) file and load it when doing manipulation. This integrates the customer program into the whole software. But the overall system response will become very slow in a few-Hz frequency due to the software rebuilding process. In the meantime, the Signal Access Module (SAM) can provide the interface where we can define our own input as well as obtain the piezo-position information for feedback control purposes.

The original system consists of the AFM, the controller and the scanner (Loop 1 in Fig. 9.5). The position command input can be facilitated by the Phantom device (red zone as Command input in Fig. 9.5) through Input1 from macro mechanism. A Linux controller was developed using the Real Time Application Interface for Linux (RTAI) architecture as shown in Loop 2. The signal input and output were achieved by A/D and D/A cards through the SAM interface. Therefore we have

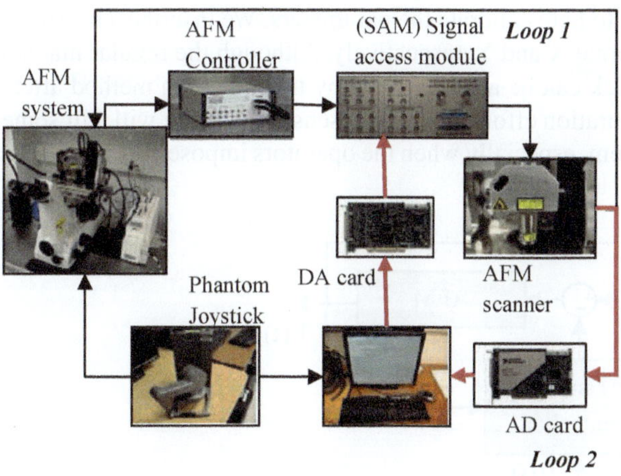

Fig. 9.5 Modified system configuration

multiple points where the control voltage can be merged into the whole system just as depicted in Fig. 9.5. The feasibility and efficiency of each scheme should be assessed whilst the outcome should be compared. The related issues of those schemes will be discussed in the following section in detail.

9.2.4 System Identification

After the hardware reconfiguration, the overall system will change its properties. We performed the system identification to build up a new dynamics model. After the modification, the scanner can be identified as a second-order system with a resonant frequency of f_0 and a Q factor of $\frac{1}{Q_0}$ which results in a damping ratio $\zeta = \frac{Q_0}{2}$. Therefore the transfer function of the AFM scanner can be obtained as

$$G(s) = \frac{\varpi^2}{s^2 + 2\zeta\varpi s + \varpi^2} = \frac{(2\pi f_0)^2}{s^2 + \frac{(2\pi f_0)}{Q_0}s + (2\pi f_0)^2}. \tag{9.1}$$

9.3 Control Scheme Design Issues

9.3.1 Feedback Issue

The intuitive way of achieving the goal would be adding the control voltage right in front of the scanner shown in Fig. 9.6. Firstly, the open-loop method has been tried. Two linear mapping equations are adopted to correlate the input voltage signal with the position of the AFM tip.

Therefore the focus would be on finding appropriate parameters for the controller and calibrating them to achieve a stable mapping. Here the AFM nanolithography experiment has been done to help calibrate the parameters. We inscribe a horizontal and a vertical line to calibrate X and Y, respectively. Although the regular manipulation tasks through joystick can be accomplished by this mapping method after a substantial amount of calibration effort, the lack of sensor feedback will still sometimes harm the whole system, especially when the operators impose abrupt motions.

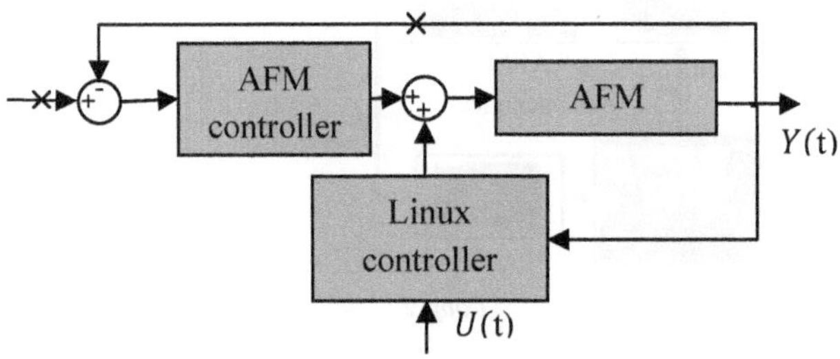

Fig. 9.6 Control scheme using the real-time Linux as main controller

9.3.2 Frequency Issue

A PI controller is designed to replace the original controller and implemented in Linux as indicated by Fig. 9.6. Since we have identified the parameters of the system by experiment, they are then adapted to the Linux real-time controller. For a general test, a step response was recorded from the oscilloscope. The stabling process takes around 0.2 sec, corresponding to a frequency around 5 Hz, which is substantially lower than expected (Fig. 9.7).

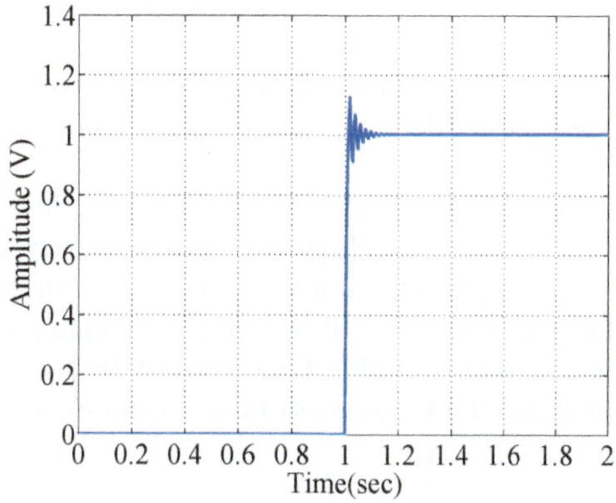

Fig. 9.7 The step response output from the scanner is observed from the oscillation scope

9.3.3 Nonlinearity Issue

The external Linux controller is able to accomplish a stable closed-loop control system. But the lack of high-frequency response is not as desirable. Since we have the original controller, which is hardware-based, implemented within the AFM system itself, it would be more efficient if we could integrate that into the new setup. By observing the whole control scheme, we revisit the signal transformation process in the following. The signal from the PI controller is (shown in Fig. 9.8A):

$$U'(t) = K_p(U(t) - Y(t)) + K_i \int (U(t) - Y(t))dt, \tag{9.2}$$

which is a sum of two signal components $U_1'(t)$ and $U_2'(t)$:

$$U_1'(t) = K_pU(t) + K_i \int U(t)dt \tag{9.3}$$

$$U_2'(t) = K_p(-Y(t)) + K_i \int (-Y(t))dt. \tag{9.4}$$

Therefore we obtain these two components separately from two controllers: the original controller and the external Linux controller shown in Fig. 9.8B. Besides,

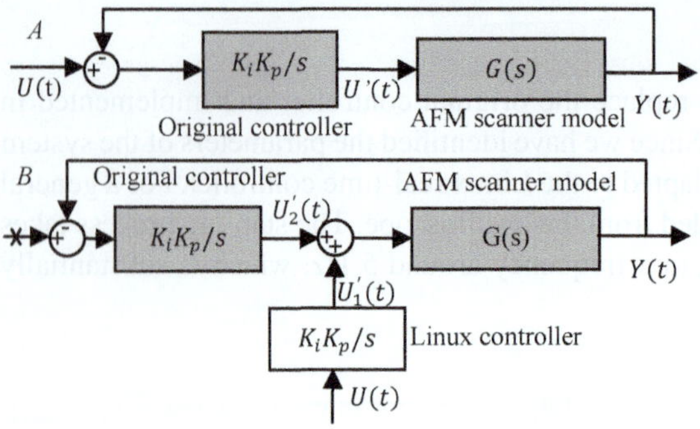

Fig. 9.8 The original control system (A) decomposed into two controller inputs to a new configuration (B)

another benefit that we will enjoy from this configuration is that the original controller, which has a high-response frequency, can be utilized.

The implementation result shows that there will be saturation in both branches of the control signal as shown in the tracking output $Y(t)$ and the control voltage from the external controller $U_2'(t)$ (Fig. 9.9). The output from the original controller and the Linux D/A card both have a saturation range from -10 V to 10 V (Scanning range: 100 μm).

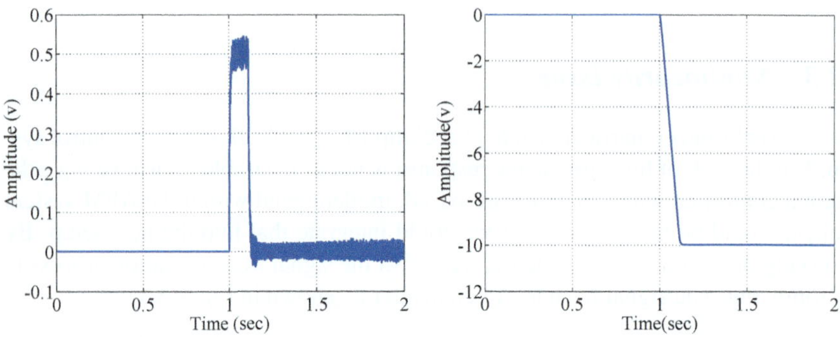

(a) Y(t) (b) $U_2'(t)$

Fig. 9.9 Saturation problem as denoted from the output $Y(t)$ and the controller output $U_2'(t)$

9.4 Design and Implementation of the Multiple-Input Single-Output System

By observing the simulation we will be able to tell that the saturation is caused by the input to the original controller. Since the input to the AFM original controller is supposed to be the difference $U_{err_1}(t)$ between the desired position and the feedback position from the scanner, it will be in quite a small range as (Configuration in Fig. 9.8A)

$$U_{err_1}(t) = U(t) - Y(t). \quad (9.5)$$

After we change the control scheme by introducing the external controller which has the input of the desired position, the feedback from the scanner becomes the only input to the original controller $U_{err_2}(t)$, which makes it very large as (Configuration in Fig. 9.8B)

$$U_{err_2}(t) = -Y(t). \quad (9.6)$$

The inputs to the system are both the scanning wave. Therefore the saturation problem can be overcome by introducing an additional input to the original controller. The aim is to bring down the input by compensation and then avoid saturation. The control scheme is shown in Fig. 9.10.

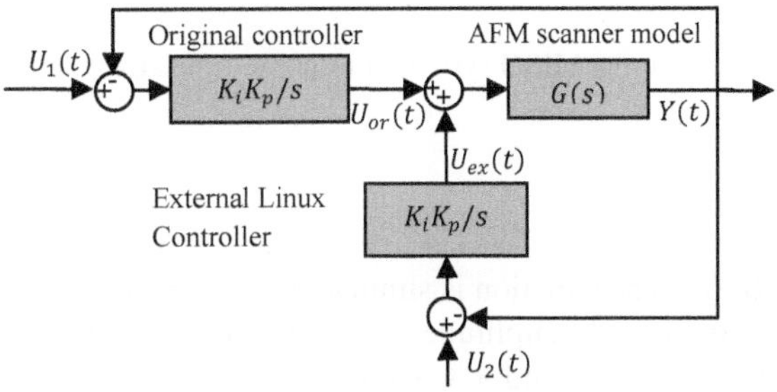

Fig. 9.10 Additional input to the AFM controller to overcome saturation

This control scheme can be analyzed in two different situations. Firstly, we assume that the input signals to both the original controller and the external controller are the same as $U_1(t) = U_2(t) = Y(t)$. Therefore the signal coming from the original and external controllers will be

$$U_{or}(t) = K_p(U_1(t) - Y(t)) + K_i \int (U_1(t) - Y(t)) \quad (9.7)$$

$$U_{ex}(t) = K_p(U_2(t) - Y(t)) + K_i \int (U_2(t) - Y(t)). \quad (9.8)$$

Since the input signal is the same, we have $U_{or}(t) = U_{ex}(t)$. In this case, the system can be regarded as a single controller scheme where the controller has proportional and integral gain doubled as compared to the original controller.

The step response to the above configuration is shown in Fig. 9.11. Compared with the original setup, the step response is in much better shape with a response time around 0.025 sec.

Secondly, the signal inputs to the two controllers are different with $U_1(t) \neq U_2(t)$. This means that at a given time t there is an offset between $U_1(t)$ and $U_2(t)$. This situation could be caused by the different sampling frequencies of the two controller branches. Although the signal source is the same, a short time delay will result in an offset between the two signals. From this point of view, the offset can be assumed to be quite small.

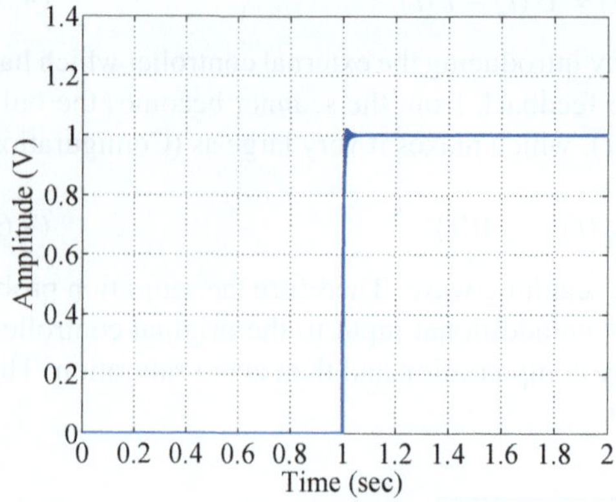

Fig. 9.11 Step response of same inputs to the MISO system; a response time around 0.025 sec can be observed

9.4.1 Step Response Analysis

Assume for now that $U_1(t) < U_2(t)$. The situation is simulated by taking $U_1(t)$ as a step input while $U_2(t)$ is a step input with amplitude 1.2. The signal outputs from both controllers are in the form of Eqs. 9.7 and 9.8. They are obviously not equal, with $U_{or}(t) < U_{ex}(t)$. From time zero to T_1, we have the following situation: $U_{or}(t) < 0$ and $U_{ex}(t) > 0$. The two controllers are working together just like the previous situation when $U_1(t) = U_2(t)$. They bring the output $Y(t)$ to the proximity of the input $U(t)$, which shows up in Fig. 9.12 as the end period of T_1. A critical point

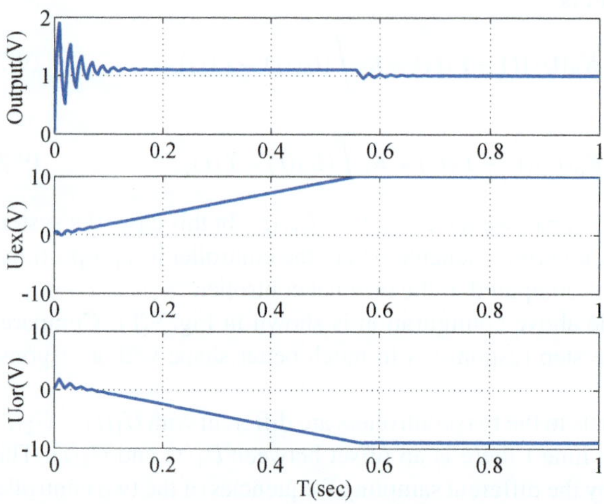

Fig. 9.12 Step response of two different inputs to the MISO system; two stabling process can be observed

is reached when the feedback signal $Y(t)$ is equal to the smaller input, which in this case is $U_1(t)$. This will make $U_{or}(t)$ cross the zero line to the negative region. Another critical point is when the feedback signal $Y(t)$ reaches the following:

$$Y(t) = \frac{U_1(t) + U_2(t)}{2}. \tag{9.9}$$

This time is set as T_1. It has the following characteristics:

$$U_{or} < 0 \tag{9.10}$$

$$U_{ex} > 0 \tag{9.11}$$

$$U_{or}(t) + U_{ex}(t) = \frac{U_1(t) + U_2(t)}{2}. \tag{9.12}$$

From T_1 until T_2 these characteristics, Eqs. 9.9 and 9.12, hold, and we call this period a temporarily stable time. The output signals from the controllers are increasing since both have constant error inputs:

$$U_{err_1}(t) = U_1(t) - \frac{U_1(t) + U_2(t)}{2} = \frac{U_1(t) - U_2(t)}{2} \tag{9.13}$$

$$U_{err_2}(t) = U_2(t) - \frac{U_1(t) + U_2(t)}{2} = \frac{-U_1(t) + U_2(t)}{2}. \tag{9.14}$$

The proportional terms remain the same while the integral errors will keep adding up. The last critical point T_2 is reached when one of the controller outputs get saturated, which in this case is $U_{ex}(t)$. From this time on, the original controller, which is active, will adjust the output $U_{ex}(T_2)$ to overcome a constant disturbance which is equal to 10 V

Finally, when the whole system is stable, the output from the scanner is tracking the final voltage from the original controller $U_1(t)$. The saturation problem is overcome since one of the controllers can be at an active state while the other is in saturation. Note that in order to achieve a higher response frequency, we want maintain the control voltage to guarantee that the stabling process stays within the first stage, which will give us the settling time which is around 0.025 sec corresponding to a frequency of 40 Hz.

9.4.2 Continuous Input

As in the continuous input, the system will always perform at the first stage illustrated above from time zero until time T_1. With most systems, the sampling frequency will be much larger than 40 Hz; hence the time it takes for a new sample to arrive is shorter than T_1. When new samples arrive, both controllers will work together to obtain a new equilibrium. Therefore, the overall system will track $\frac{U_1(t)+U_1(t)}{2}$ at all times.

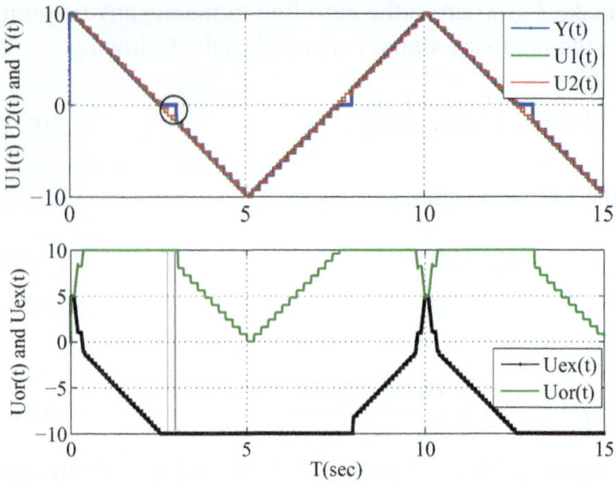

Fig. 9.13 Continuous input with a triangular wave response with a sampling time less than T_1 (circle indicates the kink when both controllers in saturation)

A scanning pattern is used as input to both the external and the original controller with sampling frequency at 10 Hz and 20 Hz respectively. The result shows that the saturation problem will not be a factor anymore, since most of the time there will be one controller that is not saturated as illustrated in Fig. 9.13. But there will be a subtly short period when both of them are saturated indicated by the grey line. There will be a small kink in those time periods as indicated by the circle. When looked further, this small kink can be eliminated by increasing the sampling frequency.

9.5 Experimental Test of Control Scheme by a Nanobiomanipulation Task

9.5.1 Materials and Methods for Sample Preparation

Keratinocyte cells were grown to confluence in DMEM medium (Gibco-Invitrogen, Carlsbad, CA, USA) supplemented with 10% fetal calf serum (Gemini Bio-products, West Sacramento, CA, USA) and 1% penicillin at 37 °C in a humidified atmosphere containing 5% CO_2, and plated onto poly-L-ornithine (PLO) (Sigma, St. Louis, MO, USA) coated glass coverslips. PLO was used to enhance primary cell adhesion to the glass coverslips. The coverslips were then washed with PBS and transferred directly to a petri dish full of PBS under the AFM scanner for imaging and nanobiomanipulation. For AFM imaging, the tapping mode was used in liquid to minimize the damage the AFM tip could do to the delicate cell membrane.

For scanning electron microscopy (SEM) imaging, the cultured cells were fixed in a mixture of 2.5% glutaraldehyde and 2.5% formaldehyde in the 0.1 M Sodium Cacodylate Buffer for 2 h at room temperature. After rinse with the buffer only solution for 15 min, the sample will go through the dehydration process within ethanol of 25%, 50%, 75%, 95% and 100% sequentially. The dehydrated sample will be critical-point-dried inside a critical-point drier (PELCO CPD2, Pelco International Redding, CA, USA). The image was taken under 20 kV acceleration voltage in a Joel SEM (JEOL, Peabody, MA, USA).

9.5.2 Structural Characterization of Cell-Cell Adhesion Structure by Different Techniques

Keratinocytes are typical human-skin cells characterized by their strong adhesion through bundles of intermediate filaments. The cells normally have a diameter around 15 μm while the intercellular junction area composed of intermediate filaments and cell membrane materials can be less than 100 nm. Optical image shows the single layer sheet of keratinocytes adherent onto glass slides. The unique cell-cell adhesion structure was characterized by a variety of techniques. SEM images were obtained and compared with AFM images with similar scan size (Fig. 9.14).

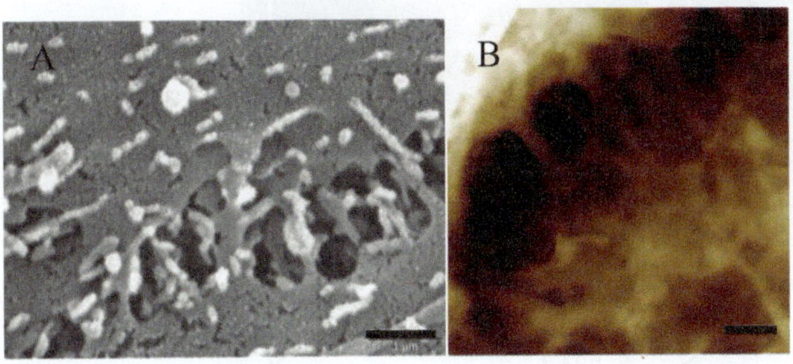

Fig. 9.14 SEM (A) and AFM (B) image of cell-cell adhesion between neighboring keratinocytes, scale bar: 1 μm

9.5.3 Experimental Setup and Results

The MISO system was tested for nanobiomanipulation tasks. To dissect those adhesion junctions, the viscoelastic nature of the material requires the high-response frequency of the control system. The motion path is denoted around the peripheral of the cell perpendicular to intermediate filaments (Fig. 9.15).

The topography image shows three keratinocytes cells connected by intercellular junction molecules, mainly intermediate filaments. The AFM images before and after the cutting clearly illustrate the topographical difference due to the cutting (Fig. 9.16). One of the intermediate filament bundles was cut off as indicated by the arrows. A detailed section image shows that there are around 100 nm in height difference, suggesting the successful dissection of the intermediate filament bundle. During the cutting process, the traveling distance of the AFM tip is around 10 μm. The voltage profile during this cutting operation is shown in Fig. 9.16. In

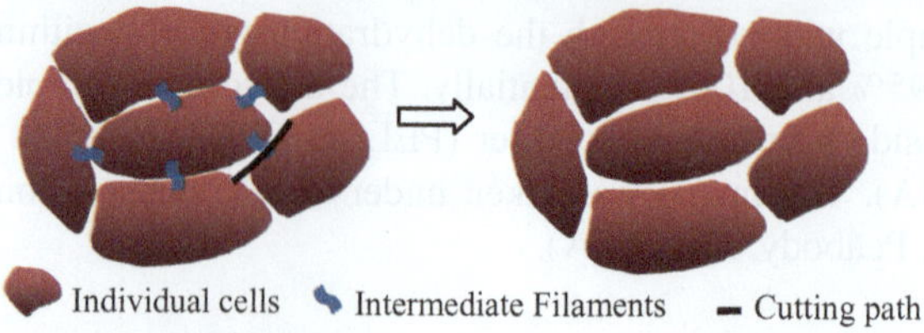

Fig. 9.15 Schematic drawing of the nanobiomanipulation tasks dissecting the intercellular junction of keratinocytes

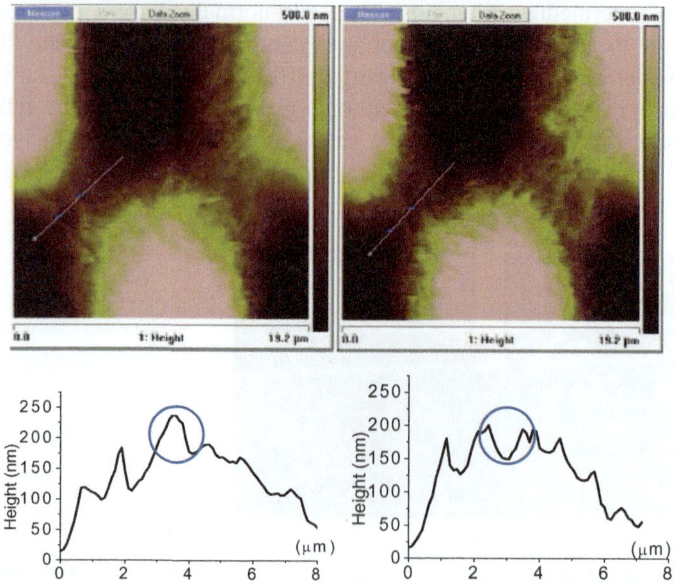

Fig. 9.16 The keratinocytes before and after cutting imaged by AFM and the cross section showing the height decrease by around 100 nm (from 0 nm to -100 nm, highlighted by the circles)

this way, the effectiveness of the MISO system has been verified. A series of other nanomanipulation experiment were also performed, such as pushing nano-particles and nanowires.

9.6 Conclusion

In the paper, several control schemes have been studied in the AFM nanomanipulation platform through using a specially designed signal access module. An external controller has been implemented in the Linux real-time framework. With the single-input system, it works well but has the low-frequency problem; we can only do manipulation at a few Hz. Then we integrated the original controller to the system for a fast response, but the saturation problem caused the nanomanipulation system work only in 5 μm range. The saturation problem has been overcome by an additional input from both the original AFM controller and the external controller. It is found that at any given time at least one of the controllers will be able to stay active and make the system stable. This MISO control not only overcomes the saturation problem but also has a high-frequency response which is around 40 Hz. When we perform manipulation tasks, this frequency is high enough for the human hand operating the haptic device. The system was finally tested with nanobiomanipulation to dissect the intercellular junctions of keratinocytes. Successful operation to cut off the intercellular material especially the intermediate filaments can be achieved.

Acknowledgements. This research work is partially supported under NSF Grants IIS-0713346 and DMI-0500372, and ONR Grants N00014-04-1-0799 and N00014-07-1-0935.

References

[1] Eigler, D.M., Schweizer, E.K.: Positioning single atoms with a scanning tunelling microscope. Nature 344, 524–526 (1990)
[2] Martin, C., Rius, G., Borriise, X., Murano, F.P.: Nanolithography on thin layers of PMMA using atomic force microscopy. Nanotechnology 16, 1016–1022 (2005)
[3] Li, G., Xi, N., Chen, H., Saeed, A., Yu, M.: Assembly of nanostructure using AFM based nanomanipulation system. In: IEEE Int. Conf. on Robotics and Automation, New Orleans, LA (2004)
[4] Requicha, A., et al.: Nanorobotic assembly of two dimensional structures. In: IEEE Int. Conf. on Robotics and Automation, Leuven, Belgium (1998)
[5] Zhang, J., Li, G., Xi, N.: Modeling and control of active end effector for the AFM based nanorobotic manipulators. In: IEEE Int. Conf. on Robotics and Automation, Barcelona, Spain (2005)
[6] Thelander, C., Samuelson, L.: AFM manipulation of carbon nanotubes: Realization of ultra-fine nanoelectrodes. Nanotechnology 13, 108–113 (2005)
[7] Guthold, M., Falvo, M.R., Matthews, W., Paulson, S., Erie, D.A.: Controlled manipulation of molecular samples with nanomanipulator. IEEE/ASME Trans. Mechatron. 5, 189–198 (2000)
[8] Kreplak, L., Bar, H., Leterier, J.F., Hermann, H., Aebi, U.: Exploring the mechanical behaviour of single intermediate filaments. J. Mol. Biol. 354, 569–577 (2005)
[9] Requicha, A.A., Arbuckle, D.J., Mokaberi, B., Yun, J.: Algorithms and software for nanomanipulation with atomic force microscopes. Int. J. Robotics Research 28, 512–522 (2009)
[10] Sitti, M., Aruk, B., Shintani, H., Hashimoto, H.: Development of a scaled teleoperation system for nanoscale interation and manipulation. In: IEEE Int. Conf. on Robotics and Automation, Seoul, Korea (2001)
[11] Aruk, B., Hashimoto, H., Sitti, M.: Man-machine interface for micro/nano manipulation with AFM probe. In: IEEE Int. Conf. Nanotechnology, Maui, HI (2001)
[12] Li, G., Xi, N., Yu, M.: Calibration of AFM based nanomanipulation system. In: IEEE Int. Conf. Robotics and Automation, New Orleans, LA (2004)

[13] Rifai, K., Rifai, O., Toumi, K.: Modeling and control of AFM based nanomanipulaton systems. In: IEEE Int. Conf. on Robtics and Automation, Barcelona, Spain (2005)
[14] Mokaberi, B., Requicha, A.A.: Compensation of scanner creep and hysteresis for AFM nanomanipulation. IEEE Trans. Automation Sci. and Eng. 5, 197–206 (2008)
[15] Kim, S., Ratchford, D.C., Li, X.: Atomic force microscope nanomanipulation with simultaneous visual guidance. ACS Nano 3, 189–198 (2009)
[16] Li, G., Xi, N., Yu, M., Fung, W.: Development of augmented reality system for AFM based nanomanipulation. IEEE Trans. Mechatronics 9, 358–366 (2004)

Chapter 10
Nanobioscience Based on Nanorobotic Manipulation

Toshio Fukuda and Masahiro Nakajima

Abstract. In this chapter, we present a summary of nanorobotic manipulations under various microscopes and their nanobioscience applications. Until now, nanomanipulation is mainly applied to the scientific exploration of mesoscopic phenomena and the construction of prototype nanodevices. Recently, the evaluation of bio-samples has gotten much attention for nanobio applications in nano-biotechnology. Single-cell analysis gets a lot of attention because of its potential for revealing unknown biological aspects of individual cells. Nanomanipulation techniques are one of the promising ways to develop nanobio-applications on the single-cell level for drug delivery, nano-therapy, nano-surgery, and so on.

10.1 Introduction

An important area in nanotechnology is the evaluation and assembly techniques of nanoscale structures. Nanostructures are basically fabricated by the top-down and bottom-up approaches as shown in Fig. 10.1. Nanomanipulation is considered as technique to evaluate and assemble them to realize a wide range of controlled devices, from the atomic to the meter scale in the near future.

Currently, we use various devices based on micro-nano technologies. Those devices are investigated based on the micro-nanomechatronic technologies used to realize high efficiency, high integration, high functionality, low energy consumption, low cost, miniaturization, and so on. The micro-nano technologies can contribute to these innovations based on mechatronics, material science, medical science, biological science, power generation, and so on. One of the promising applications of micro-nanomechatronics is in biological and medical fields. In this paper, we present current and future issues of nanomanipulation techniques for use in biological applications.

Toshio Fukuda · Masahiro Nakajima
Nagoya University, Furo-cho, Chikusa-ku, Nagoya Aichi
e-mail: {fukuda,nakajima}@mein.nagoya-u.ac.jp

E. Eleftheriou & S.O.R. Moheimani (Eds.): Cntrl. Tech. for Emerging Micro/Nanoscale Sys. LNCIS 413, pp. 169–180.
springerlink.com

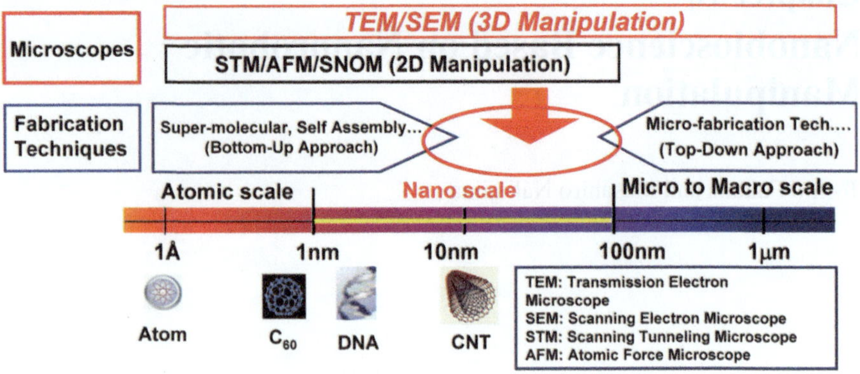

Fig. 10.1 Schematic of nanotechnology based on nanomanipulation techniques

Until now, nanomanipulation is applied to the scientific exploration of mesoscopic phenomena and the construction of prototype nanodevices. It is a fundamental technology for property characterization of nano materials, structures and mechanisms, for the fabrication of nano building blocks, and for the assembly of nanodevices (1) (Fig. 10.1). Nanoelectromechanical systems (NEMS) are expected to realize highly integrated, miniaturized, and multi-functional devices for various applications (2). To realize such high-precision system, one of the effective ways is the direct use of bottom-up-fabricated nanostructures. Recently, the evaluation of bio-samples and the precise control of single cells have gotten much attention for nanobioscience applications (3; 4). The single-cell analysis gets a lot of attention because of its potential for revealing the unknown biological aspects of individual cells. Nanomanipulation techniques are one of the promising ways to develop nanobio-applications on the single-cell level for drug delivery, nano-therapy, nanosurgery, and so on (5).

10.2 Nanorobotic Manipulation System

10.2.1 Nanorobotic Manipulation System under Various Microscopes

Nanorobotic manipulation or nanomanipulation has received much more attention because it is an effective strategy for the property characterization of individual nanoscale materials and the construction of nanoscale devices. Such devices might ultimately be the core-most part of nanotechnology. One of the attractive future applications of nanomanipulation is to realize the ultimate goal of nanotechnology: nanomanipulation. To manipulate nanoscale objects, it is necessary to observe them with a resolution higher than nanoscale. Hence, manipulators and observation systems, or microscopes in general, are necessary for nanomanipulation.

Figure 10.2 shows the strategies of nanomanipulation with various kinds of microscopes. The optical microscope (OM) is one of the most historical and basic microscope. It is suitable for manipulating various transparent samples, especially biological cells. However, its resolution is limited to ~200 nm because of the diffraction limit of optical wavelengths explained by the well-known Abbe's law (6). Hence, special techniques (using, for example, evanescent light or fluorescent light) are needed for the observation of objects on the nanoscale (7). Scanning tunnelling microscopes (STMs) and atomic force microscopes (AFMs) have functions for both observation and manipulation on the nanoscale by means of their nanoscale tip. Their high resolution makes them capable of atomic manipulation. In 1990, Eigler and Schweizer demonstrated the first nanomanipulation of atoms by means of an STM (8). Avouris et al. used an AFM to bend and translate carbon nanotubes (CNTs) on a substrate (9). They combined the technique with the inverse process, namely straightening, by pushing along a bent tube, and realized the translation of a tube to another place. Ning Xi et al. at the Michigan State University developed an AFM-based nanomanipulation system with an interactive operation system (10). The system realized real-time visual feedback during AFM-based nanomanipulation.

Normally, SPM systems are limited to observations of objects in the 2D plane with quite a smooth surface. Moreover, the observation area is limited and a long time is needed to get one image (several minutes). This limitation even increases for 3D nanomanipulation of nanostructures. On the other hand, electron microscopes (EMs) provide atomic-scale resolution with the electron beam whose wavelength is less than ~0.1 Å. EMs are mainly divided into two types, scanning electron microscopes (SEMs) and transmission electron microscopes (TEMs). For example, Yu et al. presented the tensile strength of individual CNTs inside a SEM

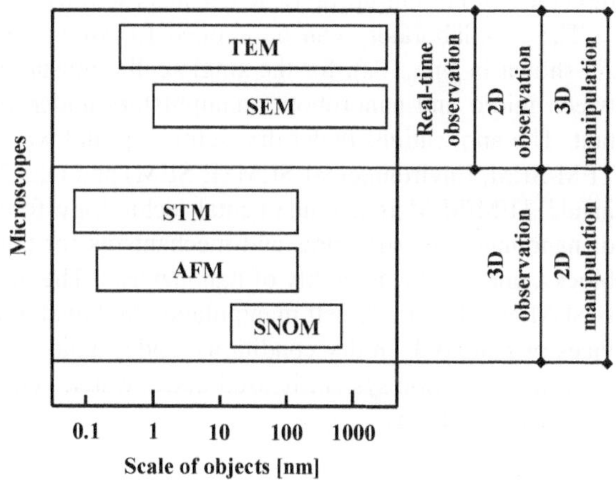

Fig. 10.2 Strategies of nanomanipulation under various microscopes

(11). However the resolution of SEM, generally $\sim$1 nm, is approximately one order of magnitude lower than that of TEM. The high-resolution transmission images of TEMs are useful for measuring and evaluating nanoscale objects. Kizuka et al. proposed a manipulation holder for use inside a high-resolution transmission electron microscope (HR-TEM) that was specially designed for atomic-level positioning resolution (12).

However, the specimen chamber and observation area of TEM are too narrow to house manipulators with complex functions. Hence, special sample preparation techniques are also needed. We proposed a hybrid nanorobotic manipulation system which is based on integrated TEM and SEM nanorobotic manipulators as core system for the nanolaboratory (13). The strategy is termed hybrid nanomanipulation so as to differentiate it from those with an exchangeable specimen holder only. The most important feature of the manipulator is that it contains several passive degrees of freedem (DOFs), which makes it possible to perform relatively complex manipulations, but still keeps the volume compact so that it can be installed inside the narrow vacuum chamber of a TEM.

10.2.2 Nanolaboratory Based on Nanorobotic Manipulations

We have proposed a "nanolaboratory" based on a nanorobotic manipulation system from 2000 (14). It is one of the systems to realize various nanoscale fabrication and assembly techniques to develop novel nanodevices for the seamless integration of technologies based on a nanorobotic manipulation system. It is readily applied to the scientific exploration of macroscopic phenomena and the construction of prototype nanodevices. It would be one of the most significant enabling technologies to realize manipulation of and fabrication with individual atoms and molecules for the assembly of devices. Recently, the investigation of NEMS has attracted much attention because of its potential for realizing highly integrated, miniaturized, and multi-functional devices for various applications. One of the effective ways is the direct use of bottom-up-fabricated nanostructures.

The nanolaboratory can be applied for single-cell analysis and manipulation. As shown in Fig. 10.3, for the single-cell nanosurgery system the integration between micro and nanorobotic manipulators under various microscopes is important. The applications under dry, semi-wet, and wet conditions can be done under TEM/SEM, Environmental-SEM (E-SEM) and OM. The nanomanipulation system inside TEM/SEM is a fundamental technology for the property characterization of nanomaterials, structures and mechanisms for the fabrication of nano building blocks, and for the assembly of nanodevices. The nanomanipulation system inside E-SEM enables single-cell manipulation and analysis for the assembly of nanodevices or nanotools in dry conditions, under nanoscale-resolution images. The OM micromanipulation system is used under water, hence biological cells can be cultured with medium.

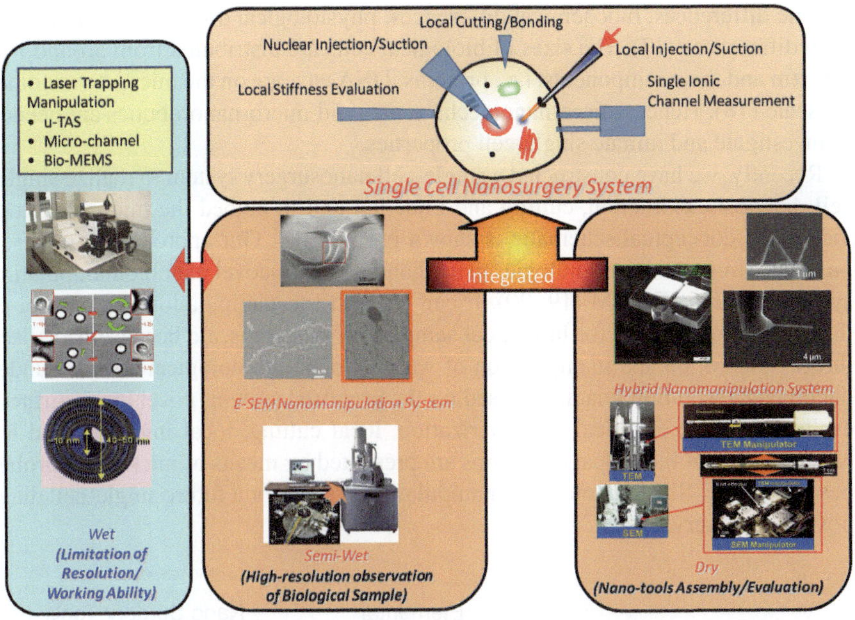

Fig. 10.3 Single-cell nanosurgery system based on micro/nanomanipulators under various microscopes (wet/semi-wet/dry conditions)

10.3 Nanobioscience Applications Based on Nanorobotic Manipulations

10.3.1 Single-Cell Analysis and Nanosurgery System Based on Nanorobotic Manipulations

Microbiology has traditionally been concerned with and focused on studies at the population level (105–107 cells) (15). On the other hand, single-cell analysis contributes to the important research of the existence of cellular heterogeneity within individual cells. This technique is important for next-generation analysis methods in biological and medical fields.

Cellular heterogeneity is widespread in bacteria and increasingly apparent in eukaryotic cells (16). Heterogeneity at the single-cell level is typically masked and is therefore unlikely to be acknowledged in conventional studies of microbial populations, which rely on data averaged across thousands or millions of cells. Bulk-scale measurements made on a heterogeneous population of cells report only average values for the population and are not capable of determining the contributions of individual cells. The types of individual differences contributing to heterogeneity within a microbial population can be divided into at least four general classes:

genetic differences, biochemical differences, physiological differences, and behavioral differences (17). The sizes of biological cells are distributed from around 1 to 100 μm and their components, i.e., proteins, DNA etc., are on the micro- to nanometer scale (18). Hence, micro-nanomechatronics and micro-nanorobotics are needed to investigate and imitate single-cell properties.

Recently, we have constructed a single-cell nanosurgery system to realize single-cell diagnosis, extraction, cutting and injection and to embed the micro-nano devices. The conceptual schematic is shown in Fig. 10.4. Our approaches are based on nanomanipulation technologies using nanotools in a novel nanobiomanipulation system inside an E-SEM (19; 20). It can be used for manipulation and control of the local environments for biological samples on the nanoscale because it realizes direct observation and manipulation of water-containing biological samples under nanometer-resolution imaging. Based on the proposed system, novel local stiffness evaluation, local electrical characterization, local cutting, local injection, and local extraction of biological organisms are presented by means of micro-nanoprobes based on the E-SEM nanorobotic manipulation system for a future single-cell diagnosis and surgery system.

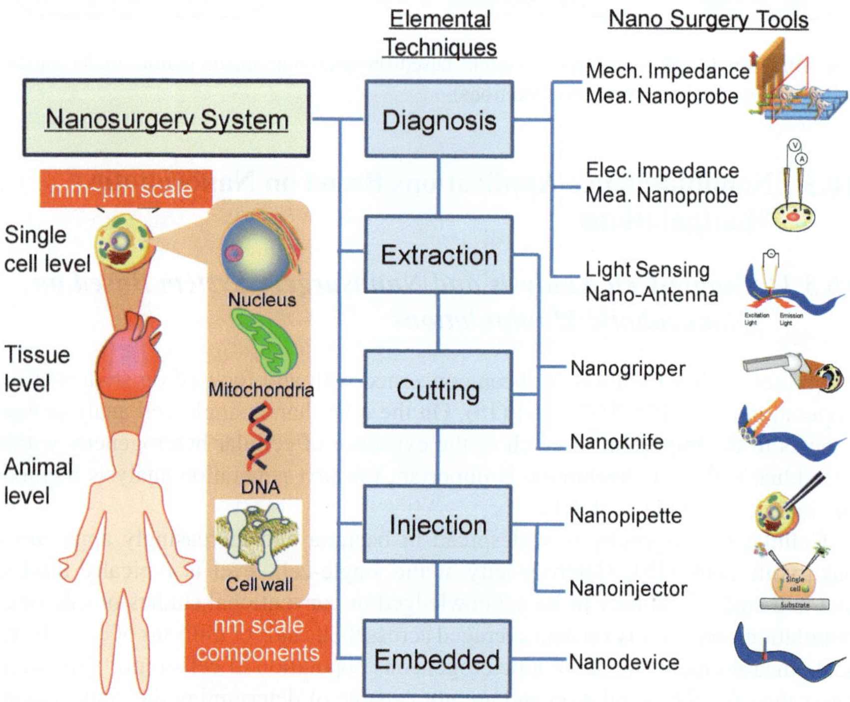

Fig. 10.4 Configuration of elemental technologies for nanosurgery system

10.3.2 E-SEM Nanonanorobotic Manipulation System

Under conventional SEMs and TEMs, the sample chambers are under high vacuum (HV) to reduce the disturbances of the electron beam for observation. To investigate water-containing samples, e.g., bio-cells, appropriate drying and dyeing treatments are needed before they can be examined. Hence, direct observations of water-containing samples are normally quite difficult using these electron microscopes. On the other hand, the E-SEM can realize the direct observation of water-containing samples with nanometer resolution (up to 3.5 nm imaging resolution) by means of a specially built secondary electron detector. The evaporation of water is controlled by the sample temperature (around 0–40°C) and the sample chamber pressure (10–2600 Pa). The unique characteristic of the E-SEM is the direct observation of hydroscopic samples with non-drying treatment. Hence, nanomanipulation inside the E-SEM is considered to be an effective tool for water-containing samples.

Our E-SEM nanorobotic manipulation system (Fig. 10.5) was built to realize effective biological manipulation and evaluation with 3 units and 7 DOFs. The sample temperature is controlled by the cooling stage unit (19; 20). It has a working area of 16 mm × 16 mm × 12 mm × 5° working area and a positioning accuracy of ~30 nm.

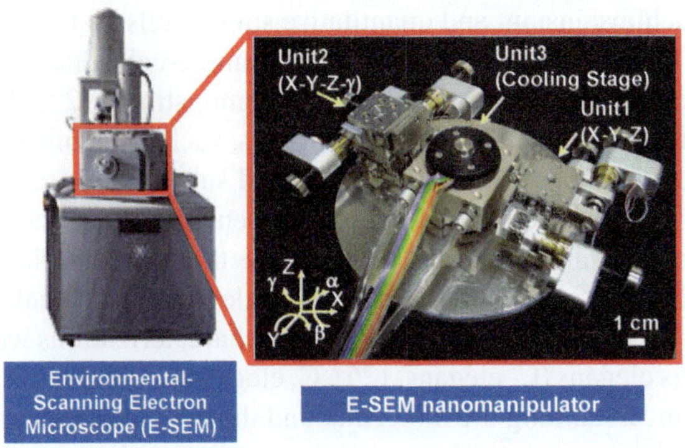

Fig. 10.5 E-SEM nanorobotic manipulation system

10.3.3 Examples for Single-Cell Manipulations Based on E-SEM Nanorobotic Manipulation System

We used the E-SEM nanomanipulation system for single-cell analysis and nanosurgery operation.

(1) In-situ measurements of the mechanical properties of individual wild-type yeast cells at different cell sizes (about 3 μm diameter), environmental conditions (600 Pa and 3 mPa), and growth phases (early log, mid log, late log and saturation) were conducted (19; 20). Figure 10.6 shows E-SEM images obtained during the single-cell stiffness measurement using the Si nanoprobe. The buckling nanoneedle was proposed as a novel force sensor for stiffness characterization of single cells, and its implementation in cell stiffness characterizations was presented. We used

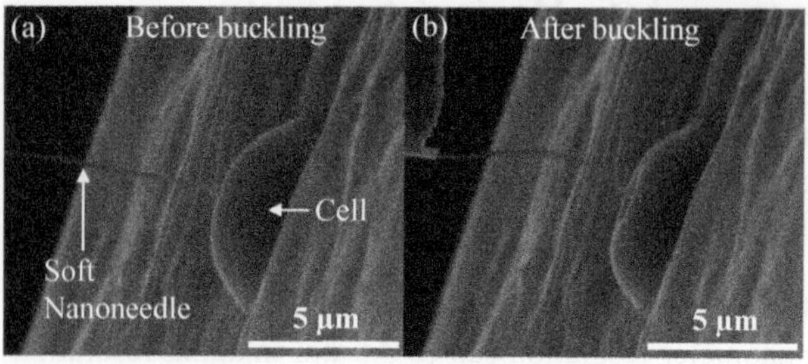

Fig. 10.6 E-SEM images taken during the single-cell stiffness measurement using the Si nanoprobe. (a) Straight and (b) buckling condition.

different rigidities to characterize the mechanical properties of single cells and to perform single-cell surgery by inserting probes. This method was applied to measure the viscoelastic properties of single cells using sharp tips, flat tips and buckling nanoneedles, in which strong conclusions could be obtained by comparing the results from each method (21).

(2) The electrical properties of single cells were measured by conducting single-pulse current measurements of the cell's cytoplasm using a dual nanoprobe to achieve instant and quantitative single-cells viability detection (22).

(3) A novel method to measure single-cell adhesion forces by means of a nanofork on a pillar-based substrate was demonstrated (23). It is expected that the adhesion forces are related to cell conditions, such as viability. It is also important for the cell culturing process between cell and substrate for regeneration engineering.

(4) Single-cell cutting was presented for yeast cells using a nano-knife (24). The nano-knife has a smaller tip angle than current diamond knives to reduce the cell compression caused by the tip angle of the nanoscale tip.

(5) Local mechanical property characterizations were presented for Caenorhabditis elegans (C. elegans) (25). C. elegans is widely investigated as model organism for understanding the molecular and the cellular mechanisms of human genes, because it has many homologue genes with human disease genes.

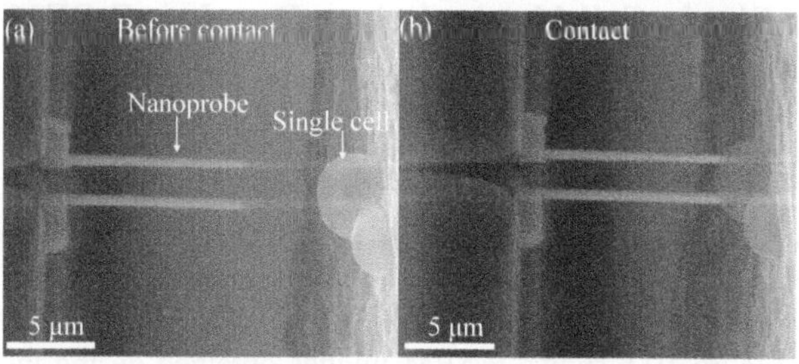

Fig. 10.7 E-SEM images obtained during the electrical probing on single cells using a nanoprobe. (a) Before and (b) during contact.

10.4 Future Direction of Nanobioscience Applications Based on Nanorobotic Manipulations – "in vitro" Realization of "in vivo" Environment for Bio-medical Applications

Seamless applications of micro-nanomechatronics are investigated such as Micro Biology, Wet Mechatronics, Micro Total Analysis System, Micro Medical Engineering, and Regenerative Medical Engineering. For such bio-medical applications, various micro-nanodevices have to be developed, such as micro-actuators, micro-ink-jet heads, micro-force sensors, micro-tactile sensors, micro-fuel batteries, micro fluidics devices, a blood vessel simulator, and so on. For example, for medical applications, we developed the patient blood-vessel simulator "EVE (Endo Vascular Educator)" to extend the frontiers of surgical simulation with novel surgical supporting or simulating technologies (26). It serves three main purposes: 1) training for catheter surgical operations, 2) evaluator for catheter surgery, and 3) simulator for catheter surgery before surgical operations. Figure 10.8 shows the EVE and related technologies from our laboratory. The simulator is fabricated and assembled using micro-fabrication technologies of the rapid prototyping technique.

We expect that future applications of nanobioscience include an "in vitro" realization of the "in vivo" environment (Fig. 10.9). For this purpose, it is important to achieve a seamless collaboration between engineering, medical, and biological fields. The investigation approach using micro/nano devices should be able measure and control inside "in vivo" biological systems, hence the possibility to reveal unknown properties in biological systems is considered to be important. Some "in vitro" construction and measurement techniques are needed to clearly reveal the biological systems in single cells and between multiple cells.

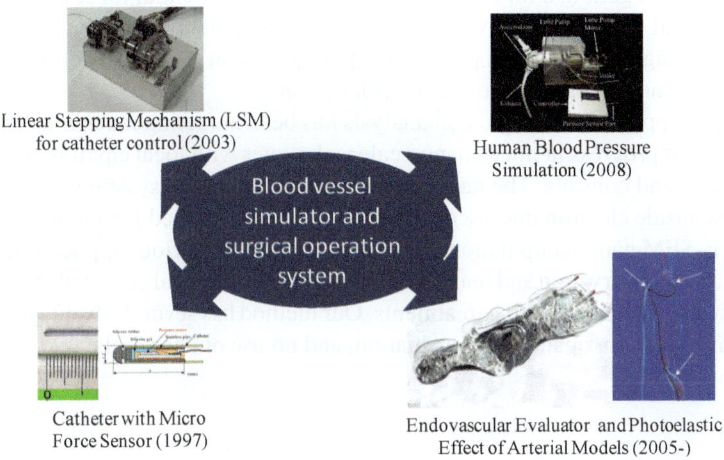

Fig. 10.8 Blood vessel simulator and surgical operation system

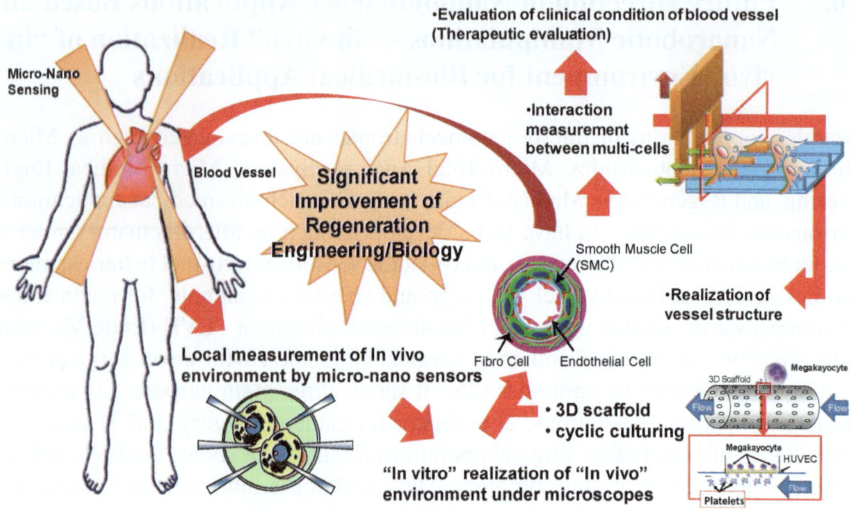

Fig. 10.9 "In vitro" realization of "in vivo" environment

10.5 Conclusion

In this chapter, we presented a summary of nanorobotic manipulations under various microscopes and their nanobioscience applications. So far, nanomanipulation is mainly applied to the scientific exploration of mesoscopic phenomena and the construction of prototype nanodevices. Recently, the evaluation of bio-samples has received much attention for nanobio applications in nano-biotechnology. Especially single-cell analysis is getting a lot of attention because of its potential for revealing unknown biological aspects of individual cells. Nanomanipulation techniques are one of the promising ways to develop nanobio applications on the single-cell level for drug delivery, nano-therapy, nano-surgery, and so on.

For biological applications, single-cell analysis has been attracting much interest based on the recent progress in micro/nano-scale techniques with local environmental measurements and controls. The nanorobotic nanomanipulation system we constructed for use inside electron microscopes was successfully applied for single-cell analysis. The E-SEM nanomanipulation system can be used for various applications involving the direct observation and manipulation of single biological cells with non-drying, non-dyeing, and non-coating treatments. Our method has several advantages, namely, instant detection, quantitative evaluation, and no use of chemicals.

References

[1] Du, E., Cui, H., Zhu, Z.: Review of nanomanipulators for nanomanufacturing. Int. J. Nanomanufacturing 1, 83–104 (2006)

[2] Craighead, H.G.: Nanoelectromechanical systems. Science 290, 1532–1535 (2000)

[3] Staples, M., Daniel, K., Sima, M.J., Langer, R.: Applications of micro- and nano-electromechanical devices to drug delivery. Pharmaceutical Research 23, 847–863 (2006)
[4] Leary, S.P., Liu, C.Y., Apuzzo, M.L.J.: Toward the emergence of nanoneurosurgery: Part III - Nanomedicine: Targeted nanotherapy, nanosurgery, and progress toward the realization of nanoneurosurgery. Neurosurgery 58, 1009–1026 (2006)
[5] Feynman, R.P.: There's plenty of room at the bottom. Caltech's Engineering and Science 23, 22–36 (1960)
[6] Lewis, A., Taha, H., Strinkovski, A., Manevitch, A., Khatchatouriants, A., Dekhter, R., Ammann, E.: Near-field optics: From subwavelength illumination to nanometric shadowing. Nature Biotechnol. 21, 1378–1386 (2003)
[7] Hell, S.W.: Far-field optical nanoscopy. Science 316, 1153–1158 (2007)
[8] Eigler, D.M., Schweizer, E.K.: Positioning single atoms with a scanning tunneling microscope. Nature 344, 524–526 (1990)
[9] Hertel, T., Martel, R., Avouris, P.: Manipulation of individual carbon nanotubes and their interaction with surfaces. J. Phys. Chem. B 102, 910–915 (1998)
[10] Li, G., Xi, N., Chen, H., Poneroy, C., Prokos, M.: Videolized atomic force microscopy for interactive nanomanipulation and nanoassembly. IEEE Trans. Nanotechnol. 4, 605–615 (2005)
[11] Yu, M.F., Lourie, O., Dyer, M.J., Moloni, K., Kelley, T.F., Ruoff, R.S.: Strength and breaking mechanism of multiwalled carbon nanotubes under tensile load. Science 287, 637–640 (2000)
[12] Kizuka, T., Yamada, K., Deguchi, S., Naruse, M., Tanaka, N.: Cross-sectional time resolved high-resolution transmission electron microscopy of atomic-scale contact and noncontact-type scannings on gold surfaces. Phys. Rev. B 55, 7398–7401 (1997)
[13] Nakajima, M., Arai, F., Fukuda, T.: In situ measurement of Young's modulus of carbon nanotube inside TEM through hybrid nanorobotic manipulation system. IEEE Trans. Nanotechnol. 5(3), 243–248 (2006)
[14] Fukuda, T., Arai, F., Dong, L.X.: Assembly of nanodevices with carbon nanotubes through nanorobotic manipulations. Proc. IEEE 91, 1803–1818 (2003)
[15] Sedgwick, H., Caron, F., Monaghan, P.B., Kolch, W., Cooper, J.M.: Lab-on-a-chip technologies for proteomic analysis from isolated cells. J. R. Soc. Interface 5, S123–S130 (2008)
[16] Ferrell, J.E., Machleder, E.M.: The biochemical basis of an all-or-none cell fate switch in xenopus oocytes. Science 280, 895–898 (1998)
[17] Avery, S.V.: Microbial cell individuality and the underlying sources of heterogeneity. Nature Reviews: Microbiology 4, 577–587 (2006)
[18] Wilson, J., Hunt, T.: Molecular Biology of the Cell. Garland Science (2002)
[19] Ahmad, M.R., Nakajima, M., Kojima, S., Homma, M., Fukuda, T.: In situ single cell mechanics characterization of yeast cells using nanoneedles inside environmental SEM. IEEE Trans. Nanotechnol. 7, 607–616 (2008)
[20] Ahmad, M.R., Nakajima, M., Kojima, S., Homma, M., Fukuda, T.: The effects of cell sizes, environmental conditions and growth phase son the strength of individual w303 yeast cells inside ESEM. IEEE Trans. Nanobioscience 7, 185–193 (2008)
[21] Ahmad, M.R., Nakajima, M., Kojima, S., Homma, M., Fukuda, T.: Nanoindentation methods to measure viscoelastic properties of single cells using sharp, flat, and buckling tips inside ESEM. IEEE Trans. Nanobioscience 9(1), 12–23 (2010)

[22] Ahmad, M.R., Nakajima, M., Kojima, S., Homma, M., Fukuda, T.: Single cell electrical characterizations using nanoprobe via ESEM-nanomanipulator system. In: Proc. 8th IEEE Int. Conf. on Nanotechnology (IEEE NANO-2009), pp. 589–592 (2009)

[23] Ahmad, M.R., Nakajima, M., Kojima, S., Homma, M., Fukuda, T.: Nanofork and line-patterned substrate for measuring single cells adhesion force inside ESEM. In: Proc. 9th IEEE Int. Conf. on Nanotechnology (IEEE NANO-2010), pp. 356–359 (2010)

[24] Shen, Y., Nakajima, M., Kojima, S., Homma, M., Fukuda, T.: Nano knife fabrication and calibration for single cell cutting inside environmental SEM. In: Proc. 2010 Int. Symposium on Micromechatronics and Human Science (MHS 2010), pp. 316–320 (2010)

[25] Nakajima, M., Ahmad, M.R., Kojima, M., Kojima, S., Homma, M., Fukuda, T.: Local stiffness measurements of C. elegans by buckling nanoprobes inside and environmental SEM. In: Proc. 2009 IEEE/RSJ Int. Conf. on Intelligent Robots and Systems (IROS 2009), pp. 1849–1854 (2009)

[26] Ikeda, S., Arai, F., Fukuda, T., Negoro, M.: An in vitro soft membranous model of individual human cerebral artery reproduced with visco-elastic behaviour. In: Proc. 2004 IEEE Int. Conf. on Robotics and Automation (ICRA 2004), pp. 2511–2516 (2004)

Chapter 11
Self-configuring CMOS Microsystems

Gary K. Fedder, Tamal Mukherjee, and Lawrence Pileggi

Abstract. The continuing evolution and maturation of techniques to integrate MEMS with CMOS is enabling creation of complex on-chip microsystems. The "More than Moore" trend in the microelectronics industry is accelerating the pace of advanced packaging, which will become increasingly exploited for MEMS integration and blur the lines between monolithic integration and heterogeneous stacked multi-chip systems. A requirement of many emerging MEMS component applications is the need for self-configuration and control in the presence of manufacturing and environmental variability. These are general attributes that hold promise of providing high manufacturing yield, resiliency and redundancy for critical applications. One class of such systems are RF circuits tightly integrated with multiple electrothermally actuated MEMS capacitors that lead to a variety of low-loss, frequency reconfigurable circuit blocks. Two exemplary self-configuring systems are micro-instrumented scanning probe arrays for tip-based nanomanufacturing and self-healing resonant mixer-filters for RF front-ends.

11.1 Introduction

Process and packaging technologies are enabling the intimate integration of CMOS circuits and systems with microelectromechanical systems (MEMS). The ability to

Gary K. Fedder
Institute for Complex Engineered Systems,
Department of Electrical and Computer Engineering and The Robotics Institute,
Carnegie Mellon University,
Pittsburgh, PA 15213
e-mail: fedder@ece.cmu.edu

Tamal Mukherjee · Lawrence Pileggi
Department of Electrical and Computer Engineering,
Carnegie Mellon University, Pittsburgh, PA 15213
e-mail: {tamal,pileggi}@ece.cmu.edu

E. Eleftheriou & S.O.R. Moheimani (Eds.): Cntrl. Tech. for Emerging Micro/Nanoscale Sys. LNCIS 413, pp. 181–200.
springerlink.com

interleave circuit functions with MEMS on a fine scale is inspiring new reconfigurable, self-configuring and self-healing systems that exploit arrays of multi-physics components. Reconfigurable microsystems offer capabilities to select MEMS components or to set parameter values of the MEMS components. Self-configurable (or adaptive) microsystems are a subset that exhibit the ability to autonomously reconfigure parameters or select components in order to operate in the presence of disturbances. Self-healing microsystems are a further subset that uses self-configuration to achieve manufacturability in the presence of component variation.

In this chapter, the general opportunity for CMOS MEMS integration will be introduced then followed by a description of a particular monolithic CMOS MEMS process developed at Carnegie Mellon. Specific examples of reconfigurable, self-configuring and self-healing microsystems will then be introduced in subsequent sections. An RF CMOS MEMS capacitor is one "building block" example component for reconfigurable microsystems, specifically RF reconfigurable circuits such as low noise amplifiers. The concept of tip-based field-emission assisted nanofabrication is discussed next, with the focus on integrated arrayed probe systems that exemplify self-configuration attributes. The last system to be described is a self-healing RF microresonator array used to create a manufacturable RF bandpass filter.

The ensuing discussion is not meant to be comprehensive but rather to inspire new uses for MEMS integration and to provide some illustrative examples of emerging self-configuring microsystems.

11.2 CMOS and MEMS Integration

The recent peak of interest in commercialization of microsystems is being motivated by the current success of MEMS in high volume applications such as motional sensing in the Nintendo Wii remote and rotational page orientation in the Apple iPhone. There are many future potential MEMS products, many of which will directly benefit from miniaturization of the electronics and MEMS either in system-in-package (SiP) or system-on-chip (SoC) solutions. Exciting applications include smart tunable passives for cognitive radio, low power miniature sensors for wireless sensor networks, lightweight, ultraminiature sensors for microrobotic air vehicles and biosensors and microfluidic devices for implantable medical microsystems.

One business sector showing increasing interest in MEMS technology is the semiconductor industry (i.e., CMOS manufacturers). The group responsible for the International Technology Roadmap for Semiconductors (ITRS) updates its 15-year forecast for the semiconductor industry every two years (1). In 2005, the group introduced the concept of "More than Moore" to describe the strategic trend in the industry toward technologies that are orthogonal to the traditional CMOS transistor scaling paradigm, as illustrated in Fig. 11.1. MEMS and other emerging technologies address new opportunities to implement radio frequency (RF), energy, sensing, actuation and biofluidic applications in systems-in-package and systems-on-chip. The latest roadmap in 2009 has moved MEMS into its own section in the RF/Analog Mixed Signal (AMS) report, signaling its growing strategic importance in the

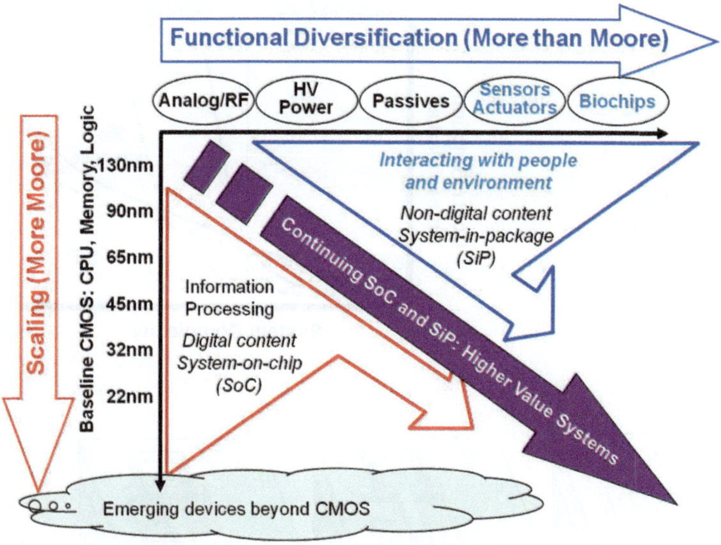

Fig. 11.1 "Moore's Law and More," adapted from (1)

semiconductor industry. Issues slowing the growth of MEMS relate to the difficulty in integrating the preponderance of different MEMS into CMOS platforms and the comparative lack of design support, packaging, performance drivers and cost drivers for MEMS.

Current options for realizing microsystem integration include System-on-Chip (SoC) and System-in-Package (SiP). Arguably, 3D integrated-circuit (3D-IC) technologies, such as chip stacking with through-hole silicon vias, can be considered an advanced form of SoC, where all integration is performed on chips that are intimately bonded together. SiP technology involves integration of chips usually by way of interposer substrates. Cost is usually the biggest driver of the decision of how to integrate and is dictated by the number of masks, the yield of the process, and the die area comparison of the MEMS and CMOS subsystems. Generally, if the MEMS die area is smaller or equal to the CMOS area, then there is a greater cost incentive to integrate. As indicated in Fig. 11.2, SiP cost roughly scales linearly with system complexity, while SoC cost increases dramatically when technology associated with advanced functionality is not manufacturable with high yield. The cost crosspoint that drives decisions to choose SoC or SiP will change with time as emerging SoC technologies, such as through-silicon vias and CMOS MEMS structures, mature. Aside from cost considerations, performance specifications motivate SoC integration when there is a need for ultra-small size, for handling large numbers of interconnects and for low parasitic capacitance to achieve ultra-low power or higher signal-to-noise microsystems.

Multiple technologies exist for co-fabricating MEMS and electronics (2). Since 1994, Fedder's group has been developing MEMS in foundry CMOS using

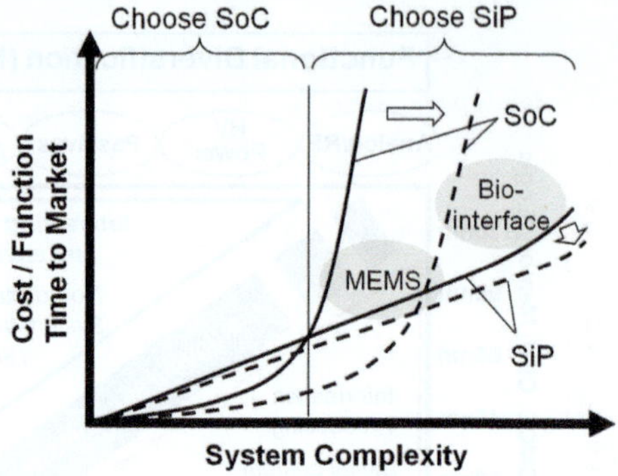

Fig. 11.2 A qualitative cost comparison between System-on-Chip (SoC) vs. System-in-Package (SiP) technology, adapted from the 2005 ITRS roadmap (1)

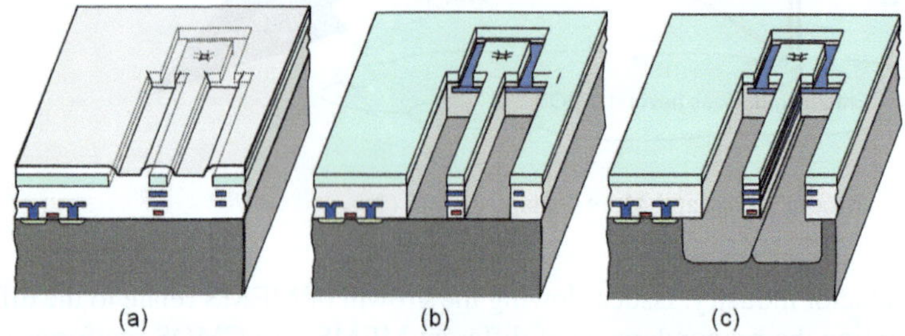

Fig. 11.3 A CMOS MEMS process flow. (a) Start with conventional foundry CMOS. (b) Reactive-ion etch of the CMOS dielectric. (c) Timed DRIE of silicon substrate followed by timed plasma etch that undercuts the silicon substrate. Adapted from (3).

derivatives of the process flow shown in Fig. 11.3 (3). The process starts with conventional foundry CMOS with MEMS features designed within the back-end-of-line metal dielectric stack. A reactive-ion etch (RIE) of the dielectric stack down to the silicon substrate defines the sidewalls of the microstructures. Subsequently, a timed deep-RIE (DRIE) of the silicon substrate followed by an isotropic plasma silicon etch undercuts and mechanically releases the microstructures. This post-CMOS process flow does not impact the foundry CMOS steps; however, its utility is restricted to processes where metal fill, normally required for advanced CMOS, can be removed from the masks in the MEMS design areas. The materials in the resulting composite microstructures are those of the CMOS metal, dielectric and transistor gate layers. The resulting devices can be designed with multiple layers of electrically isolated interconnect, as illustrated by the accelerometer shown in Fig. 11.4.

11.3 Reconfigurable RF MEMS Capacitors

Reconfigurable MEMS comprise a class of microsystem whose properties can be changed *in situ* within an application. In integrated MEMS, reconfiguration may

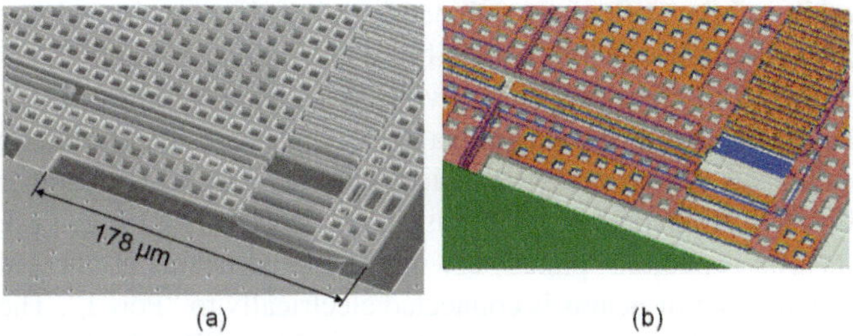

Fig. 11.4 A CMOS MEMS accelerometer illustrating CMOS interconnect internal to the microstructure. One quarter of the quad-symmetric structure is shown. (a) Scanning electron micrograph (SEM). (b) 3D emulated view generated by Semulator-3D (Coventor, Inc., Cary, NC).

be performed using electronic, mechanical, or other transduction domains, or some combination of these approaches. A simple example is setting sensor gain by switching in different values of voltage amplification. Another example is switching between an array of accelerometers, each with different gain, in order to span a larger dynamic range. In these cases, the alteration of system behavior is achieved through configuration of a multiplicity of MEMS and/or electronic components.

Future multi-channel radio is an application area where reconfigurable MEMS is expected to play a significant role. Envisioned radio-on-chip systems will benefit from RF MEMS passives that have large tuning range, low loss and high signal linearity. As illustrated in Fig. 11.5, reconfigurable MEMS passives can potentially provide these benefits in impedance matching networks, filters, low-noise amplifiers (LNA), power amplifiers (PA) and voltage controlled oscillators (VCO) within the transmit and receive chains. MEMS switches can also serve as "reconfigurable" RF components (i.e., on or off), but will not be discussed further here. RF MEMS passives used to form resonators also fall into the classification of reconfigurable components, and will be discussed in Section 11.5 for use in low power channel selection and frequency synthesis.

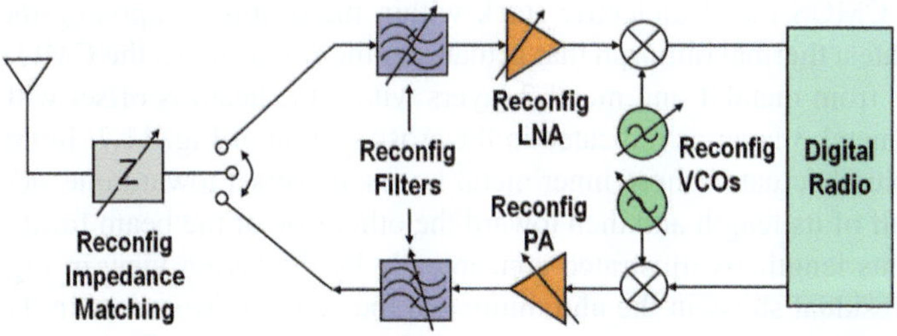

Fig. 11.5 Opportunities for MEMS component insertion into future on-chip radios

An interesting class of reconfigurable passives are RF MEMS capacitors. For example, MEMS arrayed capacitors are being developed for use in dynamic impedance matching networks to create more robust wireless links and for use in notch filters for handset tunable duplexers (4). The capacitance value is set by switching individual micromechanical parallel-plate capacitors into a low or high state corresponding to plates being apart at a fixed air gap or electrostatically clamped together. Another example of a reconfigurable RF capacitor (5; 6), using the process from Fig. 11.3, is shown in Fig. 11.6. The capacitor "plates" are interdigitated beams connected to trusses. The fixed "stator" set of beams is connected electrically to "Port 1". The moving "rotor" set of beams is electrically connected on one side to "Port 2" through a flexure that is designed for low resistance to minimize RF losses in the capacitor. The resulting quality factor at 1 GHz is 100. The other side of the rotor set of beams is connected mechanically to an electrothermal flexural actuator that moves in-plane to the wafer surface, in the x direction. The RF signal is not connected through the flexural actuator to avoid the extra capacitance on that node. This co-design of micromechanical and electrical function is critical to optimal operation. An important feature is a latching mechanism that is controlled by a second electrothermal flexural actuator that moves in the y direction.

Fig. 11.6 A RF MEMS capacitor micromachined from a foundry 0.35 μm BiCMOS process

The electrothermal flexural actuator works by thermal bimorph actuation that is designed into the CMOS metal-dielectric stack within the beams comprising the flexure (7). To create a thermal bimorph that actuates in the wafer plane, the CMOS interconnect made from metal-1 and metal-2 layers within the beam is offset with respect to the top metal-3 layer as indicated in the cross section of Fig. 11.7. In the electrothermal flexural actuator, these inner metal layers are offset toward one side of the beam for half of its length and then toward the other side of the beam for the remaining half of its length, as illustrated conceptually by the layout view in Fig. 11.8. The tensile residual stress in the aluminum and the compressive stress in the oxide causes an S-shaped displacement upon release from the substrate. Polysilicon resistors enable electrothermal heating that causes actuation in the opposite direction, since the temperature coefficient of expansion of aluminum is much larger than

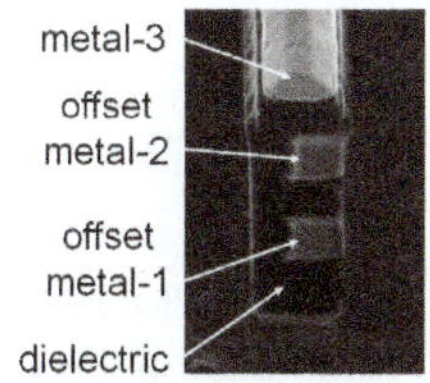

Fig. 11.7 Cross section of a CMOS MEMS beam with laterally offset metal layers to create a thermal bimorph

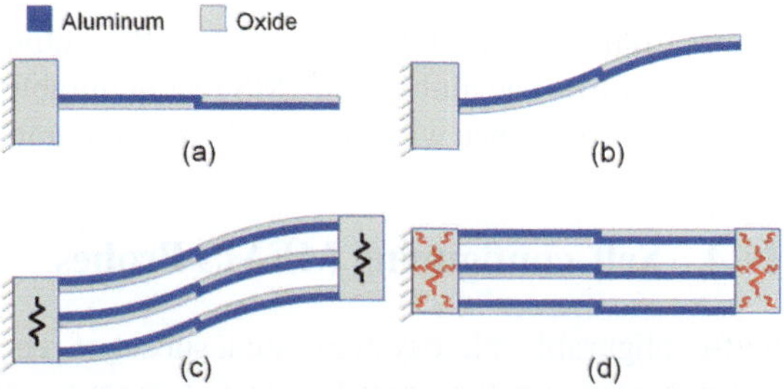

Fig. 11.8 Conceptual layout (top) view of the electrothermal flexural actuator. (a) Unreleased single beam showing offset of the inner metal layers along beam length. (b) Released single beam. (c) Ganged beams as released showing the location of heater resistors. (d) Ganged beams during electrothermal actuation.

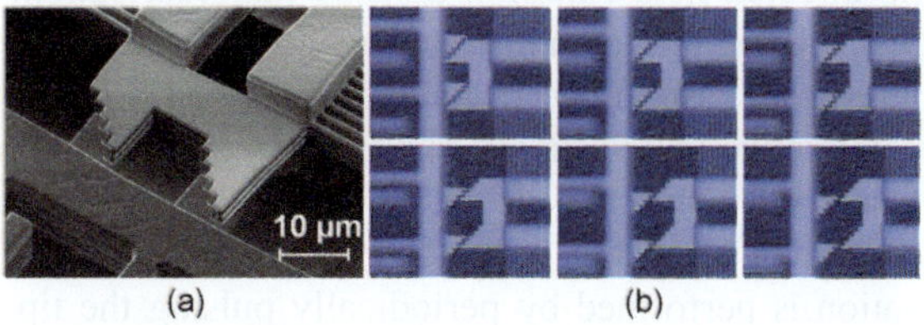

Fig. 11.9 Electrothermally actuated latch. (a) Close up. (b) Images at each position.

that of oxide. The guided-end displacement allows multiple beams to be ganged in parallel thereby creating a stiff actuator. The latching mechanism is designed to engage when no actuation power is applied. When its electrothermal actuator is powered, the latch is disengaged and the capacitor rotor beams can be reconfigured to one of six positions in 2 μm increments, as shown in Fig. 11.9. Each position corresponds to a different plate gap and capacitance. The measured maximum to minimum capacitance ratio is 8.9:1, with a maximum of 434 fF.

Multiple instances of this reconfigurable capacitor have been used in low noise amplifiers (LNA), power amplifiers and bandpass filters (8). For example, the multi-project test chip in Fig. 11.10 includes an LNA with two MEMS reconfigurable capacitors and three suspended MEMS inductors and a PA with three MEMS capacitors and four MEMS inductors. These circuits are made monolithically with the MEMS components in a BiCMOS 0.35 μm process. When the circuits are switched from their high to low capacitance states, the LNA band center moves from 2.6 to 3.2 GHz with corresponding gain of 13.5 to 18 dB (9) and the PA band center

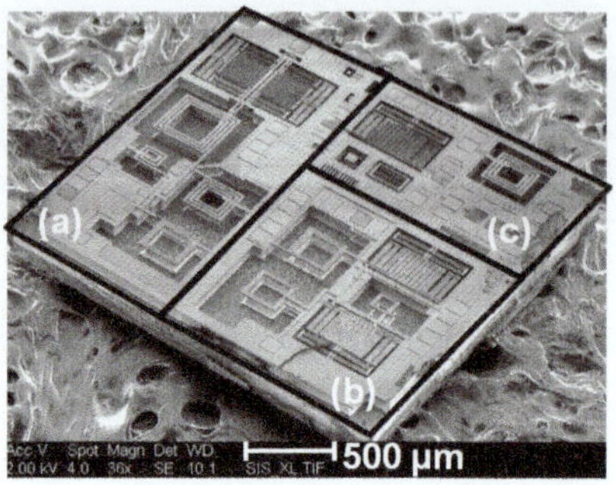

Fig. 11.10 CMOS RF MEMS multi-project chip comprising (a) a PA, (b) a LNA, (c) a stand-alone reconfigurable capacitor and a suspended inductor

moves from 3.4 to 4.3 GHz with a 19.1 dBm output power and 15.1% power aided efficiency (10). The multiple MEMS reconfigurable passives working in concert enable this performance with GHz-scale reconfiguration range on a monolithic chip.

11.4 Self-configuring MEMS Probes

Self-configurable microsystems are a subset of reconfigurable microsystems where components are autonomously selected or set to values to achieve new functionality. An example falling into this category is a parallel scanning probe system for tip-directed field-emission assisted nanofabrication. Several years ago, nanoscale lithography was demonstrated using tip-based field emission to expose patterns in e-beam resist (11). In another tip-based process, field emission was shown to be effective in selectively removing hydrogen from an atomically flat silicon surface with nanoscale resolution (12). An example pattern of desorbed hydrogen on silicon using this method is shown in Fig. 11.11 (13). The hydrogen depassivation is performed by periodically pulsing the tip at 4.5 V for sub-ms duration while performing scanning tunneling microscopy (STM) across the surface under ultra-high vacuum. The rapid write pulse does not destabilize the tunneling current servo. The STM image was created with a second pass over the area. The key

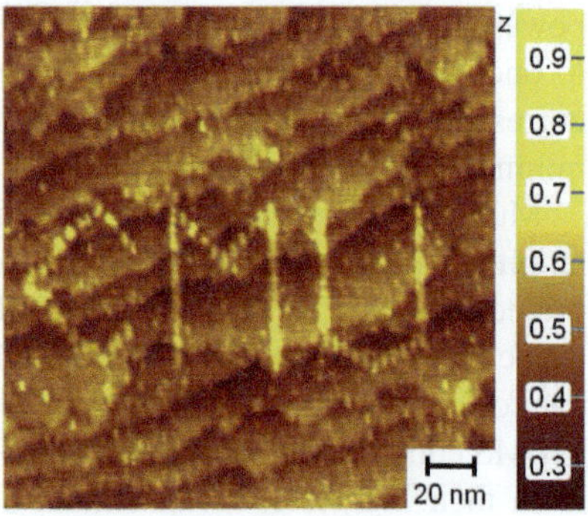

Fig. 11.11 Scanning tunneling microscope image of a hydrogen terminated ⟨100⟩ Si substrate patterned with a field emission from the scanning tip

advantage of this tip-based approach is the deterministic placement of nanoscale features with ±5 nm repeatability. The electron dose to perform writing constrains the scan speed to around 1 mm/s. Therefore, the disadvantage of tip-based writing is the slow throughput, which motivates work on parallel probe arrays.

Field emission current is a strong exponential function of tip-substrate spacing and voltage. However, the diameter of the field-emitted electron beam is dictated by the tip-substrate voltage and is roughly independent of the tip-substrate spacing, for spacing that is smaller than the tip radius. Therefore, it is possible to control both beam current and diameter independently through setting of voltage and spacing.

Depending on the tip formation process, the variation in the tip height on a probe cantilever can be much larger than the tip spacing, which is typically 1 nm to 10 nm. For example, "Spindt" tips (14) such as those shown in Fig. 11.12 have height variation of up to 100 nm (15). In order to control the tip-substrate spacing for all tips in a parallel probe array, each probe must have independent vertical-axis (i.e., out-of-plane) actuation capability. Furthermore, independent servoing is necessary since each tip sees different substrate topography at the nanoscale while scanning. CMOS MEMS probes, such as in Fig. 11.13, are being developed toward such a tip array

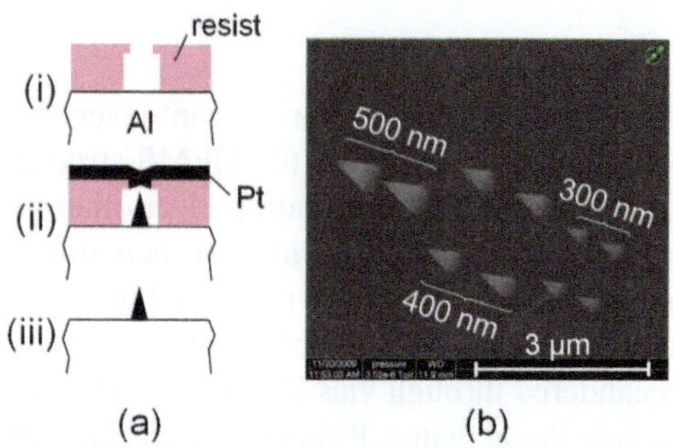

Fig. 11.12 Spindt tips. (a) Process: (i) Resist pattern, (ii) metal evaporation, (iii) resist lift-off. (b) Completed tips. The values correspond to the hole size in the e-beam resist.

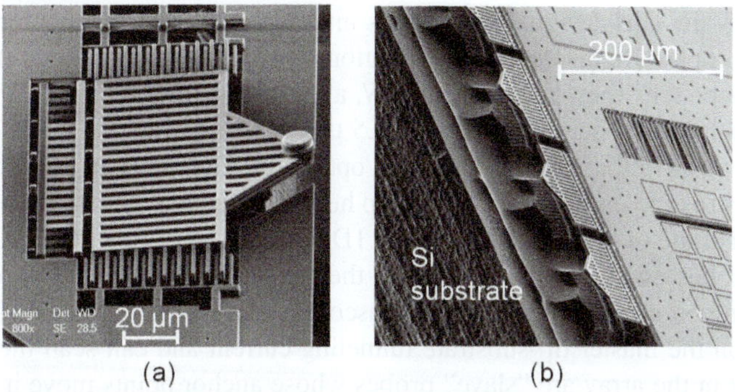

Fig. 11.13 CMOS MEMS probe platforms without tips. (a) Single probe with capacitive displacement sensing. (b) One-dimensional (1-D) probe array as viewed from the side of the chip.

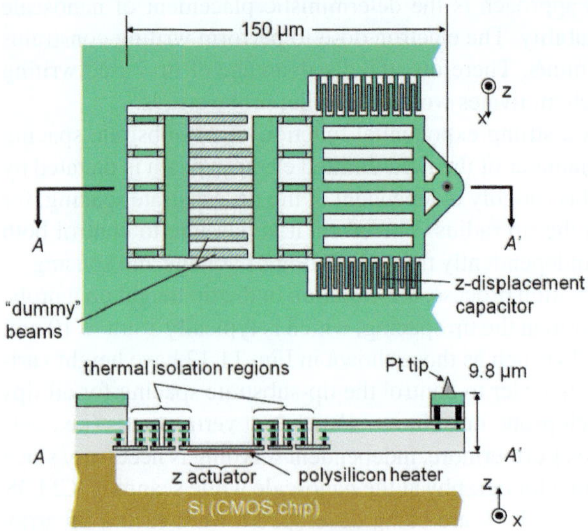

Fig. 11.14 CMOS MEMS probe schematic

servo system (15). The probe cantilevers must have a stiffness of around 50 N/m, which is relatively large for MEMS structures, to prevent electrostatic snap down to the substrate. The bimorph electrothermal actuation principle, described previously in Section 11.3 for in-plane actuators, is exploited here for vertical actuation. A schematic view of the probe in Fig. 11.13(a) is shown in Fig. 11.14. A polysilicon heater resistor is located in the thin metal/oxide bimorph actuator. Wiring is meandered through vias located on both ends of the actuator in order to thermally isolate the actuator. Running wiring directly to the anchor would shunt excessive heat and result in a low actuation sensitivity. "Dummy" heater beams are located on both sides of the actuator to help provide uniform temperature distribution when operating the actuator in air. This design also includes interdigitated comb fingers for future exploration of capacitive displacement detection. The probe in Fig. 11.13(a) has a measured actuator sensitivity of 0.25 µm/mW, a thermal cutoff frequency of 370 Hz and a maximum safe operating range of 1.5 µm before elevated temperatures in the actuator adversely affect its long-term operation. The actuated end of the probe has ample stroke to compensate for the tip height variation.

A system diagram of the on-chip CMOS MEMS 1D probe array is shown in Fig. 11.15. One of the probes on chip is designated as the "master" probe, and its tip is connected to a commercial scanning probe microscopy (SPM) system. The SPM system then servos on the master tip-substrate tunneling current and can scan the tip. The $n - 1$ probes in the array are "slave" probes whose anchor points move in rigid body motion with the SPM piezo actuation. Each slave probe has an on-chip transresistance amplifier to measure its tip-substrate current, which is fed back to the probe's electrothermal actuator. Each tip independently servos its position with

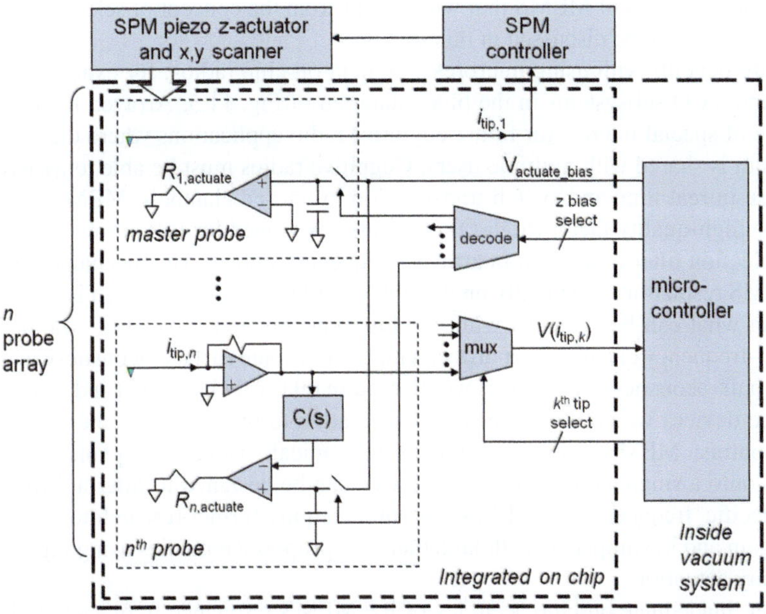

Fig. 11.15 Block diagram of an integrated MEMS probe array system for tip-directed field-emission assisted nanofabrication

respect to the substrate. Vertical bias offset due to tip height variation and to tilt alignment error of the chip to substrate is stored onto a hold circuit that adds the appropriate fixed offset to the actuator drive voltage. The target value of the tip currents is the same for all probes in the architecture with the current bias offset hidden within the compensator feedback block, C(s). Tip writing is controlled by applying the write pulse to the substrate. All probes write in parallel, however any probe can be pulled up to deselect for writing. Practical issues with the vacuum SPM system being used for this work limits the wiring to the probe head to around six connections. Therefore, the tip current signals and actuator bias voltages are multiplexed to a microcontroller to minimize the amount of electrical connections into the vacuum chamber. This arrayed probe SoC architecture provides for scalability to a large number of probes.

11.5 Self-healing RF Microresonator Systems

Self-healing microsystems are a subset of self-configurable microsystems where components are autonomously selected or set to values to achieve manufacturability or resiliency in the face of influences from the environment or intrinsic aging. The probes discussed in the prior section display one kind of self-healing behavior: the servo ability to compensate for tip height variation. Microresonators are another important class of MEMS that will benefit from the concept of self-healing microsystems, as will be discussed in this section.

There are benefits with using microresonators in on-chip radios; for example in the filter and VCO subsystems in the block diagram of Fig. 11.5. Arrayed RF resonators are of special interest for future cognitive radio applications where the signal spectrum is shared with multiple users. Cognitive radios must be able to assess spectral use in real time and switch transmission to unused channels. MEMS resonators are high-quality factor (high-Q) elements envisioned for use in low-power channel selection filter arrays and in arrayed frequency sources. The mechanical Q in RF MEMS resonators is typically on the order of 1,000 to 10,000, which is much higher than what can be achieved with on-chip inductors and capacitors. Making large multi-frequency channel filter arrays with conventional surface-acoustic-wave (SAW) or bulk-acoustic-wave devices would require relatively large area and power because the devices would require physical separation to block inter-element coupling. In contrast, MEMS enables arraying of mechanically isolated, in-plane mode resonators onto a single chip, where each device can be designed to filter a different, but specific, frequency. RF MEMS resonators require further research to attain adequate gain and low impedance, though there are proposed remedies using higher force density actuation.

However, inadequate on-chip accuracy of RF MEMS resonance frequency with high manufacturing yield remains a key problem yet to be suitably addressed. A ± 50 ppm variation (e.g., ± 50 kHz variation for a 1 GHz resonator) is typical of SAW technology and is thus a good target comparison. Micromechanical resonant frequencies are linearly dependent on characteristic dimensions of the device. For bending mode devices the most critical dimension is usually a flexural beam width while for bulk mode devices the critical dimension is the length along the wave propagation direction. These dimensions are typically micron scale for low frequency to 1 GHz resonance and become sub-micron scale for tens of GHz mechanical resonance (16). The problem of frequency variation becomes severe at high frequencies. For example, the 11.7 GHz bulk mode MEMS resonator in (16) has a characteristic dimension of 500 nm. For an assumed 10 nm dimensional process variation from device to device, the resonance frequency varies by 234 MHz, corresponding to 2% of the resonance frequency. Such relatively large frequency variation cannot be tolerated in radio applications.

While the high Q of the resonators is a benefit to performance, this property makes it difficult to tune the resonance frequency. Individual trimming of resonators is possible but very expensive. A proposed solution is to borrow from concepts that handle manufacturing offset variation in transistors to enable precision analog circuits in advanced CMOS processes. Three approaches — scaling, simple redundancy and combinatorial redundancy — are compared in Fig. 11.16 for a generic device with 1 σ variation of 1 to meet a system specification of 0.02 (17). Manufacturing yield, defined as the fraction of devices meeting specification, can be improved by simply adding together more transistors at the input stage (e.g., make them bigger). However, using this scaling or averaging approach, yield improves very gradually as $\mathrm{erfc}(a\sqrt{N})$, where N is the total number of devices in the array and

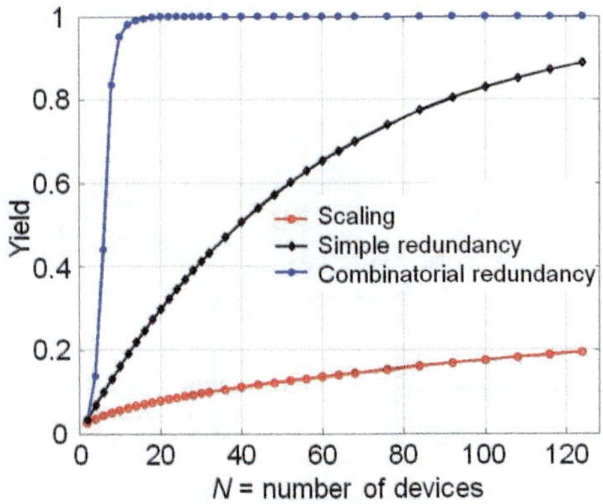

Fig. 11.16 Yield as a function of total number of devices available (N) for three design approaches: device scaling (use N of N), device redundancy (choose best 1 of N) and combinatorial device redundancy (choose best $N/2$ of N)

a is a coefficient dependent on the variation and yield specification. Redundancy is an alternative concept where the best single device meeting the specification is chosen from an array of N devices. This approach provides significantly better yield for a given number of devices. When it is possible to sum combinations of elements, an even more efficient use of devices is combinatorial redundancy, also called "statistical element selection," where the best combination of k out of N possible devices is chosen (18; 19). Since the number of potential device combinations grows nearly exponentially, the yield rapidly approaches 100% using only a very small array of devices compared to simple redundancy or scaling.

The MEMS mixer-filter receiver architecture shown in Fig. 11.17 is an initial system that our group has studied to determine if combinatorial redundancy will improve yield in the presence of resonant frequency variation. The input RF signal is routed to an array of MEMS resonant mixer-filter devices that down-converts the signal to an intermediate frequency. These devices, outlined by the dashed boxes in

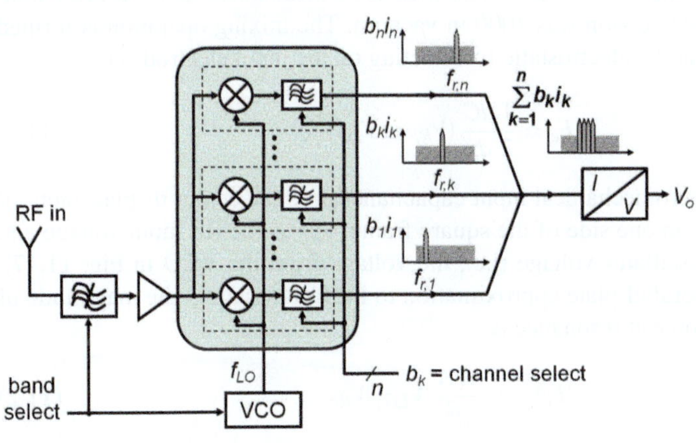

Fig. 11.17 MEMS arrayed mixer-filter receiver architecture

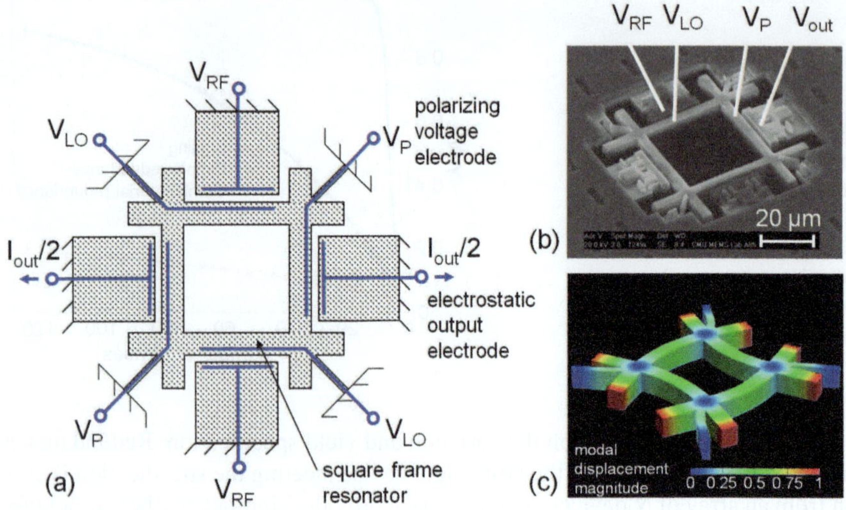

Fig. 11.18 CMOS MEMS square-frame bending mode resonator. (a) Schematic. (b) SEM. (c) Simulated mode shape.

the figure, output the demodulated signal in the form of a motional current. Only intermediate signal frequencies (IF) that lie in the narrow resonant band of the devices enabled by the channel select lines will pass. Assuming loading effects at the output are small, the motional currents can be summed by directly ganging the outputs and subsequently detected with a transimpedance amplifier.

Any electrostatic MEMS resonator can be configured as a mixer-filter. For our study, the CMOS MEMS square-frame bending mode resonator shown in Fig. 11.18 was employed (20). The device has a wine-glass mode resonance at 10 MHz, which sets the center of the IF passband. The measured mechanical Q for a prior non-optimized 6.2 MHz version was 1000 in vacuum. The mixing operation is formed by recognizing that the electrostatic force acting on the input electrodes is

$$F_e = \frac{1}{2}\frac{dC}{dx}\left(V_{RF} - V_{LO}\right)^2, \tag{11.1}$$

where C is the micromechanical input capacitance, x is the center displacement of the beam making up one side of the square frame, V_{RF} is the RF input voltage and V_{LO} is the local oscillator voltage (i.e., the voltage from the VCO in Fig. 11.17). Making a crude parallel-plate approximation to illustrate scaling, the magnitude of the electrostatic force at resonance is

$$\left|F_e\right|_{\omega_r} = \frac{\varepsilon_o A}{g^2}\left|V_{LO}\right|\left|V_{RF}\right|, \tag{11.2}$$

where ε_o is permittivity of air, A is the effective area of the sidewall parallel-plate electrodes, ω_r is radian resonant frequency and g is the effective gap spacing between electrodes. The magnitude of the motional current at resonance is

$$|i|_{\omega_r} = V_P \frac{dC}{dx}\frac{dx}{dt} = \frac{\omega_r Q \varepsilon_o^2 A^2}{kg^4} V_P |V_{LO}| |V_{RF}|, \tag{11.3}$$

where V_P is a polarization voltage and k is the effective spring constant of the mode. Note that V_P controls the gain and is ideal for use as the channel select signal, b_k, in Fig. 11.17. For the purposes of scaling, Equations (11.1) to (11.3) are general to any electrostatically driven parallel-plate RF MEMS resonator. Alternatively, the resonator can be used as a direct RF filter by replacing V_{LO} with a second polarization voltage that controls input gain.

Since a MEMS filter requires sets of different sized resonators in order to span a range of frequencies and also requires combinations of resonators to achieve gain targets, it provides a natural platform to exploit combinatorial redundancy. Frequency "bins" are created by comprising N resonators per bin, as illustrated in Fig. 11.19. The resonators comprise the mixer-filter array in Fig. 11.17 and their motional currents are summed by ganging the outputs. A combination of k of the N resonators per bin are then electrically enabled by turning on their polarization voltage. Resonators are selected that provide the filter passband with minimal ripple (or within a specification). The small random variation among them provides the tunability of the passband ripple. This approach works well for handling mismatch variation (e.g., device to device on the same die). In general, for RF filters, handling process variation (e.g., run-to-run) that is larger than mismatch variation requires extra bins to ensure resonators in the array exist with the desired resonant frequencies. The extra elements would adversely impact the area benefits of self-healing approaches. However, process variation in mixer-filters can be compensated by shifting the LO frequency to null the nominal resonant frequency offset over the

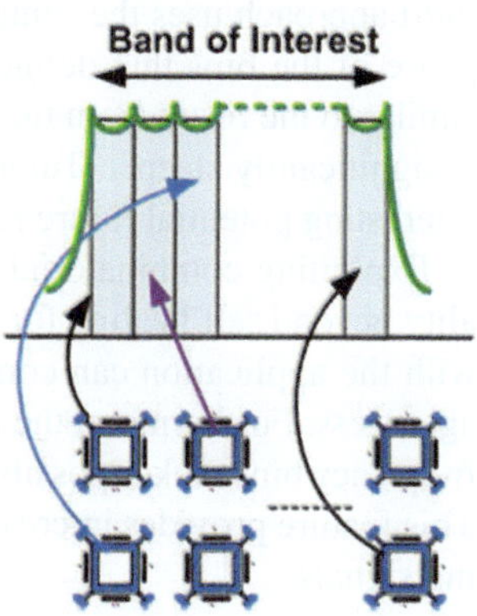

Fig. 11.19 Frequency binning concept for self-healing MEMS filter (17)

array. Therefore, extra bins are not necessary unless direct RF filters are to be implemented. Also, in future nanoscale resonators that use front-end-of-line processes, the mismatch variation is expected to be on the same order as process variation, based on the variation trends for advanced nanoscale transistors. Substantial extra bins should also not be necessary when implementing future RF arrayed systems with nanoscale devices.

Results from a filter example with a square-frame mixer-filter resonator array are shown in Fig. 11.20. The IF passband is designed from 10.0 MHz to 10.1 MHz using 34 bins spaced 3 kHz apart and with 24 resonators per bin, and assuming a mechanical Q of 2000 can be achieved. This resonator array is simulated using MEMS system-level behavioral models in schematic view in Cadence. Exact values for variation of geometry and materials parameters in MEMS processes are generally not known. Assumed 1 σ mismatch values are 13 nm for beam width and 0.5% for Young's modulus and density. The physical variation propagates though the simulation to determine mismatch variation of resonant frequency. Monte-Carlo simulation of 500 resonator samples reveals a 1 σ mismatch of 20 kHz, which is far larger than the bin spacing. Filter behavior of all resonator combinations for this array are then assessed in MATLAB using simple second-order resonator transfer functions to identify the combination of selected resonators over all bins that gives bandpass ripple within ± 1 dB. Since the number of combinations is very large (e.g., 2.7 million for 12 of 24 elements in just a single bin) and takes significant time, a random search for the best combination of resonators is limited to 10,000 trials. Three approaches for determining the profile of the number of selected resonators within each bin of 24 resonators are shown in Fig. 11.20(a). The first approach is to choose an equal number of resonators (i.e., 12) per bin. The resulting best filter response, shown in Fig. 11.20(b), has undesirable peaking at the edges of the passband. The second approach is a tailored number of resonators chosen per bin, with fewer resonators chosen near the band edge to compensate for the peaking effect. The optimal profile is determined by identifying the best filter response in MATLAB simulations with resonators having no variation. The passband is much flatter and the phase is more linear compared to the result from the first approach. The third approach uses the same tailored number of resonators per bin, but inverts the phase of the bins that define each band edge. This technique provides a flat band similar to the result from the second approach; however, the cutoff at the band edge is significantly steeper. Tuning the architecture for such MEMS filter systems is an interesting potential future research topic.

Exploiting combinatorial resonant element arrays can provide further functionality beyond self healing for improved manufacturing yield. Active selection *in situ* with the application can compensate for changes in ambient temperature or package stress. Furthermore, the additional choice of turning off all resonators in select frequency bins makes possible adaptive filters with multiple fine-grained stopbands. This feature provides interesting capabilities for adaptive radios that reject interfering signals.

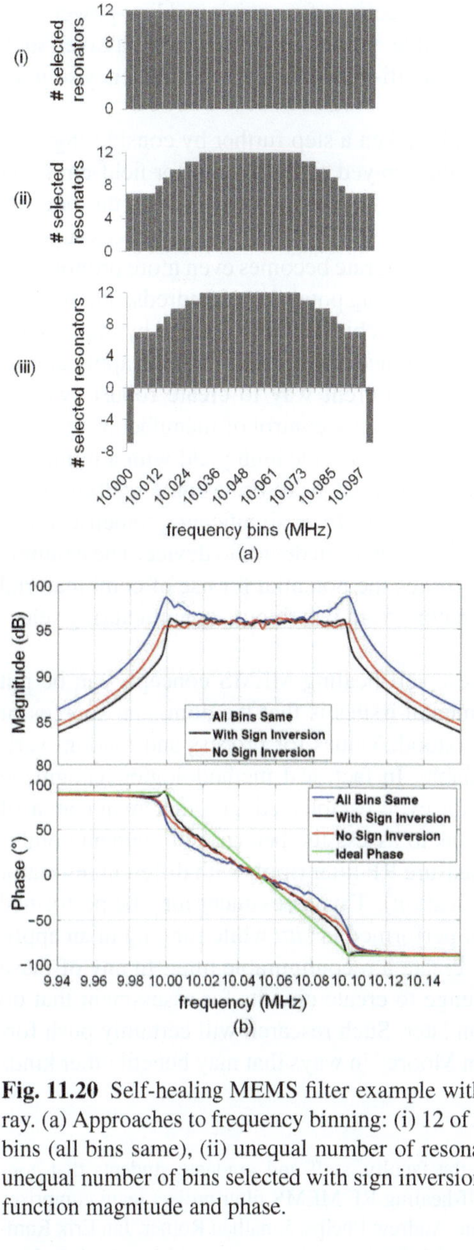

Fig. 11.20 Self-healing MEMS filter example with a square-frame mixer-filter resonator array. (a) Approaches to frequency binning: (i) 12 of 24 resonators selected for all 34 frequency bins (all bins same), (ii) unequal number of resonators selected (no sign inversion) and (iii) unequal number of bins selected with sign inversion on edge bins. (b) Resulting filter transfer function magnitude and phase.

11.6 Conclusions

Steady advances in integrated MEMS and 3D-IC technologies are providing opportunities for creating new multi-physics functionality for systems-on-chip. CMOS MEMS processes are one viable approach for monolithic integration of MEMS with electronics. In particular, RF front-end architectures for envisioned low-power cognitive radio benefit directly from integratable, reconfigurable high-Q passives such as CMOS MEMS capacitors by enabling multi-band filtering and frequency sources to be implemented on a single chip.

The concept of reconfigurability can be taken a step further by considering self-configuring systems as exemplified by the arrayed probe system for field-emission assisted nanomanufacturing. An integrated MEMS approach, such as the one described above, may be the only practical approach to accomplish the necessary parallel servo of multiple probes. The need to integrate becomes even more pronounced when one considers future 2D probe arrays having potentially hundreds, maybe even thousands, of field-emission probes on a single chip. Lessons learned by implementing this system can be applied to other envisioned arrayed nanoprobing applications.

Self-healing microsystems represent a different way to create robust systems from MEMS devices that have insufficient process control of manufacturing variation. The use of combinatorial redundancy can provide high yield with a minimum MEMS array size compared to simple redundancy or to brute-force scaling of devices. In particular, RF MEMS resonators suffer from significant geometric variation that translates into poor frequency precision from device to device. The example of the MEMS mixer-filter receiver illustrates the potential for use of combinatorial redundancy to design an array of resonators from which post-manufacturing selection tunes RF filters for high yield.

Many challenges remain before these self-healing MEMS concepts can be put into practice. One of the most fundamental issues is that the dominant sources of variation in MEMS are not well understood. Values for process and random variations in MEMS are not readily available. In fact, test methodologies recognized as standards by the community are not even established. Another major area of research is in identifying efficient means to assess the behavior of element combinations autonomously on chip (e.g., measure RF filter ripple with different resonator combinations and compare to a specification). This assessment may be performed once at startup or alternatively may be performed *in situ* while running in an application to implement compensation or create a reconfigurable filter. In any of these scenarios, it will be an exciting challenge to create circuits for assessment that do not interfere with the system operation later. Such research will certainly push forward the state-of-the-art of "More than Moore" in ways that may benefit other kinds of multi-physics systems-on-chip.

Acknowledgements. The author thanks the faculty, staff and graduate students that contributed to the research presented. The self-healing RF MEMS filter project team comprises the authors and Prof. Xin Li, Gokce Keskin, Andrew Phelps, Jonathan Rotner, Jan Erik Ramstad and Chih-Ming Sun. The tip-based nanofabrication project team comprises Prof. Fedder, Prof. James Bain, Prof. Robert F. Davis, Prof. David Ricketts, Prof. Metin Sitti, Dr. Joshua Smith, Ying Yi Dang, Weihua Hu, Onur Ozcan and Yang (Andy) Zhang. The reconfigurable RF MEMS team includes Prof. Fedder, Prof. Mukherjee, Abhishek Jajoo, John Reinke and Leon Wang. CMOS MEMS devices were fabricated with assistance from the Carnegie Mellon Nanofabrication Facility and from Suresh Santhanam. Funding for this research from the

U.S. Defense Advanced Research Projects Agency (DARPA) under awards N66001-09-1-20784 and N66001-08-1-203, the Industrial Technology Research Institute (ITRI) in Taiwan and a National Defense Science and Engineering Graduate (NDSEG) Fellowship is gratefully acknowledged.

References

[1] ITRS (2010) International Technology Roadmap for Semiconductors (ITRS), http://www.itrs.net.Cited30Sep2010
[2] Fedder, G.K., Howe, R.T., Liu, T.J.K., Quevy, E.P.: Technologies for co-fabricating MEMS and electronics. Proc. IEEE 96(2), 306–322 (2008)
[3] Fedder, G.K., Santhanam, S., Reed, M.L., et al.: Laminated high-aspect-ratio microstructures in a conventional CMOS process. Sensors & Actuators A 57(2), 103–110 (1996)
[4] De Luis, J.R., Morris, A.S., Gu, Q., De Flaviis, F.: A tunable asymmetric notch filter using RFMEMS. In: Dig. of the Int'l Microwave Symposium, pp. 1146–1149 (May 2010)
[5] Reinke, J., Jajoo, A., Wang, L., Fedder, G.K., Mukherjee, T.: CMOS-MEMS variable capacitors with low parasitic capacitance for frequency-reconfigurable RF circuits. In: Proc. IEEE Radio Frequency Integrated Circuits Conf (RFIC 2009), Boston, MA, pp. 509–512 (June 2009)
[6] Reinke, J., Fedder, G.K., Mukherjee, T.: CMOS-MEMS variable capacitors using electrothermal actuation. J. Microelectromech. Sys. 19(5), 1105–1115 (2010)
[7] Gilgunn, P.J., Liu, J., Sarkar, N., Fedder, G.K.: CMOS-MEMS lateral electrothermal actuators. J. Microelectromech. Sys. 17(1), 103–114 (2008)
[8] Jajoo, A., Wang, L., Mukherjee, T.: MEMS varactor enabled frequency-reconfigurable LNA and PA in the upper UHF band. In: Proc. IEEE International Microwave Symposium (IMS 2009), Boston, MA, pp. 1121–1124 (June 2009)
[9] Jajoo, A.: Frequency reconfigurable narrow-band low noise amplifiers using CMOS-MEMS passives for multi-band receivers. Ph.D. thesis, Dept. Electrical and Computer Eng., Carnegie Mellon University, Pittsburgh, PA (2010)
[10] Wang, L., Mukherjee, T.: A 3.4 GHz to 4.3 GHz frequency-reconfigurable class E power amplifier with an integrated CMOS-MEMS LC balun. In: Proc. IEEE Radio Frequency Integrated Circuits Symposium (RFIC 2010), Anaheim, CA, pp. 167–170 (May 2010)
[11] Wilder, K., Singh, B., Kyser, D.F., Quate, C.F.: Electron beam and scanning probe lithography: A comparison. J. Vac. Sci. Technol. B, Microelectron. Process. Phenom. 16(6), 3864–3873 (1998)
[12] Lyding, J.W., Hess, K., Abeln, G.C., et al.: Ultrahigh vacuum-scanning tunneling microscopy nanofabrication and hydrogen/deuterium desorption from silicon surfaces: Implications for complementary metal oxide semiconductor technology. Appl. Surface Sci. 130-132, 221–230 (1998)
[13] Ricketts, D.S., Bain, J.A., Luo, Y., Blanton, R.D., Mai, K., Fedder, G.K.: Enhancing CMOS using nanoelectronic devices: A perspective on hybrid integrated systems. Proc. IEEE 98(12), 2061–2075 (2010)
[14] Spindt, C.A.: A thin-film field-emission cathode. J. Appl. Phys. 39(7), 3504–3505 (1968)

[15] Zhang, Y., Santhanam, S., Liu, J., Fedder, G.K.: Active CMOS-MEMS AFM-like conductive probes for field-emission assisted nano-scale fabrication. In: Proc. 23rd IEEE Int'l Conf. on Microelectromech. Sys., Hong Kong, P.R. China, pp. 336–339 (January 2010)

[16] Weinstein, D., Bhave, S.A.: The resonant body transistor. Nanoletters 10, 1234–1237 (2010)

[17] Keskin, G.: Self-healing circuits using statistical element selection. Ph.D. Thesis, Dept. of Elec. and Computer Eng., Carnegie Mellon University, Pittsburgh, PA (2010)

[18] Li, X., Taylor, B., Chen, Y.-T., Pileggi, L.T.: Adaptive post-silicon tuning for analog circuits: Concept, analysis and optimization. In: Proc. IEEE/ACM Int'l Conf. on Computer-Aided Design, pp. 450–457 (November 2007)

[19] Keskin, G., Proesel, J., Pileggi, L.T.: Statistical modeling and post manufacturing configuration for scaled analog CMOS. In: Proc. Int'l Custom Integrated Circuits Conf. (September 2010), doi:10.1109/CICC.2010.5617625

[20] Lo, C.-C., Chen, F., Fedder, G.K.: Integrated HF CMOS-MEMS square-frame resonators with on-chip electronics and self-assembling narrow gap mechanism. In: Tech. Dig. 13th Int'l Conf. on Solid-State Sensors, Actuators, and Microsystems (Transducers 2005), Seoul, Korea, pp. 2074–2077 (June 2005)

Chapter 12
Capillary Force Actuation: A Mechatronic Perspective

Carl R. Knospe

Abstract. Capillary force actuation is a new technology holding particular promise for microelectromechanical systems. Relying upon the ability of electrical fields to alter the capillary pressure in a liquid, this technology permits larger actuation forces and strokes than conventional microactuators. Herein, the physical principles of electrowetting are discussed along with the relationships between changing apparent contact angle and capillary bridge shape. Several configurations of capillary force actuators are reviewed. For the standard configuration with two electrowetting surfaces, both equilibrium and non-equilibrium models of actuator performance are discussed. Three applications are examined where capillary force actuators may offer significant benefits.

12.1 Introduction

Actuation plays a critical role in the development of microelectromechanical systems (MEMS), converting signals from electronic components into forces acting upon the physical world. Effective microactuators are needed for many applications including micro-total-analysis systems (μ-TAS), micro optical devices, radio frequency MEMS, and microgrippers.

12.1.1 Microactuators

A variety of actuation principles are available for MEMS. These may be classified in four families: electrostatic, electromagnetic, piezoelectric, and thermal. Each

Carl R. Knospe
Mechanical and Aerospace Engineering,
University of Virginia,
Charlottesville, Virginia, USA
e-mail: knospe@virginia.edu

E. Eleftheriou & S.O.R. Moheimani (Eds.): Cntrl. Tech. for Emerging Micro/Nanoscale Sys. LNCIS 413, pp. 201–218.
springerlink.com

technology has relative advantages and disadvantages with respect to force capability, power required, bandwidth, and device manufacture (1; 9; 16):

- Electrostatic actuators are the most commonly used type for MEMS applications. These actuators are generally simple and do not require special material or difficult-to-manufacture elements (e.g., ferrous cores or conductive coils). Electrostatic actuators generally require high voltage and have large size in comparison to their limited range of motion. For small strokes normal to the lithographically-patterned plane, the parallel plate configuration, shown in Fig. 12.1a, is employed. If the required range of motion is greater than one micron, however, force production for this configuration is quite poor. For larger strokes (in-plane), a comb configuration is used, see Fig. 12.1b. Stroke here is better, but significant limitations exist for many MEMS applications. Comb drive actuators often require large structures to amplify the comb motion to that needed for the application. To achieve large forces in the lithographically-patterned plane, high-aspect-ratio micromachining must be employed.
- Electrothermal actuators employ Joule heating, either uniformly in a bi-morph or asymmetrically in a structure, to achieve motion (3; 11). The range of motion required for many applications requires large mechanical amplification factors or high temperatures ($> 250°C$), which may preclude their use. High power density is necessary to achieve quick response. Unreliable performance and even failure may result after repeated use due to degradation/debonding of electrical contacts. Problems may also occur due to overheating of compliant structures.
- Piezoelectric actuation is often used for microscale motion control. In most cases, commercial piezoelectric stacks are employed to deform compliant structures. Because piezoelectric actuators yield small strains, either large stacks or significant mechanical amplification factors (20 – 50) must be achieved to realize the desired range of motion. This greatly increases the device area. Piezoelectric actuators, like electrostatic actuators, require very high voltages for gripper operation, typically more than 100 volts. Piezoelectric actuation is difficult to integrate into MEMS fabrication.

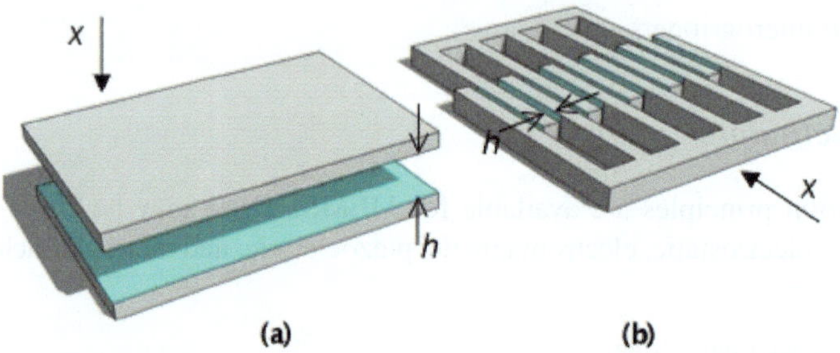

Fig. 12.1 Electrostatic actuators - (a) parallel plate; (b) comb drive. Indicated are the direction of displacement (x) and the capacitive spacing (h).

12.1.2 Capillary Force Actuation

Capillary force actuators (CFA) are a new technology for microscale actuation based upon the electrical activation of changes in capillary pressure (6; 13). Figure 12.2 shows one of several configurations of capillary force actuators, which will be referred to as the *standard configuration* in the sequel. As illustrated in the figure, a conducting liquid bridge is placed between two electrodes covered with a very thin dielectric layer. Upon application of a voltage, the fluid's contact angle upon the surface changes, a phenomenon previously observed and referred to as electrowetting on dielectric (EWOD). The change in contact angle results in a change in the bridge's capillary pressure and hence force acting upon the surfaces. Note that the dielectric layer prevents current flowing through the liquid. While electrowetting has been exploited for moving droplets laterally across a surface, its potential for creating large forces normal to a surface has previously been unrecognized.

In addition to producing significantly greater forces at low voltage, capillary force actuators possess other advantages over conventional choices for microscale actuation. The total movement achievable using a CFA is much greater than that achieved with other actuators typically employed in MEMS. Furthermore, out-of-plane forces can be readily achieved, which is beneficial to many MEMS applications as discussed below.

The actuated element in any MEMS device is supported by microfabricated flexures. For this discussion, the upper dielectric surface in Fig. 12.2 will be considered as supported by flexures and will be referred to as the platen. The lower surface will be considered fixed. Application of a constant voltage establishes a new equilibrium position of the platen where the actuator force upon the platen is balanced by the suspension's restoring force. If the actuation force increases with motion, instability of this equilibrium may result when the displacement is large. This phenomenon occurs in parallel plate electrostatic actuators, and is known as snap-in instability. This bifurcation of the equilibrium restricts the range of actuator stroke that can be regulated with open loop control. Capillary force actuators, like electrostatic, will experience snap-in of the platen when the motion is a large portion of the available stroke.

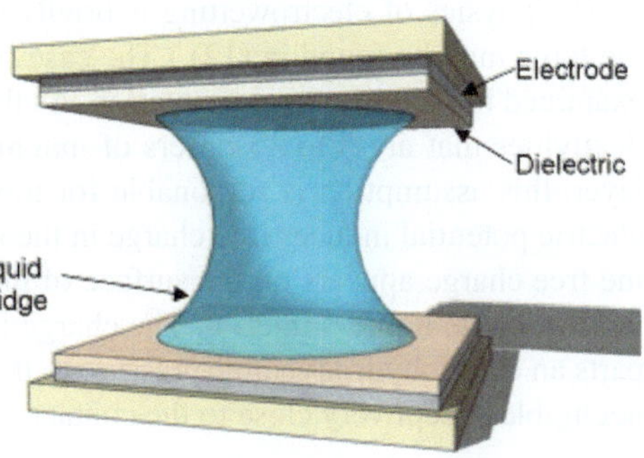

Fig. 12.2 Standard configuration of capillary force actuator (not to scale). A conducting liquid bridge spans between two dielectric-covered electrodes, one of which is supported by flexure (not shown).

12.2 Principle of Actuation

In this section the actuation mechanism is reviewed along with fundamental limitations upon its capabilities.

12.2.1 Electrowetting

Electrowetting on dielectric (EWOD) is a phenomenon where the application of an electrical potential to a conducting liquid placed on a dielectric-covered electrode causes the liquid to spread upon the dielectric surface. The apparent contact angle of the liquid on the surface, θ_v, decreases and the surface acts as increasingly hydrophilic. In a typical set-up, a sessile drop is surrounded by an ambient gas. The dielectric layer is an oxide that is covered by a thin fluoropolymer topcoat to render the surface hydrophobic and to minimize contact angle hysteresis so as to allow easy movement of the contact line upon the surface. The liquid employed is typically an aqueous solution of an inorganic salt.

Ideally, the change in apparent contact angle in electrowetting is governed by the Young–Lippmann equation:

$$\cos(\theta_v) = \cos(\theta_0) + \frac{\varepsilon_d}{2\sigma_{gl} t_d} v_d^2, \tag{12.1}$$

where θ_v is the apparent contact angle of the liquid under the applied field, θ_0 is the native contact angle (without field), σ_{gl} is the surface tension of the gas-liquid interface, ε_d is the permittivity of the dielectric, t_d is the thickness of the dielectric layer, and v_d is the voltage across the layer.

It is now recognized that electrowetting is purely an electromechanical effect, where the applied electric field causes an additional electrostatic pressure on the liquid surface near the contact line (12). The change in contact angle is due to the pressure balance on the liquid surface rather than to a change in the surface energy. As a result, the actual contact angle on the surface remains unchanged with an applied electric potential; only the apparent contact angle is altered. This change in apparent contact angle is captured by the Young–Lippmann equation (12.1).

The physics of electrowetting is briefly reviewed here. (An excellent survey of the topic may be found in (12).) The case of a perfectly conducting bridge will be examined to simplify the presentation. As the liquids typically employed have conductivities that are 5 to 15 orders of magnitude greater than that of the dielectric layer, this assumption is reasonable for this review. When voltage is applied, the electric potential induces free charge in the liquid. In the perfectly conducting case, the free charge appears on the surface of the liquid bridge and screens the electric field from the liquid's interior. The charge interacts with the electric field, and imparts an electrohydrodynamic pressure to the liquid's surface. This pressure will be negligible except very close to the contact line. The transition of the profile tangent angle from the native contact angle (θ_0) to the apparent contact angle (θ_v) occurs in this region, with dimension on the order of the dielectric layer thickness. Excluding this region, we may state that no electrical force acts upon the liquid-gas interface. As the contact line is approached on either the solid-liquid or the liquid-gas interface, the charge density and field increase rapidly (12).

12.2.2 Capillary Bridges

When an electric field is not present the equilibrium gas-liquid interface of a capillary bridge between two surfaces satisfies Laplace's equation

$$p_{gl} = \sigma_{gl}\left(\frac{1}{r_1} + \frac{1}{r_2}\right), \tag{12.2}$$

where p_{gl} is the pressure across the interface, and r_i denotes the radius of curvature of the interface along orthogonal axes. The constant mean curvature specified by (12.2) along with the axisymmetric condition of the bridge dictate that the equilibrium bridge surface will be a surface of revolution generated by a profile from the Plateau sequence of shapes (8; 13) (nodoid, unduloid, sphere, cylinder, and catenoid). The shape is determined by the liquid volume V, bridge height h, and the contact angle the liquid makes with the solid surface, θ_0. The latter is dictated by Young's equation

$$\sigma_{gs} - \sigma_{ls} - \sigma_{gl}\cos\theta_0\,,$$

where σ_{ij} denote the interfacial energies between the gas, liquid, and solid phases.

When a constant electric field is present, the apparent contact angle is dictated by Eq. (12.1). This angle is the boundary condition for the gas-liquid interface in the region where the electric field acting on the surface is negligible. Over this region, which is nearly the entire span of the bridge height, the equilibrium gas-liquid interface shape is determined by (12.2) in just the same manner as the bridge without electric field.

Because of variety of profiles in the Plateau sequence, determination of the exact solution to interface shape, and therefore capillary pressure, is quite complex - see (6) for details. Fortunately, a good approximation for the bridge shape may be obtained for bridges with a low value of the aspect ratio h/D, where D is the bridge diameter. The profile of such bridges can be approximated as a circular arc, allowing a simple relationship to be developed between apparent contact angle and pressure (6), see Fig. 12.3. First, the bridge height and the apparent contact angle determine the profile's radius of curvature, r_p, via

$$\frac{1}{r_p} = -\frac{2\cos\theta_v}{h}, \tag{12.3}$$

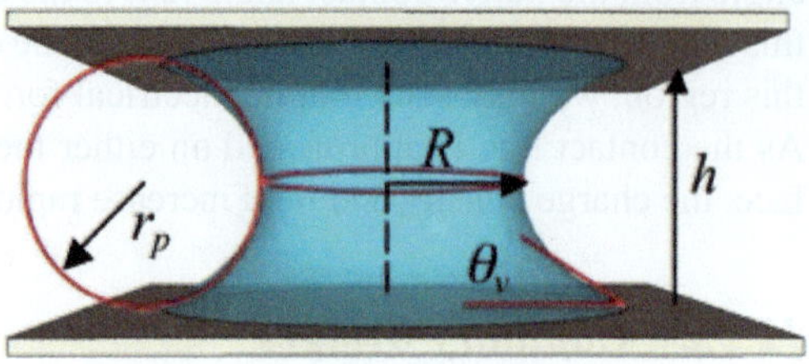

Fig. 12.3 Contact angle, waist radius, profile radius of curvature, and height for liquid bridge

where negative values correspond to bridge profiles that curve inward (bridges with a neck) and positive values correspond to profiles that curve outward (bridges with a haunch). Denote the radius of the waist of the bridge as R. We now make use of Eq. (12.2) with $r_1 = r_p$ and $r_2 = R$ to find

$$p_{gl} = \sigma_{gl}\left(\frac{1}{R} - \frac{2\cos\theta_v}{h}\right). \tag{12.4}$$

12.2.3 Equilibrium Actuator Force

The capillary pressure (12.4) acts upon the wetted area of the dielectric surface, which may be approximated as $S_{gl} = \pi R^2$. The resulting force is

$$f_p = \sigma_{gl}\pi R\left(1 - \frac{2R}{h}\cos\theta_v\right),$$

where a negative value of f_p indicates that it acts downward on the platen (upper surface) in Fig. 12.2. Combining this force with that of the surface tension, $f_\sigma = -2\sigma_{gl}\pi R$, yields the total actuation force when in equilibrium (6):

$$f = -\sigma_{gl}\pi R\left(1 + \frac{2R}{h}\cos\theta_v\right). \tag{12.5}$$

In Section 12.3.2 we examine the actuation force in the non-equilibrium case.

12.2.4 Limiting Phenomena

Actual electrowetting behavior deviates from Young–Lippmann theory. For example, Eq. (12.1) predicts that the apparent contact angle should be an even function of the applied voltage. In many cases, however, the observed contact angle displays asymmetry with respect to polarity. The Young–Lippmann equation also predicts that perfect wetting will be achieved with a finite voltage. In practice, however, charge trapping, conduction, and electrical breakdown occur in the dielectric, causing deviations in the electrowetting behavior from that predicted by Young–Lippmann. With increasing voltage the contact angle will cease to decrease, a phenomenon known as contact angle saturation (12; 18). For capillary force actuation this means reduced force production and ultimately actuator failure at higher voltages. Thus, a key research direction for the development of this actuator technology is the postponement of the onset of limiting phenomena until higher potentials through the engineering of the dielectric surface.

12.3 Actuator Dynamics

The dynamic response of capillary force actuation depends on the charging of the dielectric, the generation of electrohydrodynamic pressure upon the bridge surface and its subsequent deformation, the motion of fluid in response to the pressure changes induced, and the platen motion in response to the fluid and flexure forces. Here, we review a medium-fidelity analytical model of these dynamics.

12.3.1 Charging Dynamics

The standard configuration's electrical dynamics is that of the RC circuit shown in Fig. 12.4. The transfer function of these dynamics is given by:

$$\frac{v_d(s)}{v(s)} = \frac{R_d}{R_l R_d C_d s + R_l + 2R_d}, \tag{12.6}$$

where $v_d(s)$ and $v(s)$ are the Laplace transforms of the voltage across a single dielectric layer and the input voltage, R_d and C_d are the resistance and capacitance of a single dielectric layer, and R_l denotes the resistance of the liquid bridge (7). Since the conductivity of the bridge is many orders of magnitude greater than that of the dielectric layer, the transfer function can be simplified to

$$\frac{v_d(s)}{v(s)} = \frac{\frac{1}{2}}{\tau_e s + 1}. \tag{12.7}$$

Here τ_e denotes the electrical time constant

$$\tau_e = \frac{\varepsilon_d \rho_l h}{2t_d},$$

where ρ_l is the resistivity of the liquid.

12.3.2 Fluid Model

The fluid responds to the change in surface profile brought about via the change in apparent contact angle. For this medium-fidelity model, we will assume that the

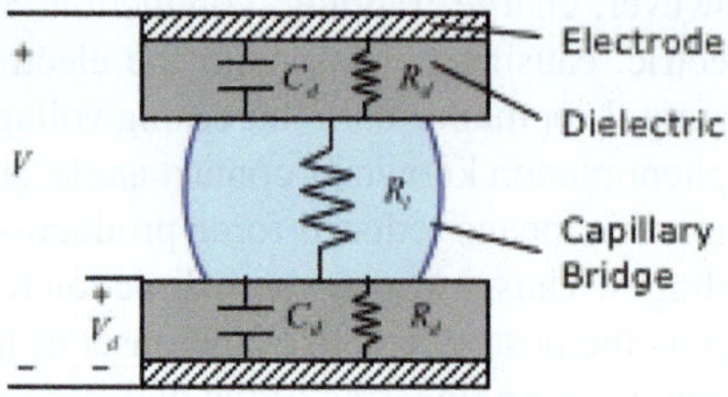

Fig. 12.4 Resistor-capacitor model of the standard configuration

time constants associated with the alteration of the surface profile are much faster than those associated with the responding radial flow. Thus, we treat the relationship between apparent contact angle and gas-liquid surface pressure by the memoryless relation (12.4). In equilibrium this pressure is equal to that exerted across the wetted surfaces as was noted in Section 12.2.3. However, in the non-equilibrium case this pressure is a boundary condition imposed in solving the fluid dynamics problem. This problem is an axisymmetric one with a no-slip condition on the dielectric surfaces, and one surface with prescribed velocity $\dot{h}$. Ignoring small convective acceleration terms, the Navier–Stokes equation in combination with incompressibility dictates a Poiseuille flow with a radial variation in pressure given by

$$\frac{dp}{dr} = \frac{6\mu\dot{h}}{h^3}r, \tag{12.8}$$

where μ is the viscosity (7). Using the pressure boundary condition at radius R, Eq. (12.4), this differential equation may be solved to analytically determine the non-equilibrium pressure variations on the dielectric surface. Integrating this pressure over the surface and adding in the surface tension contribution $f_\sigma = -2\sigma_{gl}\pi R$, yields the total capillary force acting:

$$f = -\frac{3\pi\mu R^4}{2h^3}\dot{h} - \pi R\sigma_{gl} - \frac{2\pi R^2\sigma_{gl}}{h}\cos\theta_v\,. \tag{12.9}$$

Relating the bridge radius R to the bridge volume V via the cylindrical bridge approximation $V = \pi R h^2$ yields

$$f = -\frac{3V^2\mu}{2\pi h^5}\dot{h} - \sigma_{gl}\sqrt{\frac{\pi V}{h}} - \frac{2\sigma_{gl}V}{h^2}\cos\theta_v\,. \tag{12.10}$$

12.3.3 Non-equilibrium Actuator Model

The platen will be modeled as a mass supported by linear flexures allowing motion along a single axis aligned with the bridge axis of symmetry and normal to the electrode surfaces. The equation of motion is

$$m\ddot{h} + k(h - h^*) = f, \tag{12.11}$$

where m and k are the platen mass and flexure stiffness, respectively, and h^* is the nominal bridge height (i.e., without applied fluid forces f). The entire dynamical model then consists of Eqs. (12.1), (12.7), (12.10) and (12.11) (7).

12.3.4 Equilibria

As shown in (7), the model developed has two equilibrium solutions for the bridge height when the applied voltage is sufficiently small. We label these as h_1 and h_2, with $h_1 < h_2$. Equilibrium h_1 is an unstable equilibrium branch, while h_2 is locally stable. As illustrated in Fig. 12.5, as the voltage is increased (equivalently, $\cos\theta_v$ increased), the stable and unstable equilibria will merge and equilibrium solutions will not exist for sufficiently large voltages. This bifurcation will occur when h_2 reaches a critical bridge, labeled $\tilde{h}$. Experimentally, this bifurcation would appear as a sudden snap-in of the platen as the voltage applied was incrementally increased. This behavior allows open-loop control of the platen motion only over bridge heights h greater than $\tilde{h}$. Analysis shows that $\tilde{h} < \frac{2}{3}h^*$ (7). Electrostatic actuators have a similar behavior in this regard. However, for electrostatic actuators the nominal spacing h^* is much less than for capillary force actuators.

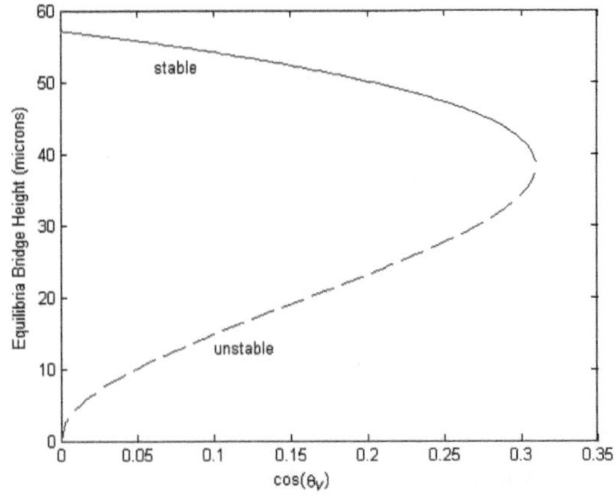

Fig. 12.5 Bridge height stable and unstable equilibrium solutions as a function of $\cos\theta_v$

12.4 Alternative Configurations

The standard configuration of capillary force actuators is not desirable for certain applications as it requires both surfaces to contain electrodes. In this section alternative configurations that have only one active surface are discussed.

12.4.1 Wetting Passive Surface Configuration

The first alternative configuration to be considered is the parallel plate case with one passive surface with fixed contact angle θ_1 and one active surface with variable apparent contact angle θ_v. The active surface would contain both electrodes arranged with one at the bridge's center and the other at its periphery. This arrangement does not significantly affect the electrowetting on the active surface. In addition to bridge volume and height, the contact angle of the liquid bridge on the passive surface is a design parameter to be considered. Here we will examine the effect that this boundary condition has upon the actuation force and compare the efficiency of force production to that of the standard configuration.

Using the analytical expressions introduced in (13) a one-dimensional root-finding problem may be formed and solved to determine precisely the bridge shape and the capillary force produced. Alternatively, an analytical approximation of the force in the low aspect ratio case may be derived by considering the bridge as cylinder. This gives considerable insight, and comparison to the exact solution indicates its accuracy. The total potential energy of such a bridge is given by

$$E = \sigma_{gl}\left(S_{gl} - (\cos\theta_1 + \cos\theta_v)\, S_{ls}\right) ,$$

where S_{ls} denotes the liquid-solid interface area. The force may be derived by taking a derivative of this energy with respect to a virtual displacement of one of the surfaces:

$$F_c = -\sigma_{gl}\frac{\partial S_{gl}}{\partial h} + \sigma_{gl}\left(\cos\theta_1 + \cos\theta_v\right)\frac{\partial S_{ls}}{\partial h} .$$

Considering the bridge with low aspect ratio as a cylinder with radius r and height h, we may obtain for a fixed volume bridge:

$$\frac{\partial S_{gl}}{\partial h} = \pi r \qquad \frac{\partial S_{ls}}{\partial h} = -\pi\frac{r^2}{h} .$$

Substitution of these results yields

$$F_c = -\sigma_{gl}\pi r\left\{1 + \left(\cos\theta_1 + \cos\theta_v\right)\left(\frac{r}{h}\right)\right\} .$$

Now consider the change in force with an infinitesimal variation in the cosine of active surface contact angle:

$$\frac{\partial F_c}{\partial \cos\theta_v} = -\sigma_{gl}\pi r\left(\frac{r}{h}\right) .$$

This analysis demonstrates the independence of the actuator effectiveness from the passive surface contact angle. A comparison of this result with that of the standard configuration indicates that the change in force with a change in $\cos\theta_v$ is half that of the standard configuration. This conclusion, achieved via the simplified formulation, has been verified using the exact solution (13). For that analysis, we considered three different values of the passive surface contact angle (60°, 90°, 120°). The contact angle on the active surface was varied over a range of values. The height and volume of the bridge in all the cases were the same and equal to 100 μm and $7.85 \times 10^7\,\mu m^3$, respectively. The capillary force is plotted in Fig. 12.6 as a function of the cosine of the active contact angle for the three fixed values of the passive surface contact angle. A comparison of these results with those for the standard configuration (same volume and bridge height) confirms that the actuator effectiveness (change in force in response to a change in $\cos\theta_v$) is nearly half that for the standard configuration, $\approx 500\,\mu$N vs. 1040 μN.

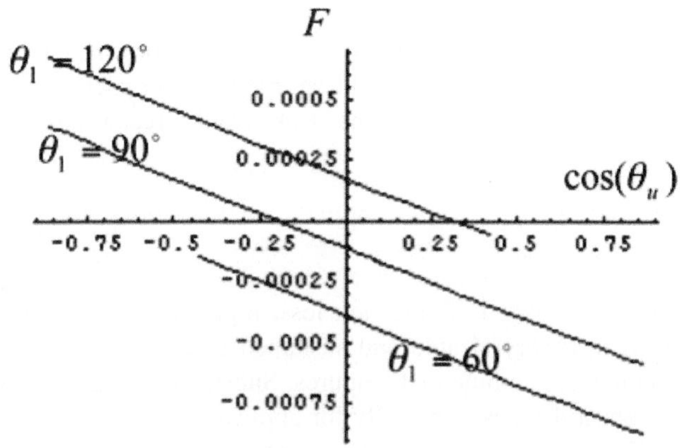

Fig. 12.6 Actuator force for wetting passive surface configuration as a function of the active surface contact angle (θ_u) for three different values of the passive surface contact angle (θ_1)

12.4.2 *Pinned Passive Surface Configuration*

We now examine another configuration with one active surface and one passive surface. Rather than the fixed contact angle boundary condition just examined, this configuration has a pinned circular contact line on the passive surface. This boundary condition could be achieved via a step in surface geometry, see Fig. 12.7. There are several design variables that enter into an analysis of this problem including bridge volume, height, and the pinned contact line radius. A detailed analysis of the problem is presented in (13). Here we consider an example to illustrate the potential of this configuration. The bridge considered has the same height and volume as in the previous example (100 μm and $7.85 \times 10^7\,\mu m^3$) and the pinned radius is 500 μm. The effectiveness of capillary force actuation $\left(\frac{\partial F_c}{\partial \cos\theta_v}\right)$ was found to be 864 μN. This is about 80% of the effectiveness calculated for the standard configuration and more than 1.6 times the effectiveness when the passive surface has a fixed contact angle boundary condition. However, the range of stroke in this pinned case will be more limited than that in the wetting passive surface configuration.

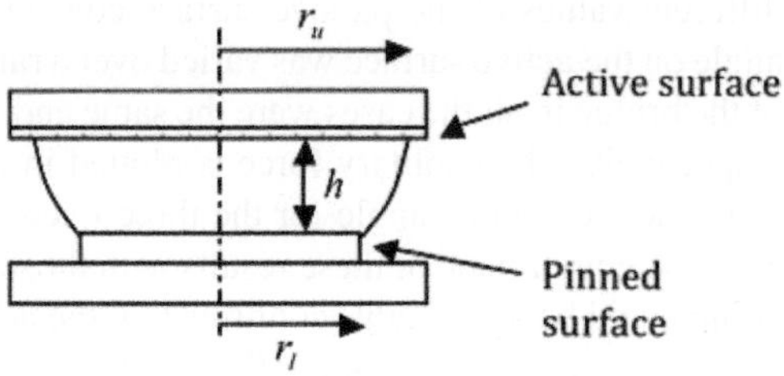

Fig. 12.7 Capillary force actuator with pinned contact line on the passive surface

12.5 Potential Applications

Several applications that may benefit from the adoption of capillary force actuation are examined here. The actuator designs that would be employed in each of the three cases examined would be quite different as the strokes and forces necessary vary widely.

12.5.1 RF MEMS

The promise of RF MEMS, including low insertion loss, high isolation, and reconfigurability, has resulted in a very sizable and sustained research and development effort, including numerous commercial ventures. Substantial progress has been made in developing robust devices, especially for applications requiring low to medium RF power ($<$10 watts) (14; 15). High RF power switches, however, are currently not feasible due to the combination of the high contact force and large contact area needed to achieve low resistance / current density when the switch is closed and the large gap ($>$3 μm) necessary for adequate isolation when the switch is open. To a significant extent, both of these shortcomings are due to a deficiency in MEMS actuation. As such, the development of improved actuation technology is an important research direction for achieving high power RF MEMS.

Many factors determine the suitability of a microactuator for high power contact switches. The actuator must generate sufficient contact force to produce low contact-resistance ($< 1\,\Omega$). Small dimples (with diameters less than 1 μm) are typically fabricated in the metal contact area to provide the electrical connection when the switch is closed. Elastic / plastic deformation of these bumps occurs under high force, increasing the contact area and reducing contact resistance. The metal employed also influences the resistance. Important factors here are the hardness of the metal and its tendency to form a surface layer from contaminants. Without adequate contact area, heat generated in the contacts by the RF power will result in eventual switch failure. Key to avoiding this is the provision of a sufficient number of dimples to transfer the RF power and an adequate contact force in each dimple (4). Previous switch designs have demonstrated current handling of 300–400 mA (15) and even as much as 1 A. The RF current on a 50 Ω transmission line associated with 100 W is 1.414 Arms, while the voltage is 71.7 Vrms. The main challenges in achieving higher current handling in a dc-contact switch are achieving an actuation force on the order of 1 mN and creating a thermal design capable of sinking over 1 W heat generation in the contact without exceeding the maximum temperature for reliable operation.

When the switch is opened, there must be sufficient stiffness in the support to break the adhesion forces between the metal contacts. It has been estimated that the adhesion forces are 40% of the total contact force achieved when the switch was closed (14). Since large contact forces are needed for high RF power, stiff suspensions will be necessary to return the switch to the up-state. The actuator must provide sufficient force to overcome this stiffness when closing the switch. The quality of isolation achieved by the switch when open is determined by the up-state capacitance and the RF frequency. Larger gaps reduce the up-state capacitance and improve isolation. High isolation (50–60 dB) at frequencies above 5 GHz can only be achieved with gaps larger than 3 μm.

The size of a switch is important for realizing fully functional RF MEMS systems, such as radar switching matrices. In RF MEMS switches to date the actuator occupies the preponderance of the switch footprint. Since an actuator's size is strongly related to its force capability and range of motion, high RF power switches based on conventional microactuators will not be compact. Furthermore, it is generally recognized that MEMS actuators that provide lateral switch movement (i.e., within the lithographically-patterned plane) are larger than those with vertical motion (i.e., normal to the plane) (14).

The voltage and current used for actuation are important factors in selecting an actuation technology. In general, it is desirable that both be low to simplify integration of the actuator with the overall system design. If current flows through the actuator during the down-state, low resistance bias lines must be used. Actuator DC power consumption in this case may be an important consideration for certain applications, from both a power loss and a thermal perspective. Finally, the impact of actuator design on the switch's thermal response to the RF power needs to be considered. The switch configuration should allow adequate conduction of heat away from the t-line.

Electrostatic actuators are by far the most commonly used technology for RF MEMS. They can be easily fabricated with a process highly compatible with standard IC processing. Another significant advantage is that they have nearly zero power consumption. However, they require either high voltages (>100 V) or very large size to achieve the considerable forces necessary for good isolation and reliable switch operation at high RF power. Electrothermal actuators employ thermal expansion to achieve switch motion (15). They typically require low voltages and are capable of achieving large contact forces. However, electrothermal actuators tend to be large unless the gap to be closed is quite small. They also suffer from high DC power consumption in the actuated state. To remedy this, an electrostatic actuator is sometimes used in combination with electrothermal actuation to permit hold-down with low power consumption. This has the disadvantage of significantly increasing the device size if the electrostatic actuator is to provide large contact forces. Electrothermal actuators must be biased with low resistance lines and therefore considerable care must be used to avoid coupling with the microwave t-lines.

Capillary force actuators have a number of advantages for this application. First, the larger force capability and stroke of CFA are beneficial respectively to achieving substantial contact force when the switch is closed and high up-state isolation when it is open. Second, actuation forces are normal to the lithographically-patterned plane. This orientation is better suited for RF MEMS than lateral actuation. Normal orientation will typically result in a more compact device. Furthermore, conventional micromachining techniques may be used in this case to engineer the contact surfaces to improve contact quality (e.g., as in (4)). In contrast, when in-plane actuation is employed, the presence of sidewall slopes and the difficulty of providing a good metal layer on these surfaces present obstacles to achieving low resistance and reliable metal-to-metal contact. Finally, in-plane motion requires high-aspect-ratio structures to achieve the large actuation forces necessary for high current handling switches.

12.5.2 Microgrippers

Key to the success of today's microdevices was the development of an economic route to fabrication via photolithography. The current trend in microsystems, however, is toward the integration of components that are incompatible with standard microfabrication tools. Such components might have a surface chemistry engineered for detecting a pathogen, or an elasticity suitable for a leak-tight valve seat. Clearly, complementary approaches to microfabrication are needed.

One approach to this microfabrication challenge is the assembly of complex, three-dimensional microsystems from heterogenous microcomponents. Such microsystems are often referred to as hybrid MEMS. To be cost effective, the assembly of hybrid MEMS must be carried out in an automated and massively parallel fashion. While pick-and-place operations are extremely common in the manufacturing of macroscale objects ($> 1\,mm^3$), they are exceedingly difficult for objects in the size range from 1 to 100 microns. At these scales, adhesion forces become much greater than inertial forces, making it difficult to grasp and release objects in a reliable and predictable manner.

Many end-effectors for executing pick-and-place operations have been investigated for microassembly. Microgrippers, while not as simple mechanically as single finger manipulators, are capable of handling a much greater range of microscale parts with greater grasping stability and dexterity. For assembly, a microgripper should have sufficient tweezing range of motion and force, finger surfaces of adequate size and appropriate geometry, low power/voltage, compact size, and ease of manufacture. The actuation technology used is a critical factor in determining whether these requirements are met.

In general, the finger surfaces of a gripper should be on the same scale as the manipulated object (say 10–100 μm). This has important consequences for microfabrication. If the fingers are designed to have movement within the lithographically-patterned plane, high-aspect-ratio micromachining techniques must be used, such as deep reactive ion etching (DRIE) or X-ray lithography (LIGA). Sloped sidewalls would be unacceptable for finger surfaces. On the other hand, if actuation force and finger motion are normal to the lithographically-patterned plane, fingers with adequate area can be manufactured by a variety of conventional micromachining techniques. In this case, finger surface chemistry and morphology may be engineered without difficulty. If thick structures are needed, metal electroplating into photoresist molds may be used. While the sidewalls of these structures may be sloped due to the high aspect ratio, gripper function would not be compromised since actuation and tweezing motion are out of the lithographically-patterned plane.

Electrostatic actuators are often advocated for this application, however, their size is quite large in comparison to their limited range of finger motion (5). Comb drive actuators often require long gripper fingers to amplify the comb motion to that needed for the finger ($\sim 100\ \mu m$). To achieve wide fingers, high-aspect-ratio micromachining must be employed for comb drive designs. As a result, it is difficult to tailor finger surface geometry, morphology, and chemistry as desired for the task of pick-up and release. While surface tailoring is much easier for parallel plate designs, the actuation stroke in this case is far too small for the application.

It is desirable for the entire microgripper (including actuator and any motion amplification structures necessary) to be as compact as possible. Ideally, the gripper's volume should be comparable to that of the manipulated object, as is the case with the human hand. Compact size enhances the gripper's ability to perform assembly tasks without encountering physical obstruction from previously assembled components, other manipulators, or fixtures. Compactness also lends itself to greater dexterity for two-fingered manipulators (e.g., through the use of a "wrist"). Finally, compact size is critical to the visual feedback necessary to cope with the uncertainties inherent to microassembly. Microscope optics will have small field of view, small depth of field, and small working distance (< 1 cm). Separate illumination should be provided for each microscope. The microgripper and micromanipulator should not interfere with the light paths from the illumination sources to the object, or with those from the object to the microscopes. Small gripper size is critical to avoiding occlusion. Current microgrippers are not compact, largely because of deficiencies in the actuation technologies used. For example, the electrostatic comb microgripper reported in (2) has a tweezing gap of 100 μm. The actuator has dimensions of $2600\ \mu m \times 4000\ \mu m \times 50\ \mu m$. The volumetric utilization (i.e., the ratio of object volume to gripper volume) of this device is therefore less than 0.5%.

Capillary force actuation possesses several advantages over conventional technologies for microgripper actuation. The total movement achievable using a CFA is much greater than that achieved with other actuators typically employed in MEMS. Because of its greater effectiveness, the voltage necessary for an actuation task is also significantly lower. CFA microgrippers may be batch manufactured by a combination of conventional micromachining (for structures) and inkjet printing (for liquid bridges). Since the actuation force is directed out-of-plane, finger surfaces may be fabricated with significant width (30–100 μm) without requiring high-aspect-ratio micromachining. This orientation also permits the lithographic fabrication of complex finger planform geometries. Furthermore, physical and chemical modification of the fingertip surfaces may be readily carried out so as to reduce surface adhesion.

12.5.3 Micro-total Analysis Systems

Recently, there has been an intense interest in the miniaturization of chemical and biological analysis onto microdevices, often referred to as micro-total-analysis systems (μTAS) or Lab-on-a-Chip (LOC) (10; 17). This holds many promises including smaller samples, quicker analysis, and parallel processing capability. Critical to this technology is the ability to manage the flow of nanoliter samples in a network of microchannels through flow control devices such as valves and pumps. Actuation plays a key role here. Most valves suffer from limitations associated with their actuation principle, including slow response speed, large power requirements, and necessitated off-chip equipment. For example, one common design employs an elastomeric membrane as the valve moving-element, with actuation of this membrane accomplished pneumatically using an external air pump. Similar designs are common for flow generation, including peristaltic and diaphragm pumps (17). To reduce device size, minimize power consumption, and enhance portability, development of a totally integrated microfluidic device, with actuation microfabricated on the substrate, is highly desirable. Unfortunately, most easily manufactured MEMS actuators, such as electrostatic, are incapable of achieving the required range of motion (30–100 μm) and the necessary force for μTAS valve control, falling short of the specifications by over a factor of ten.

Capillary force actuators are quite promising for this application because of their large stroke and force capabilities at limited voltages. A configuration of CFA where an elastomeric membrane is actuated, described in (13), is particularly well-suited for this application.

12.6 Conclusions

Alteration of the capillary pressure within a liquid bridge via electrowetting provides a new means for microscale actuation. Analysis demonstrates that the forces produced can greatly exceed those of other microactuators when similar voltages are employed. Furthermore, the range of stroke achievable is significantly greater than with other actuation technologies. Actuation is normal to the lithographically-patterned plane, which is desirable in many MEMS applications.

Both equilibrium and non-equilibrium models of capillary force actuation are reviewed. The non-equilibrium model integrates electrical, electrowetting, fluid dynamic, and structural component models. In terms of the capillary force generated, the equilibrium and non-equilibrium models differ only in a highly nonlinear damping term due to the bridge's viscosity. The non-equilibrium model treats the realization of the equilibrium bridge profile as instantaneous with the change in apparent contact angle.

Several potential applications of capillary force actuation are discussed. These illustrate both the need for new microscale actuation technologies and the advantages capillary force actuators have over conventional approaches.

Acknowledgements. The author wishes to thank Ali Nezamoddini, Hossein Haj-Hariri, Michael Reed, N. Scott Barker, and Giovanni Zangari for many helpful discussions. This research was supported in part by National Science Foundation Grant ECCS 0801908 "Dynamics of Capillary Force Actuators".

References

[1] Bell, D., Lu, T., Fleck, N., Spearing, S.: MEMS actuators and sensors: Observations on their performance and selection for purpose. Journal of Micromechanics and Microengineering 15, 153–164 (2005)

[2] Beyeler, F., Neild, A., Oberti, S., Bell, D., Sun, Y., Dual, B., Nelson, B.: Monolithically fabricated microgripper with integrated force sensor for manipulating microobjects and biological cells aligned in an ultrasonic field. IEEE Journal of Microelectromechanical Systems 16(1), 7–15 (2007)

[3] Bordatchev, E., Zeman, M., Knopf, G.: Electro-thermo-dynamic performance of a microgripping system. In: Proc. IEEE International Conference on Mechatronics and Automation, pp. 1848–1853 (2005)

[4] Chow, L., Volakis, J., Saitou, K., Kurabayashi, K.: Lifetime extension of RF MEMS direct contact switches in hot switching operations by ball grid array dimple design. IEEE Electron Device Letters 28(6), 479–481 (2007)

[5] Kim, C.-J., Pisano, A., Muller, R., Lim, M.: Polysilicon microgripper. In: IEEE Solid-State Sensor and Actuator Workshop, June 4-7, pp. 48–51 (1990)

[6] Knospe, C., Nezamoddini, S.A.: Capillary force actuation. Journal of Micro - Nano Mechatronics 5(3-4), 57–68 (2010), doi:10.1007/s12213-009-0023-4

[7] Knospe, C., Haj-Hariri, H.: Modeling and dynamic analysis of capillary force actuators. In: 5th IFAC Symposium on Mechatronic Systems, Cambridge, MA, September 13-15 (2010)

[8] Kralchevsky, P., Nagayama, K.: Particles at Fluid Interfaces and Membranes, pp. 469–502. Elsevier, Amsterdam (2001)

[9] Madou, M.: Fundamentals of Microfabrication: The Science of Miniturization. CRC Press, Boca Raton (2001)

[10] Manz, A., Becker, H. (eds.): Microsystem Technology in Chemistry and Life Sciences. Springer, Heidelberg (1999)

[11] Mayyas, M., Zhang, P., Lee, W., Shiakolas, P., Popa, D.: Design tradeoffs for electrothermal microgrippers. In: 2007 IEEE International Conference on Robotics and Automation, pp. 907–912 (2007)

[12] Mugele, F., Baret, J.-C.: Electrowetting: from basics to applications. Journal of Physics: Condensed Matter 17(2), 705–774 (2005)

[13] Nezamoddini, S.A.: Capillary Force Actuators, PhD Dissertation, Mechanical and Aerospace Engineering, University of Virginia (August 2008)

[14] Rebeiz, G.: RF MEMS: Theory, Design, and Technology. Wiley & Sons, Chichester (2003)

[15] Streeter, R., Hall, C., Wood, R., Mahadevan, R.: VHF high power tunable RF bandpass filter using microelectromechanical (MEM) microrelays. International Journal RF Microwave CAE 11(5), 261–275 (2001)
[16] Thielicke, E., Obermeier, E.: Microactuators and their technologies. Mechatronics 10, 431–455 (2000)
[17] Unger, M., Chou, H., Thorsen, T., Scherer, A., Quake, S.: Monolithic microfabricated valves and pumps by multilayer soft lithography. Science 288, 113–116 (2000)
[18] Vallet, M., Vallade, M., Berge, B.: Limiting phenomena for the spreading of water on polymer films by electrowetting. European Physics Journal 11(4), 583–591 (1999)

Chapter 13
Control and Estimation in Force Feedback Sensors

Karl Johan Åström

Abstract. The principle of force feedback is to design a sensor as a feedback system where the action of the measured variable is compensated by a force generated by feedback. Force feedback has contributed significantly to improve the quality of sensors. A characteristic feature of MEMS devices is that they are subject to random disturbances, such as Brownian motion, Johnson–Nyquist noise, and tunneling noise. Variations in device parameters must also be accounted for. In this paper we will discuss principles for designing control and estimation algorithms for instruments based on MEMS devices with force feedback. The fact that the final goal is to design an instrument gives an interesting formulation of the problem. It is shown that shaping of the frequency response can be separated from attenuation of disturbance. The principles have been applied to design of a tunneling accelerometer. The experimental work was done at Professor Turner's laboratory at the University of California, Santa Barbara.

13.1 Introduction

Force balance is a useful principle for sensor design. Traditionally the sensor output is the feedback signal. Control design is then a regulation problem, where the objective is to keep the error small. In this article it is shown that there are significant advantages to using also the error signal. Sensor design then becomes a richer control problem. We present a control architecture where the frequency response of the sensor can be specified. Analysis reveals that there are significant advantages by co-design of sensor and controller. It is also shown that a controller can be designed so that the sensor transfer function does not depend on the error as long as the dynamics can be approximated linearly. A characterization of the resolution of the sensor

Karl Johan Åström
Department of Automatic Control LTH,
Lund University
e-mail: kja@control.lth.se

E. Eleftheriou & S.O.R. Moheimani (Eds.): Cntrl. Tech. for Emerging Micro/Nanoscale Sys. LNCIS 413, pp. 219–233.
springerlink.com

in terms of the noise sources in the system is also given. The methodology has been applied to accelerometers where mass deflection is measured by quantum tunneling.

13.2 Force Balance

Force balance is a principle that has been used to improve the performance of sensors for a long time. Even if the name force feedback refers to force measurements the principle has been used in many other contexts from process control (1) to biology (2). It is a nice illustration of how feedback can be used to obtain good systems from bad components (3). We illustrate the principle for an accelerometer. A basic accelerometer is simply a spring-mass system. An acceleration of the base results in an elongation of the spring. This elongation is measured and amplified to give the acceleration signal. The precision and linearity of the instrument depend on all elements in the chain, the spring, the mass and the amplifiers.

The accelerometer based on force feedback has an actuator that exerts a force on the mass. The sensor signal is amplified and fed back to the actuator to keep the mass at a constant position. The output of the instrument is the actuator signal. Force feedback has several advantages. The linearity and precision of the instrument depend only on the calibration of the actuator. The linearity of the sensor and the amplifier are irrelevant because the sensor is only used to detect deviation from the reference position. Sensor precision depends primarily on the precision of the actuator. Since the feedback drives the mass towards the center position the linearity of the sensor is not so important.

Force feedback also has other advantages which can be understood from an analysis of the simple accelerometer. Let the mass be m and the spring constant k; the bandwidth of the instrument is $\omega_B = \sqrt{k/m}$. The sensitivity k_S is proportional to m/k, and the product $\omega_B^2 k_S$ is thus a constant. High bandwidth and high sensitivity are thus conflicting requirements. Note in particular that a doubling of the bandwidth decreases the sensitivity by a factor of 4. The constraint is eliminated when force feedback is used. The bandwidth of a sensor with force feedback depends primarily on the tightness of the feedback loop. The sensitivity is large since the current through the actuator is typically high.

Force feedback is an nice illustration of the fact that feedback can resolve design compromises. Force feedback has been applied to many sensors, and it has typically given orders of magnitude improvement in precision, linearity and bandwidth of sensors.

13.3 Design of Force Feedback Systems

Naively the problem of control design for a force balance system is to find a feedback which applies a signal to the actuator which counteracts the applied force. The signal that drives the actuator is the sensor output, perhaps with some filtering. Traditionally the error signal is not used because it has no effect in steady state and the error is typically small. The error signal does, however, contain useful information that can be used to improve performance. Since the fundamental problem is to measure a physical variable, it is natural to consider it as a joint control and estimation problem. In this way it is possible to combine actuator and sensor signals with a mathematical model of the system.

Assume that the sensor can be modeled by a linear system

$$\frac{dx}{dt} = Ax + Bu + B_w w \qquad y = Cx, \tag{13.1}$$

where w is the signal that we want to measure, u is the the control signal, and y is the error signal that we try to keep small. The states are defined in such a way that the equilibrium corresponds to $x = 0$. From the viewpoint of control the signal w, which we want to measure, can be regarded as a disturbance. The control problem is thus to find a system which gives a good estimate of the disturbance w and at the same time keeps the error y small.

A nice feature of sensor development is that it is often possible to design the sensor hardware and the control system. It is thus a prime example of co-design of process and control. It is highly desirable to design the sensor so that the vectors B_w and B are parallel because it is then easy to reduce the effect of the disturbance w. Physically this means that the variable we want to measure, w, and the control signal u should enter the system at the same point. A system with this property is said to have a *co-located* configuration. There are considerable advantages in designing the instrument to have this property.

To have a complete model we must also have a model for the dynamics of the signal we want to measure. A simple model is to assume that w is constant but unknown. Such a signal can be modeled by

$$\frac{dw}{dt} = 0. \tag{13.2}$$

The assumption that disturbances are constant but unknown is not too special, the manner in which it is used in the control design actually captures disturbances that are changing at rates close to the bandwidth of the control signal. It is possible to introduce other models of the signal we want to estimate. Since the signal model will influence how the signals are processed in the sensor it is possible to tune a sensor to particular signal behaviors that are known a-priori. A more general model is to assume that the signal is generated by

$$\frac{dz}{dt} = A_z z, \qquad w = C_z z. \tag{13.3}$$

Matrix A_z typically has eigenvalues on the imaginary axis. For example, to model disturbances that are composed of a periodic component with frequency ω_d and a drift, matrix A_z has the form

$$A_z = \begin{bmatrix} 0 & \omega_d & 0 & 0 \\ -\omega_d & 0 & 0 & 0 \\ 0 & 0 & 0 & 1 \\ 0 & 0 & 0 & 0 \end{bmatrix}.$$

Note that the same method was used to model disturbances in Section 13.5.

A straightforward way to design the system is to use use a controller based on feedback from the estimated state, which can be generated by a Kalman filter (3). The state w is naturally not reachable (the external forces cannot be influenced by control), but it is observable if matrix A does not have an eigenvalue at the origin. The control law then becomes

$$\begin{aligned} \frac{d\hat{x}}{dt} &= A\hat{x} + B_w C_z \hat{z} + Bu + L_x(y - C\hat{x}) \\ \frac{d\hat{z}}{dt} &= A_z \hat{z} + L_z(y - C\hat{x}) \\ u &= -K_x\hat{x} + K_z\hat{z}. \end{aligned} \tag{13.4}$$

From the point of view of instrument design the external acceleration is the interesting signal, but from the point of view of control the external acceleration w is a disturbance that drives the state away from the zero position. The term $K_x\hat{x}$ is a feedback from the estimated sensor states and the term $K_w\hat{z}$ is a feedback from an estimated disturbance state. It can also be interpreted as a feedforward from an estimated disturbance.

Control design is essentially a matter of finding suitable values of the feedback gains K and K_w and the observer gains L_x and L_z. We will show that the transfer function from w to its estimate is independent of the feedback gains K and K_w but that it depends on the observer gains L and L_z. The only requirement on the gains K and K_w is that they must be chosen so that the deviations remain so small that the linear approximation is valid.

The control law (13.4) can be written as

$$\frac{d}{dt}\begin{bmatrix}\hat{x}\\ \hat{z}\end{bmatrix} = \begin{bmatrix} A - BK_x - L_xC & (B_wC_z - BK_z)C \\ -L_zC & A_z \end{bmatrix}\begin{bmatrix}\hat{x}\\ \hat{w}\end{bmatrix} + \begin{bmatrix}L_x\\ L_z\end{bmatrix} y, \quad u = -\begin{bmatrix}K_x & K_z\end{bmatrix}\begin{bmatrix}\hat{x}\\ \hat{z}\end{bmatrix},$$

and the controller has the transfer function

$$C(s) = \begin{bmatrix}K_x & K_z\end{bmatrix}\begin{bmatrix} sI - A + BK_x + L_xC & -(B_wC_z - BK_z)C \\ L_zC & sI - A_z \end{bmatrix}^{-1}\begin{bmatrix}L_x\\ L_z\end{bmatrix}.$$

If the instrument is designed to have a co-located configuration the feedback gain K_z can be chosen so that $B_wC_z = BK_z$. The transfer function then becomes

$$C(s) = K_z(sI - A_z)^{-1}L_z + (K_x + K_z(sI - A_z)^{-1}L_zC)(sI - A + BK_x + L_xC)^{-1}L_x. \tag{13.5}$$

Note that the controller transfer function has zeros at the eigenvalues of matrix A_z in the disturbance model. For the special case $A_z = 0$ and $B_z = 1$, the transfer function is

$$C(s) = \frac{K_z L_z}{s} - \frac{K_z L_z C}{s}[sI - A + BK_x + L_x C]^{-1} L_x.$$

The controller thus has integral action. Furthermore if $K_x = 0$ the controller transfer function becomes

$$C(s) = \frac{K_z L_z}{s}\left(1 - C(sI - A + L_x C)^{-1} L_x\right).$$

This transfer function has poles at $s = 0$ and at the eigenvalues of $A - L_x C$ and zeros at the eigenvalues of A. The controller can be interpreted as an integrating controller with a notch filter that cancels the system dynamics.

13.4 Determining the Observer Gains

There are many ways to determine the observer gains. One possibility is to determine the gains to give the sensor the desired transfer function, another is to find gains that give the best estimates. We will first discuss how the gains can be chosen to shape the frequency response.

The sensor transfer function describes how the estimate $\hat{w}$ is related to the physical variable w, which we want to measure. This transfer function is a property of the closed-loop system obtained when the feedback given by (13.4) is applied to the system (13.1). Consider the dynamics of $\hat{z}$ and $\tilde{x} = x - \hat{x}$.

$$\begin{aligned}
\frac{d\hat{z}}{dt} &= A_z \hat{z} + L_z(y - C\hat{x}) = A_z + L_z C\tilde{x} \\
\frac{d\tilde{x}}{dt} &= A\tilde{x} + B_w \tilde{w} - L_x(y - C\hat{x}) = (A - L_x C)\tilde{x} + B_w \tilde{w}.
\end{aligned}$$

Taking Laplace transforms gives

$$\begin{aligned}
\hat{W} &= C_z \hat{Z} = C_z(sI - A_z)^{-1} L_z C\tilde{X} \\
\tilde{X} &= (sI - A + L_x C)^{-1} B_w \tilde{W} = (sI - A + L_x V)^{-1} B_w (W - \hat{W}),
\end{aligned}$$

where we used the notation $\mathscr{L}x = X$. The transfer function then becomes

$$\begin{aligned}
G_{\hat{w}w} &= \left(I + F(s)\right)^{-1} F(s) \\
F(s) &= C_z(sI - A_z)^{-1} L_z C(sI - A + L_x C)^{-1} B_w.
\end{aligned} \tag{13.6}$$

The transfer function depends on the sensor model (A, B, C), the disturbance model (A_z), and the filter gains (L_x, L_w) but *not* on the feedback gains (K_x, K_z). The architecture of the controller (13.4) thus gives a nice separation of the design problem. The filter gains can be chosen to give the desired transfer function of the sensor and the controller gains can be chosen for other purposes.

Also note that the transfer function $F(s)$ has the eigenvalues of A_z as poles. If A_z has an eigenvalue at the origin the function $F(s)$ has a pole at the origin and the transfer function $G_{\hat{w}w}$ has unit zero-frequency gain.

For the special case of unknown but constant accelerations we have $A_z = 0, B_z = 1$ and $C_z = 1$, and the transfer function (13.6) is

$$G_{\hat{w}w} = \frac{L_z C(sI - A + L_x C)^{-1} B_w}{s + L_z C(sI - A + L_x C)^{-1} B_w}. \tag{13.7}$$

We will now discuss how to choose the gains K_x and K_w so that the error is small. Note that it follows from the previous discussion that the sensor transfer function does not depend on K_x and K_w. The important issue is to make sure that the signals are not so large so that the system is not driven outside the range where it is linear.

The dynamics of the system is given by (13.1). Inserting the feedback law (13.4) we find

$$\begin{aligned}\frac{dx}{dt} &= Ax + B_w w - BK_x\hat{x} - BK_z\hat{z}\\ &= (A - BK_x)x + (B_w C_z - BK_z)z + BK_x\tilde{x} + B_w K_z\tilde{z},\end{aligned}$$

where the state x of the instrument denotes deviations from the ideal position. It is desirable to choose feedback gains so that x is small. To achieve this, the driving terms of the right-hand side should be small. The terms proportional to $\tilde{x}$ and $\tilde{w}$ will be small if the estimator is well designed. The effect of the external signal w (the signal we want to measure) is small if $B_w C_z - BK_w$ is small. If the instrument is designed to have a co-located configuration the matrices B_w and B span the same space and the feedback gain K_z can then be chosen so that $B_w C_z = BK_z$, and the term $(B_w C_z - BK_z)z$ then vanishes. The effect of all disturbances is reduced if the feedback gain K is chosen so that the eigenvalues of $A - BK_x$ are far to the left. Large K_x will however increase the effect of the term $BK_x\tilde{x}$. If information about the disturbances is available the gains K_x and K_z can be chosen to find the optimal compromise.

The filter gains can also be chosen to give the best estimates. To do so we must know the character of the noise sources and the stochastic properties of the acceleration we want to measure. The properties of the noise sources can often be found from physics as is illustrated in Section 13.7, but it may be more difficult to find the characteristics of the signal we want to measure, one exception being signal monitoring when the signal processing can be tuned to the particular task.

Assume that the disturbances are modeled as white noise in the right-hand sides of (13.1) and (13.3). The process state x and the disturbance state z are combined to a new state vector which we call x with some abuse of notation. Furthermore we introduce the matrices

$$A_a = \begin{bmatrix} A & B_w C_z \\ 0 & A_w \end{bmatrix}, \quad B_a = \begin{bmatrix} B \\ 0 \end{bmatrix}, \quad C_y = \begin{bmatrix} C & 0 \end{bmatrix}, \quad C_w = \begin{bmatrix} 0 & C_z \end{bmatrix}. \tag{13.8}$$

Matrix C_y picks out the output and matrix C_w picks out the acceleration. The method is illustrated by an example. The augmented system can then be modeled by

$$dx = A_a x dt + B_a u dt + dv, \qquad dy = C_y x dt + de, \tag{13.9}$$

where x is the augmented state and v and e are Wiener processes with incremental covariances $R_x dt$ and $R_y dt$. The estimator of the state is given by

$$d\hat{x} = A_a \hat{x} dt + B_a u dt + L(dy - C_y \hat{x} dt), \qquad L = \begin{bmatrix} L_x \\ L_z \end{bmatrix}. \tag{13.10}$$

The steady-state variance of the estimates is given by

$$(A_a - LC_y)P + P(A_a - LC_y)^T + R_x + LR_y L^T. \tag{13.11}$$

The filter gains that minimize the variance are given by the Kalman filter equations (3)

$$L = \begin{bmatrix} L_x \\ L_z \end{bmatrix} = PC_y^T R_y^{-1}. \qquad L = P_0 C_y^T R_y^{-1}, \tag{13.12}$$

where matrix P_0 is given by

$$A_a P_0 + P_0 A_a + R_x - P_0 C_y^T R_y^{-1} C_y P_0 = 0.$$

By comparing matrix P of (13.11) with P_0, we can find out how much variance is lost by shaping the transfer function as compared with optimal filtering.

The variance of the measured signal is $\sigma_{\hat{w}}^2 = C_w P C_w^T$, where C_w is the vector that picks out w from the augmented state. For the comparison with the resolution of a simpler scheme where the control signal is used as an estimate of the scaled acceleration, we observe that $u = K_x \hat{x} K_w \hat{w}$ and the acceleration estimate is given by

$$\bar{w} = \hat{w} + (K_x / K_w)\hat{x},$$

which has the variance

$$C_w P C_w^T + \frac{1}{K_w^2} K_x C_x P C_x^T K_x^T.$$

The second term represents how much we gain by exploiting the error signal as compared to a simpler force feedback system.

13.5 A Tunneling Accelerometer

In a tunneling accelerometer the position of the mass is sensed using the quantum tunnel effect. A small tip is placed on the proof mass, and a bias voltage is applied between the tip and the base (4; 5; 6; 7). The mass is provided with a force actuator and a feedback system which attempts to keep the tunneling current constant.

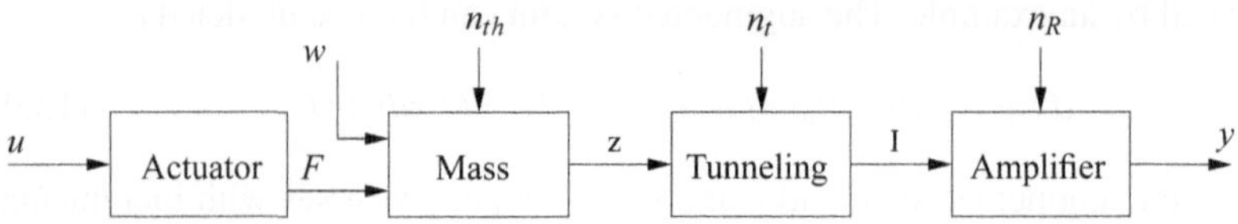

Fig. 13.1 Block diagram of the accelerometer system. The signal w is the acceleration, and F is the force applied to the mass by the comb drive.

There are four blocks representing the comb-drive actuator, the spring-mass, system, the tunneling tip and the amplifier, see Fig. 13.1. The input u is the voltage applied to the drive electronics of the comb drive and the output y is the output of the amplifier for the tunneling detector. There are three major noise sources thermal actuation of the mass which is represented by the force n_{th} in the block diagram, tunneling noise n_t and resistor noise n_R. There are parameter variations between the chips.

It is straightforward to model the system. The essential nonlinearities are in the comb-drive actuator and in the tunneling sensor. The major disturbances are the forces generated by the acceleration we would like to measure. There are random disturbances. The mass exhibits Brownian motion due to air molecules that collide with the mass, there is also resistor and tunneling noise. Most coefficients can be obtained from physics. The resonant frequency ω_0 and the Q value can be obtained by simple experiments. Sensor noise characteristics are obtained from physics.

Nominal values of are given in Table 13.1.

Table 13.1 Nominal values of the system parameters

Boltzmann's constant	k_B	1.3807×10^{-23} J/K
Charge of electron	q_0	1.602×10^{-19} C
Dielectric constant	ε	8.854×10^{-12} F/m
Tunneling constant	α	1.025 Å^{-1}eV^{-05}
Tunneling barrier	ϕ	0.05 eV
Temperature	T	293 K
Mass	m	4.92 μg
Resonant frequency	f_0	4.2 kHz
Q-value	Q	10
Number of comb fingers	N	540
Comb finger height	h	18 μm
Spacing	d	2.65 μm
Height	h	18 μm
Actuator gain	k_a	9.2×10^{-7} N/V
Tunneling gain	k_t	4 A/m
Preamp resistance	R	10.2 MΩ
Voltage gain	k_v	2
Sensor gain	$k_s = k_t k_v R$	81.6 MV/m

The force F generated by the comb-drive is

$$F = N\frac{h\varepsilon}{d}(V_e + u)^2, \quad (13.13)$$

where N is the number of moving fingers, h is the length of the fingers, d is the spacing between the fingers and ε is the dielectric constant of the medium surrounding the drive (6). Linearizing the model (13.13) around the equilibrium $u = 0$ gives

$$F = k_a u, \qquad k_a = \frac{2Nh\varepsilon V_e}{d}, \quad (13.14)$$

where k_a is the actuator gain. With $V_e = 14.2$ V we get $k_a = 0.2 \times 10^{-7}$ N/V.

The mass and its suspension are modeled as a spring-mass system

$$m\frac{d^2z}{dt^2} + c\frac{dz}{dt} + kz = F + mw + n_{th} = k_a u + mw + n_{th}, \quad (13.15)$$

where m is the mass, c is the damping coefficient, k is the spring constant of the flexure, F is the force applied by the comb drive and w the acceleration of the frame.

The disturbances n_{th} are the force generated by the air molecules that collide with the mass causing Brownian motion. The force can be represented as white noise with the spectral density $4ck_BT$ [N^2/Hz] (single sided), where $k_B = 1.38 \times 10^{-23}$ [J/K] is Boltzmann's constant and T is the absolute temperature.

Tunneling begins when the gap is on the order of 1 nm or 10 Å. The tunneling current then increases exponentially with decreasing gap. The following approximate model is given in (6)

$$I = kV_b e^{-\alpha z\sqrt{\phi}}, \quad (13.16)$$

where V_b is the bias current, k is a constant, z is the tunneling gap, α is the tunneling constant and ϕ is the effective tunneling barrier. Linearizing the model (13.16) gives

$$I = k_t z + n_t, \qquad k_t = \alpha\sqrt{\phi}I_0. \quad (13.17)$$

We have chosen the sign convention so that k_t is positive. The parameter k_t [A/m] is the linearized tunneling gain and n_t is the tunneling noise, which has shot noise character. The tunneling gain k_t is proportional to the tunneling current. Tunneling noise is white with the spectral density $\Phi_t = 2q_0I_0$ [A^2/Hz], where $q_0 = 1.6 \times 10^{-19}$ [C] is the charge of the electron and I_0 is the tunneling current. The corresponding incremental covariance is $r_t\,dt$ with $r_t = \Phi_t/2$ [A^{2}s].

The tunneling current, which is a few nano amperes, is amplified electronically. The first stage is a low noise operational amplifier with a feedback resistance R, the second step is a voltage amplifier with gain k_v. It is advantageous to have the resistance as large as possible. The largest value is limited by offset voltages of the opamp and the input capacitance; it also depends on the type of amplifier. A simple model is

$$V = k_v(RI + n_R). \quad (13.18)$$

The gain from current to voltage is $k_v R$ [V/A]. Using the nominal parameters in Table 13.1 we have $R = 10.2\,\mathrm{M}\Omega$ and $k_v = 2$. The amplifier gain is thus 2.04×10^7 V/A. A tunneling current of 2 nA gives an output voltage of 40 mA.

Resistor noise is modeled as Johnson–Nyquist noise which has the spectral density $4k_B TR$ [V^2/Hz] (single sided). The corresponding incremental covariance is $r_R dt$, where $r_R = 2k_B TR$ [$\mathrm{V}^2\mathrm{s}$]). The noise in the first stage dominates and we will neglect the resistor noise in the second stage. The preamplifier used in the experiments is more complicated with compensating capacitors. A more complicated dynamic model is given in (7). The simple model given here suffices if the bandwidth of the instrument is not too high.

Finally we will assume that the acceleration to be measured can be modeled as a Wiener process with incremental covariance $r_w dt$.

Summarizing we find that the accelerometer can be modeled as a stochastic differential equation

$$dx = A_a x dt + B_a u dt + dv, \qquad dy = C_a x dt + de, \tag{13.19}$$

where the state is $x = (z;\, \dot{z};\, w)$ and the matrices are given by

$$\begin{aligned}
A_a &= \begin{bmatrix} 0 & 1 & 0 \\ -k/m & -c/m & -1 \\ 0 & 0 & 0 \end{bmatrix}, \qquad B_a = \begin{bmatrix} 0 \\ k_a/m \\ 0 \end{bmatrix} \\
C_y &= \begin{bmatrix} k_s \; 0 \; 0 \end{bmatrix}, \qquad C_w = \begin{bmatrix} 0 \; 0 \; 1 \end{bmatrix} \\
R_x &= E dv dv^T = \mathrm{diag}(0,\, 2ck_B T/m^2,\, r_w) \\
R_y &= E(de)^2 = k_v^2 (2k_B TR + R^2 q_0 I_0),
\end{aligned} \tag{13.20}$$

where R_x and R_y are the incremental covariance of v and e.

Since we have neglected the dynamics of the amplifier, the measurement noise in (13.20) is a combination of tunneling noise and resistor noise. A more detailed dynamics model is given in (7).

13.6 The Controller

Comparing with the general model in Section 13.3 we have

$$\begin{aligned}
A &= \begin{bmatrix} 0 & 1 \\ -k/m & -c/m \end{bmatrix}, \quad B = \begin{bmatrix} 0 \\ k_a/m \end{bmatrix}, \quad B_w = \begin{bmatrix} 0 \\ 1 \end{bmatrix}, \quad C = \begin{bmatrix} k_s \; 0 \end{bmatrix} \\
D &= 0, \quad A_z = 1, \quad C_z = 1, \quad L_x = \begin{bmatrix} l_1 \\ l_2 \end{bmatrix}, \quad L_z = l_3.
\end{aligned} \tag{13.21}$$

The disturbance state z is thus the acceleration w. The system in Fig. 13.1 has a co-located structure because the comb drive was modeled as a static model (13.13). The decoupling condition $B_w C_z = BK_z$ is satisfied by choosing the controller gain $K_z = k_3 = m/k_a$. The controller (13.4) becomes

$$
\begin{aligned}
d\hat{x}_1 &= \hat{x}_2\,dt + l_1(dy - k_s\hat{x}_1\,dt)\\
d\hat{x}_2 &= -(k/m)\hat{x}_1\,dt - (c/m)\hat{x}_2\,dt - k_1\hat{x}_1\,dt - k_2\hat{x}_2\,dt + l_2(dy - k_s\hat{x}_1\,dt)\\
d\hat{x}_3 &= l_3(dy - k_s\hat{x}_1\,dt)\\
u &= -\frac{m}{k_a}(k_1\hat{x}_1 + k_2\hat{x}_2 + \hat{x}_3).
\end{aligned}
\tag{13.22}
$$

Note that $\hat{x}_3 = \hat{w}$ is the estimated acceleration. A block diagram of the controller represented by the equations (13.22) is shown in Fig. 13.2. Note that a model of the spring-mass system is a central part of the controller. The feedback gains k_1 and k_2

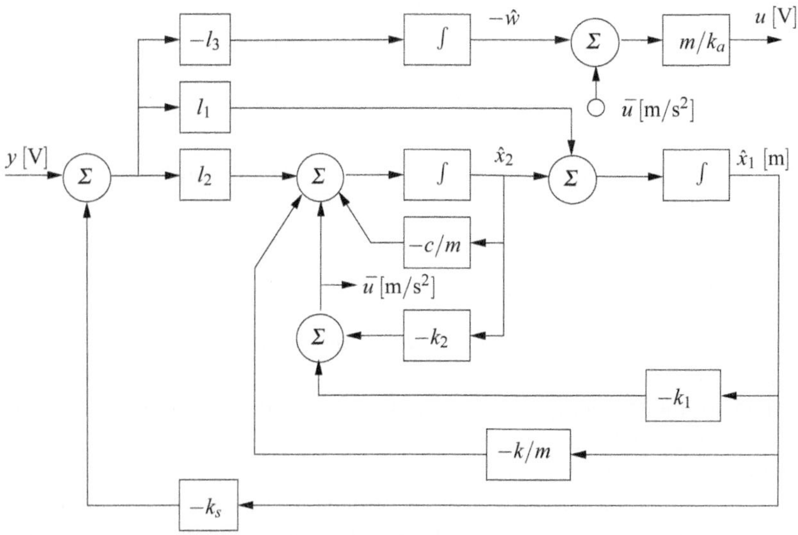

Fig. 13.2 Block diagram of the controller. The signal $\hat{w}$ is the estimated acceleration. The signal $\bar{u} = -k_1\hat{x}_1 - k_2\hat{x}_2$ has been introduced to avoid clutter.

determine how well the mass is kept to its center position. The gain k_1 influences the stiffness and the gain k_2 influences damping. The controller gains will not influence the sensor transfer function as long as the deviations are so small that the linear approximation is valid. If the feedback gains are zero we have $\bar{u} = 0$ and the controller can be interpreted as a second-order notch filter in cascade with an integrator. The notch filter attenuates the resonant mode of the mass.

The filter gains l_1 and l_2 and l_3 are chosen to shape the frequency response of the sensor. The transfer function of the sensor is given by (13.6). Inserting the matrices (13.21) into (13.6) gives

$$
\begin{aligned}
F(s) &= l_3 \begin{bmatrix} k_s & 0 \end{bmatrix} \begin{bmatrix} s + k_s l_1 & -1 \\ k_s l_2 + k/m & s + c/m \end{bmatrix}^{-1} \begin{bmatrix} 0 \\ 1 \end{bmatrix}\\
&= \frac{k_s}{s^2 + s(k_s l_1 + c/m) + k_s(l_1 c/m + l_2) + k/m},
\end{aligned}
$$

where $k_s = k_t k_v R$. The transfer function of the sensor then becomes

$$G_{\hat{w}w}(s) = \frac{l_3 k_s}{s^3 + s^2(k_s l_1 + c/m) + s\big(k_s(l_1 c/m + l_2) + k/m\big) + l_3 k_s}. \tag{13.23}$$

The transfer function has unit zero-frequency gain and the transfer function can be shaped by choosing the filter gains. Assume, for example, that we want the denominator

$$(s + \alpha_c \omega_c)(s^2 + 2\zeta_c \omega_c s + \omega_c^2).$$

The filter gains then becomes

$$\begin{aligned} l_1 &= \frac{(\alpha_c + 2\zeta_c)\omega_c - c/m}{k_v k_t R} = \frac{(\alpha_c + 2\zeta_c)\omega_c - 2\zeta_0\omega_0}{k_s} \\ l_2 &= \frac{(1 + 2\alpha_c\zeta_c)\omega_c^2 - k/m - \big((\alpha_c + 2\zeta_c)\omega_c - c/m\big)c/m}{k_v k_t R} \\ &= \frac{(1 + 2\alpha_c\zeta_c)\omega_c^2 - \omega_0^2 - 2\zeta_0\omega_0\big((\alpha_c + 2\zeta_c)\omega_c - 2\zeta_0\omega_0\big)}{k_s} \\ l_3 &= \frac{\alpha_c\omega_c^3}{k_v k_t R} = \frac{\alpha_c\omega_c^3}{k_s}, \end{aligned} \tag{13.24}$$

where $\omega_0 = \sqrt{k/m}$ and $\zeta_0 = c/(2m\omega_0)$. Note that the filter gain l_k increases with the k-th power of ω_c.

The shape of the frequency response is given by the parameters α_c and ζ_c, and the bandwidth is given by ω_c.

If the power spectral density of the acceleration is known we can instead choose filter gains that minimize the variance of the estimate $\hat{w}$. In this way it is possible to design a sensor that is highly tuned to a particular situation. In other cases it is more natural to use the filter gains to give the desired frequency response of the sensor.

13.7 Experiments

Many experiments were performed, the first experiments (7) used MATLAB, SIMULINK and the xPC target from MathWorks. The sampling rate was limited to 5 kHz because of hardware restrictions. An integrating controller was used. There were significant problems in measuring the tunneling signal. The theoretical models were helpful to guide improvements. Later the electronics were rebuilt and the xPC target was replaced by CompactRIO from National Instruments, which has a field programmable gate array (FPGA), a PowerPC processor board and plug-in cards for analog IO with 16-bit resolution (10). A controller with the structure shown in Fig. 13.2 was implemented in the FPGA. It is important to scale Eqs. (13.22) describing the controller in order to obtain a robust implementation. It is particularly important if the controller is implemented using fix point calculations in the FPGA.

The tunneling tip is very sensitive. The accelerometer is therefore designed so that the tip distance is normally about a micrometer. This distance, which is maintained by the springs that support the mass, allows normal handling. At that distance there is no tunneling current. To use the sensor it must be initialized by reducing the tunneling gap to about a nanometer so that tunneling starts. The control system for that has to be done in a gentle way to avoid that the sensor is spoiled by crashing the tip to the mass. The control system has to accomplish two tasks: To move the tip safely to a position where tunneling starts and then to regulate the tip to maintain a constant tunneling current.

Initialization is difficult because tunneling does not occur until the gap is very small (about a nanometer). Initialization therefore has to be done in open loop. A safe way to initialize the system is to use an integrating controller with low gain. The gap will decrease linearly at a controlled rate by limiting the output rate of the integrator. To make sure that the tip is not damaged, the gain of the integrating controller can then be increased for tighter control. This initialization scheme can easily be combined with the controller structure in Fig, 13.2, which becomes an integrating controller if the gains in the blocks labeled l_1, l_2 and k_s are set to zero. The set point is typically given as a desired tunneling current. The controller is then smoothly switched to the regulation controller.

Figure 13.3 shows an experiment with initialization and regulation. Initially the drive signal u to the comb drive increases practically linearly because of the integrating controller whose input is saturated. The output of the amplifier is practically zero, the fluctuations are dominated by thermal noise in the resistor R. Tunneling starts around time $t = 2.4$ and increases almost exponentially. The gain l_3 is increased when the tunneling current is close to its desired value and the integrating controller then regulates the tunneling current. The fluctuations in tunneling current increase when tunneling starts primarily because the mass picks up Brownian motion. The magnitude of the fluctuations increases with increasing tunneling current because of the gain increase; it is also partially hidden because of the exponential

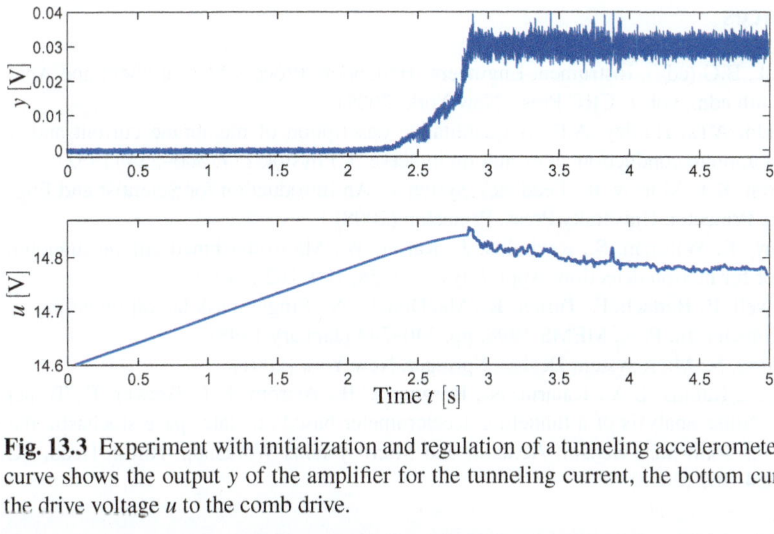

Fig. 13.3 Experiment with initialization and regulation of a tunneling accelerometer. The top curve shows the output y of the amplifier for the tunneling current, the bottom curve shows the drive voltage u to the comb drive.

increase of the current. The standard deviation in steady state regulation in Fig. 13.3 is about 3 mV, which corresponds to a position deviation of about 0.3 Å.

Figure 13.3 shows that the the control signal drifts in steady state. Analysis of long data records (9) shows that the drift can be modeled as $1/f$ noise. The drift is probably due to imperfections in the input stage of the operational amplifier.

13.8 Summary

In this paper we have developed a flexible design methodology for the design of sensors based on force feedback. The design problem is formulated as an estimation problem. A key feature is that both the feedback signal and the error signal are used to form the estimate of the measured variable. The method permits shaping of the sensor transfer function and analysis of the performance of the sensor. The design leads to an architecture where the sensor transfer function depends on the observer gains but not on the feedback gains. The analysis also gives insight into desirable features of instrument design, a key property being the co-location property that makes it possible to keep the control error small.

Acknowledgements. This work was carried out in Professor Kimberly Turner's laboratory at the University of California, Santa Barbara. I am grateful to Kimberly and her students for many stimulating discussions and interesting experiences in hunting noise sources in the lab. Laura Oropeza-Ramon designed and made the chips; she also did the first experiments. Nitin Kataria, under supervision of Professor Forest Brewer, designed the input amplifiers, and also programmed the FPGA. Chris Burgner did many of the experiments and the data analysis. Zi Yie simulated the systems and participated in experiments. I would also like to thank Jeannie Falcon, Javier Guiterrez and Brian Mac Cleery at National Instruments for valuable support on LabVIEW and the CompactRIO platforms.

References

[1] Liptak, B.G (ed .): Instrument Engineers' Handbook. Process Measurement and Analysis, 4th edn., vol. 1. CRC Press, New York (2009)
[2] Hodkin, A.L., Huxley, A.F.: A quantitative description of membrane current and its application to conduction an excitation in nerve. J. Physiol. 117, 500–554 (1952)
[3] Åström, K.J., Murray, R.: Feedback Systems - An Introduction for Scientist and Engineers. Princeton University Press, Princeton (2008)
[4] Kenny, T., Waltman, S., Reynolds, J., Kaiser, W.: Micro-machined silicon tunneling sensor for motion detection. Appl. Phys. Lett. 58, 100–102 (1991)
[5] Hartwell, P., Bertsch, F., Turner, K., MacDonald, N.: Single mask lateral tunneling accelerometer. In: Proc. MEMS 1998, pp. 340–344 (January 1998)
[6] Senturia, S.: Microsystem Design. Springer, New York (2001)
[7] Oropeza-Ramos, L.A., Kataria, N., Burgner, C.B., Åström, K.J., Brewer, F., Turner, K.L.: Noise analysis of a tunneling accelerometer based on state space stochastic theory. In: Solid-State Sensor, Actuator and Microsystems Workshop, Hilton Head, SC, pp. 364–367 (2008)

[8] Åström, K.J.: Introduction to Stochastic Control Theory. Dover, New York (2006)
[9] Burgner, C.B., Yie, Z., Kataria, N., Oropeza, L., Åström, K.J., Brewer, F., Turner, K.L.: Digital control of a tunneling accelerometer. In: IEEE Sensors 2009, Christchurch, New Zealand, pp. 1824–1827 (2009)
[10] Yie, Z., Kataria, N., Burgner, C., Åström, K.J., Brewer, F., Turner, K.L.: Control design for a force balance sensor. In: American Control Conf. 2009, St. Louis, MO, pp. 4208–4214 (2009)

Chapter 14
H2 Guaranteed Cost Control in Track-Following Servos

Richard Conway, Jianbin Nie, and Roberto Horowitz

Abstract. This chapter presents two new control synthesis approaches for dual-stage track-following servo systems. Both approaches are based on $\mathscr{H}_2$ guaranteed cost analysis, in which an upper bound on the worst-case $\mathscr{H}_2$ performance of a discrete-time system with gain-bounded unstructured causal LTI uncertainty is determined by solving either a semi-definite program (SDP) or several Riccati equations. We review the results of a paper on $\mathscr{H}_2$ guaranteed cost analysis and a paper on optimal full information $\mathscr{H}_2$ guaranteed cost control and then use these results to develop two output feedback control synthesis approaches. The first approach is based entirely on the solution of SDPs whereas the second approach exploits Riccati equation structure to reduce the number and complexity of the SDPs that need to be solved. Throughout the paper, we apply the analysis and control techniques to a hard disk drive model with a PZT-actuated suspension and demonstrate that the approaches that exploit Riccati equation structure are faster and at least as accurate as their SDP counterparts.

14.1 Introduction

For several decades now, the areal storage density of hard disk drives (HDDs) has been doubling roughly every 18 months, in accordance with Kryder's law. As the storage density is pushed higher, the concentric tracks on the disk which contain data must be pushed closer together, which requires much more accurate control of the read/write head. Currently available hard drives can store 2 TB of data on a 3.5" drive with three platters. This corresponds to an areal data density of 600 gigabit/in^2. The current goal of the magnetic recording industry is to achieve an areal storage density of 4 terabit/in^2. It is expected that the track width required to achieve this

Richard Conway · Jianbin Nie · Roberto Horowitz
Department of Mechanical Engineering, University of California, Berkeley
e-mail: rconway345@berkeley.edu, njbin@berkeley.edu
horowitz@me.berkeley.edu

E. Eleftheriou & S.O.R. Moheimani (Eds.): Cntrl. Tech. for Emerging Micro/Nanoscale Sys. LNCIS 413, pp. 235–270.
springerlink.com

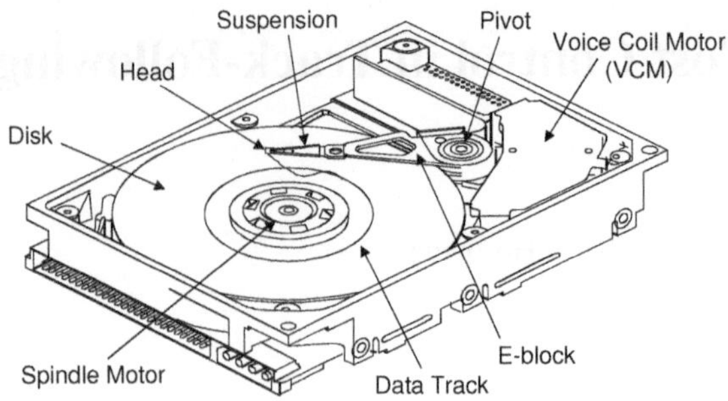

Fig. 14.1 Schematic of a typical HDD

data density is 25 nm. To achieve this specification for track-following control, in which the read/write head is maintained as close to the center of a given data track as possible, the 3σ value of the closed-loop position error signal (PES) should be less than 2.5 nm.

To help achieve this goal, the use of a secondary actuator has been proposed to give increased precision in read/write head positioning. There are three classes of secondary actuators: actuated suspensions (6), actuated sliders (7), and actuated heads (17). Each of these proposed secondary actuator classes corresponds to a different actuator location in Fig. 14.1. In the actuated head configuration, a microactuator (MA) actuates the read/write head with respect to the slider mounted at the tip of the suspension. In the actuated slider configuration, an MA directly actuates the head/slider assembly with respect to the suspension. For both of these configurations, it is difficult to design an MA which can be easily incorporated into the manufacture and assembly of a HDD on a large scale. In the actuated suspension configuration, the MA actuates the suspension with respect to the E-block. This secondary actuator scheme is the least difficult to design and has been incorporated into some consumer products. We will use this secondary actuation scheme in this paper.

Since there tend to be large variations in HDD dynamics due to variations in manufacture and assembly, it is not enough to achieve the desired level of performance for a single plant; the controller must guarantee the desired level of performance for a large set of HDDs. Thus, we are interested in finding a controller which gives robust performance over a set of HDDs. One framework for solving this problem is guaranteed cost control. This methodology is a control design methodology whose objectives involve worst-case quadratic time domain costs over a modeled set of parametric uncertainty. Both the state feedback synthesis problem and the output feedback synthesis problem can be solved for discrete-time systems by using semidefinite programs (SDPs)—convex optimization involving linear matrix inequalities (LMIs)—as is done in (14) and (18), respectively.

As mentioned earlier, the relevant performance metric in a HDD is the standard deviation of the PES. Since the squared $\mathscr{H}_2$ norm of a system can be interpreted as the sum of variances of the system outputs under the assumption that the system is driven by independent white zero mean Gaussian signals with unit covariance, the $\mathscr{H}_2$ norm is a useful performance metric for HDDs.

This chapter reviews the basic results of $\mathscr{H}_2$ guaranteed cost analysis (2), which use the techniques of guaranteed cost control to yield an upper bound on the worst-case $\mathscr{H}_2$ performance of a system with dynamic unstructured uncertainty. Using this characterization of performance, we then review the solution of the corresponding full information control problem (3)—a generalization of the state feedback control problem—and use the results to generate two heuristics for solving the output feedback control problem.

For all of the problems in this chapter, we consider two approaches: an approach based on solving SDPs and an approach based on solving Riccati equations. Parallel to reviewing and developing the relevant theory, we apply these techniques to the design and analysis of HDD track-following controllers. For all numerical experiments in this paper, we use a 2.2 GHz Intel Core 2 Duo processor with 2 GB RAM running MATLAB 7.4.0 (with multithreaded computation disabled) under 32-bit Windows Vista. We solve SDPs two ways in this paper: using SeDuMi (16) with YALMIP (12) and using the `mincx` command in the Robust Control Toolbox for MATLAB without YALMIP. Through these HDD track-following control examples, we demonstrate the computational advantages of the approach based on Riccati equation solutions.

14.1.1 Preliminaries

In this chapter, we will denote the spectral norm (i.e. the maximum singular value) and the Frobenius norm of a matrix M respectively as $\|M\|$ and $\|M\|_F$. We will say that M is Schur if all of its eigenvalues lie strictly inside the unit disk in the complex plane. The operator "diag" takes several matrices and stacks them diagonally:

$$\mathrm{diag}[M_1,\ldots,M_n] = \begin{bmatrix} M_1 & & 0 \\ & \ddots & \\ 0 & & M_n \end{bmatrix}. \tag{14.1}$$

Positive definiteness (resp. semi-definiteness) of a symmetric matrix X will be denoted by $X \succ 0$ (resp. $X \succeq 0$), and a $\bullet$ in a symmetric matrix will represent a block which follows from symmetry.

For given (A,B,Q,R,S), where $Q = Q^T$ and $R = R^T$, we define the functions

$$\mathscr{R}_{(A,B,Q,R,S)}(P) := A^T PA + Q - (A^T PB + S)(B^T PB + R)^{-1}(B^T PA + S^T) \tag{14.2a}$$

$$\mathscr{K}_{(A,B,Q,R,S)}(P) := -(B^T PB + R)^{-1}(B^T PA + S^T)\,. \tag{14.2b}$$

We will make the notation more compact in the remainder of the paper by respectively denoting these functions as $\mathscr{R}_\phi(P)$ and $\mathscr{K}_\phi(P)$ where ϕ is an appropriately defined 5-tuple. Note that the equation $\mathscr{R}_\phi(P) = P$ is a discrete algebraic Riccati equation (DARE). If $\mathscr{R}_\phi(P) = P = P^T$ and $A + B\mathscr{K}_\phi(P)$ is Schur, then P is called a stabilizing solution of the DARE. Throughout the paper, we will implicitly use the property that if a DARE has a stabilizing solution, it is unique (11).

A matrix pair (A,B) will be called d-stabilizable if $\exists K$ such that $A+BK$ is Schur. A matrix pair (A,C) will be called d-detectable if $\exists L$ such that $A+LC$ is Schur. For a given stable and causal LTI system G, its $\mathscr{H}_2$ and $\mathscr{H}_\infty$ norms will respectively be denoted as $\|G\|_2$ and $\|G\|_\infty$. For two causal LTI systems G_1 and G_2, we denote the lower linear fractional transformation (LFT) of G_1 by G_2 (shown in Fig. 14.2(a)) as $\mathscr{F}_l(G_1, G_2)$. We will denote the upper LFT of G_1 by G_2 (shown in Fig. 14.2(b)) as $\mathscr{F}_u(G_1, G_2)$.

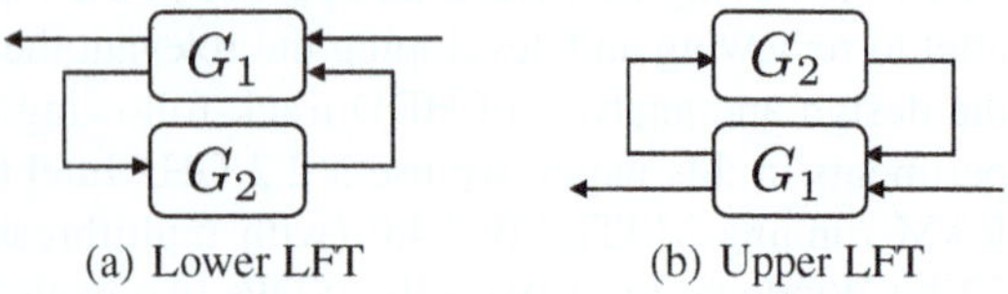

Fig. 14.2 Linear fractional transformations (LFTs)

14.2 Hard Disk Drive Model

In this section, we present the HDD model we will be using throughout this chapter. The HDD we are considering has the PZT-actuated suspension shown in Fig. 14.3, which is a Vector model suspension provided to us by Hutchinson Technology Inc. In our setup, we use a laser Doppler vibrometer (LDV) to measure the absolute radial displacement of the slider. The control circuits include a Texas Instrument TMS320C6713 DSP board and an in-house made analog board with a 12-bit ADC, a 12-bit DAC, a voltage amplifier to drive the MA, and a current amplifier to drive the voice coil motor. The DSP sampling period is 1.4×10^{-5} s and the controller delay, which includes the ADC and DAC conversion delay and the DSP computation delay, is 6 µs. A hole was cut through the case of the drive to allow the LDV laser to shine on the slider. It should be noted that these modifications affect the response of the drive and may have detrimentally affected the attainable performance of the servo system.

The block diagram of our HDD setup is shown in Fig. 14.4 and the relevant signals and their units are listed in Table 14.1. In this block diagram, we treat the dynamics from the two control inputs to the head displacement as a single block

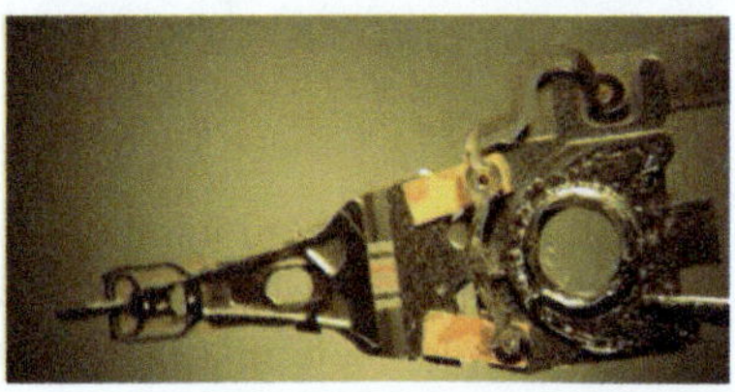

Fig. 14.3 Picture of the Vector model PZT-actuated suspension used in our experimental setup

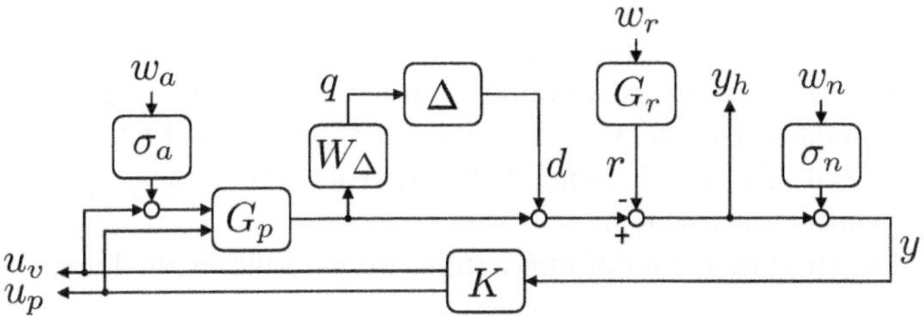

Fig. 14.4 HDD block diagram

Table 14.1 HDD signals

Signal	Description	Units
r	Disturbances on the head position (8)	nm
u_p	PZT actuator control signal	V
u_v	Voice coil motor control signal	V
w_a	Airflow disturbances	(normalized)
w_n	PES sensor noise	(normalized)
w_r	Disturbances on the head position	(normalized)
y	PES	nm
y_h	Head displacement relative to the track center	nm

Table 14.2 Model parameters for G_p

mode, i	$\begin{bmatrix} a_{i,1} \\ a_{i,2} \end{bmatrix}$	B_i
1	$\begin{bmatrix} 1.99607 \\ -0.996286 \end{bmatrix}$	$\begin{bmatrix} 3.982 & 0 \\ -6.262 & 0 \end{bmatrix}$
2	$\begin{bmatrix} 1.474 \\ -0.9680 \end{bmatrix}$	$\begin{bmatrix} -1.415 & 1.263 \\ 1.368 & -1.659 \end{bmatrix}$
3	$\begin{bmatrix} 1.381 \\ -0.9762 \end{bmatrix}$	$\begin{bmatrix} -0.8049 & 2.338 \\ 2.082 & -0.01339 \end{bmatrix}$
4	$\begin{bmatrix} 0.5387 \\ -0.9632 \end{bmatrix}$	$\begin{bmatrix} 0.1366 & -4.740 \\ -1.520 & 25.62 \end{bmatrix}$
5	$\begin{bmatrix} 0.04209 \\ -0.9353 \end{bmatrix}$	$\begin{bmatrix} -0.06772 & 9.273 \\ -0.1826 & 105.8 \end{bmatrix}$
6	$\begin{bmatrix} -1.653 \\ -0.9527 \end{bmatrix}$	$\begin{bmatrix} -0.09272 & -1.618 \\ -0.1086 & -0.2218 \end{bmatrix}$

to take into account the knowledge that both actuators can excite the same vibration modes in the suspension. Exploiting this knowledge allows us to form a model which does not have redundant states resulting from including two copies of the suspension vibration modes. The block Δ is an unknown stable causal LTI system which satisfies $\|\Delta\|_\infty \leq 1$. This block, along with W_Δ, characterizes the output dynamic multiplicative uncertainty on G_p.

To construct a discrete-time model of our system, we used the methodology of (13). To find the model of G_p, we first obtained frequency responses of our system from u_v to y and u_p to y. Using weighted least squares, we separately fit a continuous-time model to each of these frequency responses, which we then combined into a single model and used common mode identification (4) to eliminate redundant copies of the suspension vibration modes. We then discretized this model with the 6 μs delay on each of its two inputs to yield the model for G_p. This model is given by

$$G_p(z) = \begin{bmatrix} -0.6858 & 20.94 \end{bmatrix} z^{-1} + \sum_{i=1}^{6} \begin{bmatrix} 1 & 0 \end{bmatrix} \left(zI - \begin{bmatrix} a_{i,1} & 1 \\ a_{i,2} & 0 \end{bmatrix} \right)^{-1} B_i \tag{14.3}$$

where the model parameters are as listed in Table 14.2.

Because there are two poles at $z = 0$—one for each input channel—the state-space model of G_p has 14 states. The poles at $z = 0$ were introduced by the discretization of the continuous-time input delay. The six vibration modes in (14.3) are ordered from lowest to highest resonance frequency. The Bode plot of this model is shown in Fig. 14.5.

The weighting for the dynamic multiplicative uncertainty of G_p, given by

$$W_\Delta = \frac{0.9733 - z}{z - 0.465}, \tag{14.4}$$

was chosen so that the uncertain model enveloped the experimental frequency response of G_p. The Bode magnitude plot of W_Δ is shown in Fig. 14.6(a). Since Δ is a SISO system, upper and lower bounds on the magnitude of each input/output pair in G_p can be easily computed one frequency at a time. Doing so yields the upper and lower bounds on the Bode magnitude plots of G_p shown in Fig. 14.7. The values $\sigma_a = 0.04854$ and $\sigma_n = 1.3$ were determined by matching the power spectrum density of the open loop slider motion respectively at low and high frequency. The disturbances on the head position are characterized by

$$\begin{aligned} G_r(z) = {} & \begin{bmatrix} 1 & 0 \end{bmatrix} \left(zI - \begin{bmatrix} 1.964 & 1 \\ -0.975 & 0 \end{bmatrix} \right)^{-1} \begin{bmatrix} -0.2574 \\ 0.25 \end{bmatrix} \\ & + \begin{bmatrix} 1 & 1 \end{bmatrix} \left(zI - \begin{bmatrix} 0.9956 & -0.0745 \\ 0 & 0.9956 \end{bmatrix} \right)^{-1} \begin{bmatrix} -0.9533 \\ 0.919 \end{bmatrix}. \end{aligned} \tag{14.5}$$

Figure 14.6(b) shows the Bode magnitude plot of G_r. In addition to capturing the effect of disturbances on the head position, this model of G_r also captures the

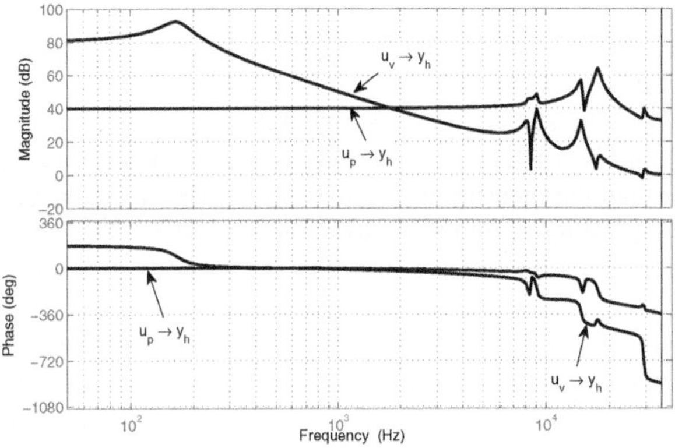

Fig. 14.5 Bode plot of G_p

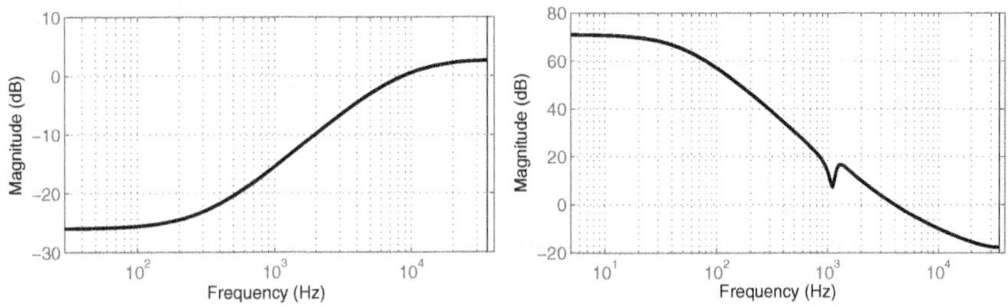

(a) W_Δ (b) G_r

Fig. 14.6 Bode magnitude plots of W_Δ and G_r

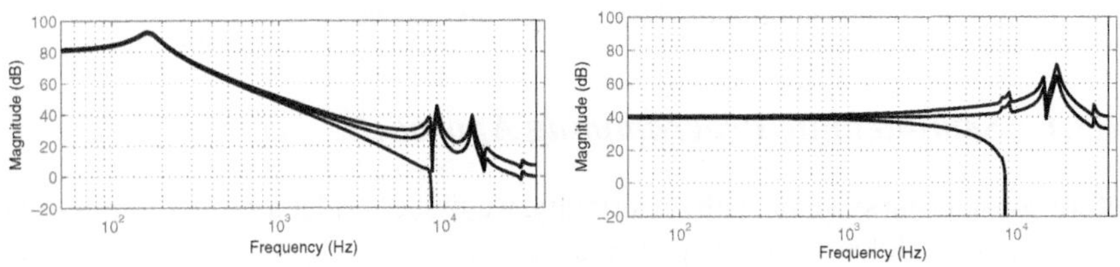

(a) $u_v \to y_h$ (b) $u_p \to y_h$

Fig. 14.7 Nominal Bode magnitude plots for G_p along with pointwise upper and lower bounds over modeled uncertainty

low-frequency drift in the LDV position measurements resulting from integration of velocity measurements. The second-order mode near 1 kHz in this model captures the effect of disk modes between 1 kHz and 3 kHz.

These disturbances, although realistic for our experimental setup, are larger than the disturbances typically found in a HDD. First of all, the measurement noise of the LDV is somewhat larger than the measurement noise of the PES. Moreover, as we previously mentioned, the LDV has a significant low-frequency drift. These two factors along with the mechanical modifications of the drive significantly deteriorate the achievable level of closed-loop performance.

With some manipulation, the blocks in Fig. 14.4 can be grouped to form the LFT representation in Fig. 14.8. In this form G_H has 19 states. For the remainder of this chapter, we will use the balanced realization of G_H for analysis and control design.

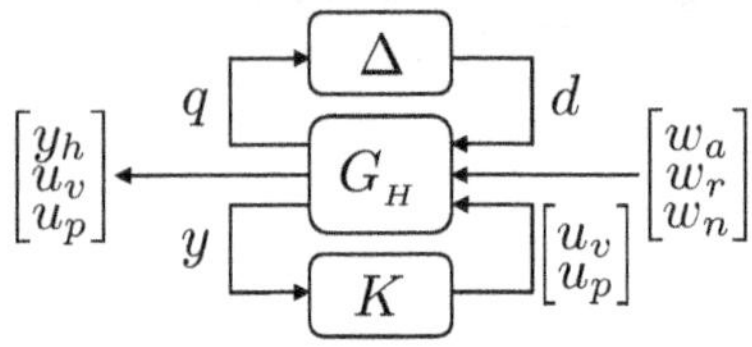

Fig. 14.8 LFT representation of HDD model

14.3 H2 Guaranteed Cost Analysis

In this section, we review the results of (2) on $\mathscr{H}_2$ guaranteed cost analysis. In particular, we first present an SDP for determining the $\mathscr{H}_2$ guaranteed cost performance of a given system and then show that this convex optimization can be efficiently solved using nonlinear convex optimization involving Riccati equation solutions. For the sake of brevity and clarity of presentation, we do not present the proofs here; interested readers should read the paper cited above.

14.3.1 Semi-definite Programming Approach

Before considering systems with uncertainty, we first consider a given discrete-time LTI system $\tilde{G}$ with known state-space realization

$$\tilde{G} \sim \left[\begin{array}{c|c} A_{\tilde{G}} & B_{\tilde{G}} \\ \hline C_{\tilde{G}} & D_{\tilde{G}} \end{array}\right]. \tag{14.6}$$

A well-known characterization of the $\mathscr{H}_2$ norm is that $\|\tilde{G}\|_2^2 < \gamma$ if and only if there exist $P \succ 0$ and W such that

$$\mathrm{tr}\{W\} < \gamma \tag{14.7a}$$

$$W \succ B_{\tilde{G}}^T P B_{\tilde{G}} + D_{\tilde{G}}^T D_{\tilde{G}} \tag{14.7b}$$

$$P \succ A_{\tilde{G}}^T P A_{\tilde{G}} + C_{\tilde{G}}^T C_{\tilde{G}}\,. \tag{14.7c}$$

In this context, however, it is beneficial for us to consider an alternate characterization which says that $\|\tilde{G}\|_2^2 < \gamma$ if and only if there exist $P \succ 0, W, V$ such that

$$\mathrm{tr}\{W\} < \gamma \tag{14.8a}$$

$$\begin{bmatrix} P & V^T \\ V & W \end{bmatrix} \succ \begin{bmatrix} A_{\tilde{G}} & B_{\tilde{G}} \\ C_{\tilde{G}} & D_{\tilde{G}} \end{bmatrix}^T \begin{bmatrix} P & 0 \\ 0 & I \end{bmatrix} \begin{bmatrix} A_{\tilde{G}} & B_{\tilde{G}} \\ C_{\tilde{G}} & D_{\tilde{G}} \end{bmatrix}. \tag{14.8b}$$

It should be noted that eliminating V from (14.8b) using the matrix elimination technique (see, e.g., (1)) yields (14.7b)–(14.7c). This alternate characterization is more suitable for two reasons. First, as will be discussed later in this section, it will allow us to consider a richer set of system uncertainty models. Second, it will allow us to use the matrix variable elimination technique to derive an optimal control scheme in Sect. 14.4, which will be important in our approaches to the output feedback problem developed in Sect. 14.5.

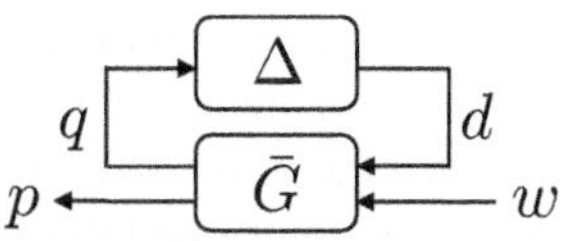

Fig. 14.9 LFT representation of uncertain system

We now turn our attention to analyzing the $\mathscr{H}_2$ performance of the system interconnection shown in Fig. 14.9 where $\bar{G}$ has the state-space realization

$$\bar{G} \sim \left[\begin{array}{c|cc} \bar{A} & \bar{B}_1 & \bar{B}_2 \\ \hline \bar{C}_1 & \bar{D}_{11} & \bar{D}_{12} \\ \bar{C}_2 & \bar{D}_{21} & \bar{D}_{22} \end{array}\right] \tag{14.9}$$

and Δ is a real matrix satisfying $\|\Delta\| \le 1$. Closing the loop yields

$$\begin{aligned} \mathscr{F}_u(\bar{G},\Delta) &\sim \left[\begin{array}{c|c} \bar{A}+\bar{B}_1(I-\Delta\bar{D}_{11})^{-1}\Delta\bar{C}_1 & \bar{B}_2+\bar{B}_1(I-\Delta\bar{D}_{11})^{-1}\Delta\bar{D}_{12} \\ \hline \bar{C}_2+\bar{D}_{21}(I-\Delta\bar{D}_{11})^{-1}\Delta\bar{C}_1 & \bar{D}_{22}+\bar{D}_{21}(I-\Delta\bar{D}_{11})^{-1}\Delta\bar{D}_{12} \end{array}\right] \\ &=: \left[\begin{array}{c|c} A_\Delta & B_\Delta \\ \hline C_\Delta & D_\Delta \end{array}\right]. \end{aligned} \tag{14.10}$$

We are thus interested in determining if $\|\mathscr{F}_u(\bar{G},\Delta)\|_2^2 < \gamma$, $\forall \|\Delta\| \le 1$. Using the characterization of the $\mathscr{H}_2$ norm given by (14.8), we would like to know if there exists $P \succ 0, W, V$ such that $\mathrm{tr}\{W\} < \gamma$ and

$$\begin{bmatrix} P & V^T \\ V & W \end{bmatrix} \succ \begin{bmatrix} A_\Delta & B_\Delta \\ C_\Delta & D_\Delta \end{bmatrix}^T \begin{bmatrix} P & 0 \\ 0 & I \end{bmatrix} \begin{bmatrix} A_\Delta & B_\Delta \\ C_\Delta & D_\Delta \end{bmatrix}, \qquad \forall \|\Delta\| \le 1. \tag{14.11}$$

It should be noted that, although the state-space matrices are a function of Δ, the analysis variables (P, W, and V) are not. Applying the S-procedure (see, e.g., (1)) to

(14.11) yields the equivalent condition that there exists $\tau > 0, P \succ 0, W, V$ such that $\mathrm{tr}\{W\} < \gamma$ and

$$\mathcal{M}(\tau,P,W,V) := \begin{bmatrix} P & 0 & V^T \\ 0 & \tau I & 0 \\ V & 0 & W \end{bmatrix} - \begin{bmatrix} \bar{A} & \bar{B}_1 & \bar{B}_2 \\ \bar{C}_1 & \bar{D}_{11} & \bar{D}_{12} \\ \bar{C}_2 & \bar{D}_{21} & \bar{D}_{22} \end{bmatrix}^T \begin{bmatrix} P & 0 & 0 \\ 0 & \tau I & 0 \\ 0 & 0 & I \end{bmatrix} \begin{bmatrix} \bar{A} & \bar{B}_1 & \bar{B}_2 \\ \bar{C}_1 & \bar{D}_{11} & \bar{D}_{12} \\ \bar{C}_2 & \bar{D}_{21} & \bar{D}_{22} \end{bmatrix} \succ 0 . \tag{14.12}$$

By Schur complements, this is equivalent to the existence of τ, P, W, V such that

$$\bar{\mathcal{M}}(\tau,P,W,V) := \begin{bmatrix} P & \bullet & \bullet & \bullet & \bullet & \bullet \\ 0 & \tau I & \bullet & \bullet & \bullet & \bullet \\ V & 0 & W & \bullet & \bullet & \bullet \\ P\bar{A} & P\bar{B}_1 & P\bar{B}_2 & P & \bullet & \bullet \\ \tau\bar{C}_1 & \tau\bar{D}_{11} & \tau\bar{D}_{12} & 0 & \tau I & \bullet \\ \bar{C}_2 & \bar{D}_{21} & \bar{D}_{22} & 0 & 0 & I \end{bmatrix} \succ 0 . \tag{14.13}$$

These two new conditions, which are equivalent to each other, remove the dependence of the matrices on Δ at the expense of introducing an extra scalar parameter, τ. Since the matrices in the two new conditions do not depend on Δ, they give us computationally tractable means to verify that γ is an upper bound on the worst-case $\mathcal{H}_2$ performance of $\mathcal{F}_u(\bar{G}, \Delta)$ when Δ is a real matrix satisfying $\|\Delta\| \leq 1$.

Rewriting these feasibility problems as an optimization problems to find the smallest upper bound of this type yields

$$\inf_{\tau>0, P\succ 0, W, V} \mathrm{tr}\{W\} \quad \text{s.t.} \quad \mathcal{M}(\tau,P,W,V) \succ 0 \tag{14.14}$$

$$\inf_{\tau, P, W, V} \mathrm{tr}\{W\} \quad \text{s.t.} \quad \bar{\mathcal{M}}(\tau,P,W,V) \succ 0 . \tag{14.15}$$

We will refer to the square root of the value of these optimization problems as the $\mathcal{H}_2$ guaranteed cost of $\bar{G}$. Of these two optimizations, (14.14) is more useful for determining the $\mathcal{H}_2$ guaranteed cost of a given system because its matrix inequalities are smaller in dimension and the optimization problem has fewer dual variables. However, as we will see in Sects. 14.4 and 14.5, (14.15) will be more suitable for control design. Relaxing the strict inequalities in either of these optimization problems to non-strict inequalities results in a SDP. Thus, a reasonable way to solve the $\mathcal{H}_2$ guaranteed cost analysis problem is to relax (14.14) to a SDP then solve the SDP using an appropriate solver.

From the derivation, it is obvious that the $\mathcal{H}_2$ guaranteed cost is an upper bound on the worst-case $\mathcal{H}_2$ performance of the interconnection in Fig. 14.9 when Δ is a real matrix satisfying $\|\Delta\| \leq 1$. What is not immediately apparent, however, is that the $\mathcal{H}_2$ guaranteed cost is also an upper bound on the worst-case $\mathcal{H}_2$ performance of the interconnection in Fig. 14.9 when Δ is only known to be a causal LTI system satisfying $\|\Delta\|_\infty \leq 1$.

14.3.2 Riccati Equation Approach

We begin this subsection by noting that (14.14) can equivalently be expressed as

$$\inf_{\tau>0} J_\tau(\bar{G}) \tag{14.16}$$

where

$$J_\tau(\bar{G}) := \inf_{P\succ 0, W, V} \operatorname{tr}\{W\} \quad \text{s.t.} \quad \mathscr{M}(\tau, P, W, V) \succ 0\,. \tag{14.17}$$

It can be shown that $J_\tau(\bar{G})$ is well-defined, i.e. for a fixed value of $\tau > 0$, $J_\tau(\bar{G})$ is independent of the realization of $\bar{G}$.

Whenever the optimization problem (14.17) is infeasible for a particular value of τ, we will say that $J_\tau(\bar{G}) = \infty$. Defining

$$\begin{bmatrix} \bar{Q} & \bar{S} \end{bmatrix} := \bar{C}_2^T \begin{bmatrix} \bar{C}_2 & \bar{D}_{21} \end{bmatrix} + \tau \bar{C}_1^T \begin{bmatrix} \bar{C}_1 & \bar{D}_{11} \end{bmatrix} \tag{14.18a}$$

$$\begin{bmatrix} \bar{Q}_W & \bar{S}_W \end{bmatrix} := \bar{D}_{22}^T \begin{bmatrix} \bar{D}_{22} & \bar{D}_{21} \end{bmatrix} + \tau \bar{D}_{12}^T \begin{bmatrix} \bar{D}_{12} & \bar{D}_{11} \end{bmatrix} \tag{14.18b}$$

$$\bar{R} := \bar{D}_{21}^T \bar{D}_{21} + \tau(\bar{D}_{11}^T \bar{D}_{11} - I) \tag{14.18c}$$

$$\bar{\phi} := (\bar{A}, \bar{B}_1, \bar{Q}, \bar{R}, \bar{S}) \tag{14.18d}$$

$$\bar{\psi} := (\bar{B}_2, \bar{B}_1, \bar{Q}_W, \bar{R}, \bar{S}_W) \tag{14.18e}$$

it was shown (2) that $J_\tau(\bar{G}) \neq \infty$ if and only if the DARE $\mathscr{R}_{\bar{\phi}}(P) = P$ has a stabilizing solution, P_0, such that $\bar{B}_1^T P_0 \bar{B}_1 + \bar{R} \prec 0$. In this case,

$$J_\tau(\bar{G}) = \operatorname{tr}\{\mathscr{R}_{\bar{\psi}}(P_0)\}\,. \tag{14.19}$$

It was also shown that if $J_\tau(\bar{G}) \neq \infty$ for $\tau = \tau_0$, then $J_\tau \neq \infty$ for all $\tau > \tau_0$. Moreover, there exists a value of τ such that $J_\tau(\bar{G}) \neq \infty$ if and only if the interconnection in Fig. 14.9 is robustly stable over $\|\Delta\|_\infty \leq 1$, i.e. the $\mathscr{H}_\infty$ norm of $\bar{G}$ from d to q is less than 1. This condition is in turn equivalent to the DARE $\mathscr{R}_\rho(P) = P$ having a stabilizing solution P_0 such that $\bar{B}_1^T P_0 \bar{B}_1 + \bar{D}_{11}^T \bar{D}_{11} - I \prec 0$ where

$$\rho := (\bar{A},\, \bar{B}_1,\, \bar{C}_1^T \bar{C}_1,\, \bar{D}_{11}^T \bar{D}_{11} - I,\, \bar{C}_1^T \bar{D}_{11})\,. \tag{14.20}$$

Thus, once we have verified that the optimization problem (14.16) is feasible by using the DARE $\mathscr{R}_\rho(P) = P$, we can always find values of τ for which $J_\tau(\bar{G}) \neq \infty$ simply by making τ increasingly large.

With this in mind, we would like to know how $J_\tau(\bar{G})$ varies as τ varies. First, since minimizing a convex function of several variables over a subset of those variables produces a convex function of the remaining variables, we see that $J_\tau(\bar{G})$ is a convex nonlinear function of τ. Second, since the stabilizing solution of a DARE is analytic in its parameters (5) and $J_\tau(\bar{G})$ is an analytic function of the stabilizing DARE solution, $J_\tau(\bar{G})$ is an analytic function of τ. We will therefore find the global optimal value of $J_\tau(\bar{G})$ by finding a value of τ such that $(\mathrm{d}/\mathrm{d}\tau)(J_\tau(\bar{G})) = 0$.

Although the most straightforward way to find $(\mathrm{d}/\mathrm{d}\tau)(J_\tau(\bar{G}))$ is by directly taking the derivative of the relevant equations with respect to τ, as was done in (2), we will pursue a slightly different approach here which is more computationally efficient. First, we define $\varepsilon := \tau^{-1}$ and

$$\hat{\phi} := (\bar{A},\, \bar{B}_1,\, \varepsilon\bar{Q},\, \varepsilon\bar{R},\, \varepsilon\bar{S}) \tag{14.21a}$$

$$\hat{\psi} := (\bar{B}_2,\, \bar{B}_1,\, \varepsilon\bar{Q}_W,\, \varepsilon\bar{R},\, \varepsilon\bar{S}_W)\,. \tag{14.21b}$$

Multiplying the DARE by ε yields $\mathscr{R}_{\hat{\phi}}(\varepsilon P_0) = \varepsilon P_0$. Also note that $\varepsilon\mathscr{R}_{\bar{\psi}}(P_0) = \mathscr{R}_{\hat{\psi}}(\varepsilon P_0)$. Taking derivatives of these two equations yields, after some algebra, that

$$\begin{aligned}\frac{\mathrm{d}}{\mathrm{d}\varepsilon}(\varepsilon P_0) &= \left(\bar{A}+\bar{B}_1\mathscr{K}_{\bar{\phi}}(P_0)\right)^T \frac{\mathrm{d}}{\mathrm{d}\varepsilon}(\varepsilon P_0)\left(\bar{A}+\bar{B}_1\mathscr{K}_{\bar{\phi}}(P_0)\right)\\ &\quad+\left(\bar{C}_2+\bar{D}_{21}\mathscr{K}_{\bar{\phi}}(P_0)\right)^T\left(\bar{C}_2+\bar{D}_{21}\mathscr{K}_{\bar{\phi}}(P_0)\right)\end{aligned} \tag{14.22a}$$

$$\begin{aligned}\frac{\mathrm{d}}{\mathrm{d}\varepsilon}\left(\varepsilon\mathscr{R}_{\bar{\psi}}(P_0)\right) &= \left(\bar{B}_2+\bar{B}_1\mathscr{K}_{\bar{\psi}}(P_0)\right)^T \frac{\mathrm{d}}{\mathrm{d}\varepsilon}(\varepsilon P_0)\left(\bar{B}_2+\bar{B}_1\mathscr{K}_{\bar{\psi}}(P_0)\right)\\ &\quad+\left(\bar{D}_{22}+\bar{D}_{21}\mathscr{K}_{\bar{\psi}}(P_0)\right)^T\left(\bar{D}_{22}+\bar{D}_{21}\mathscr{K}_{\bar{\psi}}(P_0)\right)\,.\end{aligned} \tag{14.22b}$$

The first of these equations is a discrete Lyapunov equation for $(\mathrm{d}/\mathrm{d}\varepsilon)(\varepsilon P_0)$. Since $\bar{A}+\bar{B}_1\mathscr{K}_{\bar{\phi}}(P_0)$ is stable (by the definition of a stabilizing solution of a DARE), we see that there exists upper triangular $\bar{U}$ such that $\bar{U}^T\bar{U} = (\mathrm{d}/\mathrm{d}\varepsilon)(\varepsilon P_0)$ and we can directly solve for $\bar{U}$ using the `dlyapchol` function in MATLAB. Using this, we express

$$\mathrm{tr}\left\{\frac{\mathrm{d}}{\mathrm{d}\varepsilon}\left(\varepsilon\mathscr{R}_{\bar{\psi}}(P_0)\right)\right\} = \|\bar{U}\left(\bar{B}_2+\bar{B}_1\mathscr{K}_{\bar{\psi}}(P_0)\right)\|_F^2 + \|\bar{D}_{22}+\bar{D}_{21}\mathscr{K}_{\bar{\psi}}(P_0)\|_F^2\,. \tag{14.23}$$

Using the chain rule, we see that

$$\frac{\mathrm{d}}{\mathrm{d}\varepsilon}\left(\varepsilon\mathscr{R}_{\bar{\psi}}(P_0)\right) = \mathscr{R}_{\bar{\psi}}(P_0) + \varepsilon\frac{\mathrm{d}}{\mathrm{d}\varepsilon}\left(\mathscr{R}_{\bar{\psi}}(P_0)\right)\,. \tag{14.24}$$

Thus, taking the derivative of (14.19) and applying the chain rule to the right-hand side yields

$$\frac{\mathrm{d}}{\mathrm{d}\tau}\left(J_\tau(\bar{G})\right) = -\tau^{-2}\mathrm{tr}\left\{\frac{\mathrm{d}}{\mathrm{d}\varepsilon}\left(\mathscr{R}_{\bar{\psi}}(P_0)\right)\right\} = \tau^{-1}\mathrm{tr}\left\{\mathscr{R}_{\bar{\psi}}(P_0) - \frac{\mathrm{d}}{\mathrm{d}\varepsilon}\left(\varepsilon\mathscr{R}_{\bar{\psi}}(P_0)\right)\right\}\,. \tag{14.25}$$

Therefore, we have that

$$\frac{\mathrm{d}}{\mathrm{d}\tau}\left(J_\tau(\bar{G})\right) = \frac{1}{\tau}\left(J_\tau(\bar{G}) - \|\bar{U}\left(\bar{B}_2+\bar{B}_1\mathscr{K}_{\bar{\psi}}(P_0)\right)\|_F^2 - \|\bar{D}_{22}+\bar{D}_{21}\mathscr{K}_{\bar{\psi}}(P_0)\|_F^2\right)\,. \tag{14.26}$$

In comparison to formulas given in (2), using (14.26) to compute the cost derivative is more computationally efficient because it requires fewer matrix multiplications and additions.

The value and derivative of $J_\tau(\bar{G})$ is also useful for generating a lower bound on the optimal value of (14.16). Consider Fig. 14.10, which shows a representative

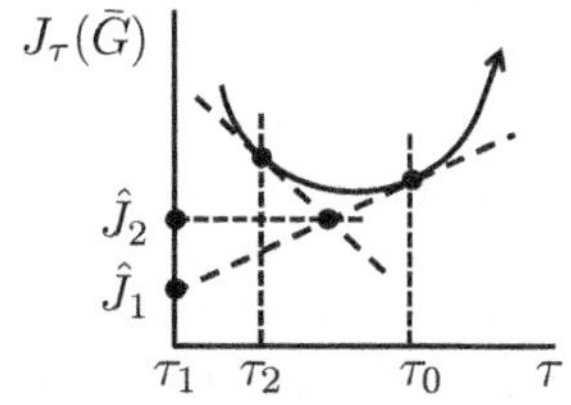

Fig. 14.10 Illustration of lower bound computation in $\mathscr{H}_2$ guaranteed cost analysis

graph of $J_\tau(\bar{G})$ in which τ_0 is known to be an upper bound on the minimizing value of τ. By convexity, if τ_1 is known to be a lower bound on the minimizing value of τ, the value and derivative of $J_\tau(\bar{G})$ at τ_0 gives us the lower bound $\hat{J}_1$. If instead, the value and derivative of $J_\tau(\bar{G})$ at τ_0 and τ_2 is known, we have the lower bound $\hat{J}_2$. These lower bounds are respectively given by

$$\hat{J}_1 = J_{\tau_0} - m_0(\tau_0 - \tau_1) \tag{14.27a}$$

$$\hat{J}_2 = \frac{m_2[m_0(\tau_0 - \tau_2) - (J_{\tau_0} - J_{\tau_2})]}{m_0 - m_2} + J_{\tau_2} \tag{14.27b}$$

where J_{τ_i} is interpreted as $J_\tau(\bar{G})$ evaluated at $\tau = \tau_i$ and m_i is $(\mathrm{d}/\mathrm{d}\tau)(J_\tau(\bar{G}))$ evaluated at $\tau = \tau_i$. It should be noted that the second of these lower bounds is less conservative when it is applicable.

With these results in place, we can easily solve (14.16) using the following methodology:

1. **Check Feasibility:** Check that the DARE $\mathscr{R}_\rho(P) = P$ has a stabilizing solution P_0 such that $\bar{B}_1^T P_0 \bar{B}_1 + \bar{D}_{11}^T \bar{D}_{11} - I \prec 0$.
2. **Find Initial Interval:** Choose $\alpha > 1$. As previously noted, the DARE $\mathscr{R}_{\bar{\phi}}(P) = P$ will have a stabilizing solution with the required properties for $\tau = \alpha^k$, for large enough k. Starting from $k = 0$, iterate over increasing k until a value of $\tau = \alpha^k$ is found such that the DARE has a stabilizing solution P_0 that satisfies $\bar{B}_1^T P_0 \bar{B}_1 + \bar{R} \prec 0$ and $(\mathrm{d}/\mathrm{d}\tau)(J_\tau(\bar{G})) > 0$. Denote this value of τ by τ_u. Note that this corresponds to an upper bound on the optimal value of τ. If $\tau_u = 1$, then 0 is a lower bound on the optimal value of τ, otherwise τ_u/α is a lower bound.
3. **Bisection:** Solve the equation $(\mathrm{d}/\mathrm{d}\tau)(J_\tau(\bar{G})) = 0$ over τ using bisection. Use (14.26) for evaluating $(\mathrm{d}/\mathrm{d}\tau)(J_\tau(\bar{G}))$. Whenever the DARE $\mathscr{R}_{\bar{\phi}}(P) = P$ does not have a stabilizing solution with the required properties for a given value of τ, this corresponds to a lower bound on the optimal value of τ.

In our implementation, we use $\alpha = 100$. Also, except when the lower bound on the optimal value of τ is 0, we use the geometric mean instead of the arithmetic mean to better deal with large intervals in which the optimal value of τ could lie. We use two stopping criteria in our implementation. If we define the relative error as $\nu := 1 - \underline{J}/J_\tau(\bar{G})$ where $\underline{J}$ is the lower bound computed by (14.27), we terminate the algorithm when either $\nu < 10^{-10}$ or 30 iterations have been executed in steps 2–3.

14.3.3 Application to Hard Disk Drives

So far in this section, we have developed two methodologies for analyzing the robust performance of a system with dynamic uncertainty—one based on the solution of a SDP and another based on nonlinear convex optimization involving Riccati equation solutions. We will now use these two methodologies to examine the performance of a closed-loop HDD system for a specified controller.

For the controller, we choose the controller returned by the MATLAB Robust Control Toolbox function `hinfsyn` applied to the optimization problem

$$\inf_{K} \left\| \mathscr{F}_l(\hat{G}_H, K) \right\|_\infty \tag{14.28}$$

where $\hat{G}_H := \mathrm{diag}[0.001, 0.001, 1, 1, 1]G_H$. The controller returned by `hinfsyn`, K^o, contained 19 states and achieved $\|\mathscr{F}_l(\hat{G}_H, K^o)\|_\infty = 0.3831$. For this controller, the nominal $\mathscr{H}_2$ norm of the interconnection in Fig. 14.8 (i.e. with $\Delta = 0$) was 11.6212.

We used three approaches to find the $\mathscr{H}_2$ guaranteed cost of the system $G_{cl} := \mathscr{F}_l(G_H, K^o)$: solving the SDP (14.14) using the `mincx` command, solving (14.14) using SeDuMi and YALMIP, and solving (14.16) using the methodology at the end of Sect. 14.3.2. We will refer to the latter of these approaches as the DARE approach. Using these three approaches to analyze the performance of this 38th-order system yielded the results listed in Table 14.3. Although all three approaches yielded similar values of the $\mathscr{H}_2$ guaranteed cost and the corresponding value of τ, the DARE approach was more than 100 times faster for this system than the other two approaches.

We now look more closely at the results of applying the DARE approach to this problem. In particular, we are interested in the values of $J_\tau(G_{cl})$ and $(\mathrm{d}/\mathrm{d}\tau)(J_\tau(G_{cl}))$ as functions of τ. Figure 14.11 shows a plot of these two quantities for 50 linearly spaced points in the interval $[2,6]$ along with an estimate of $(\mathrm{d}/\mathrm{d}\tau)(J_\tau(G_{cl}))$ obtained by applying the central difference approximation to $J_\tau(G_{cl})$. As we would expect, the curves are smooth, $(\mathrm{d}/\mathrm{d}\tau)(J_\tau(G_{cl}))$ is monotonic non-decreasing, and the computed values of $(\mathrm{d}/\mathrm{d}\tau)(J_\tau(G_{cl}))$ agree with the central difference approximations. It is interesting to note that as τ becomes large, $(\mathrm{d}/\mathrm{d}\tau)(J_\tau(G_{cl}))$ becomes constant and J_τ varies linearly with τ.

The $\mathscr{H}_2$ guaranteed costs computed so far reflect a combination of the "sizes" of the signals y_h, u_v, and u_p. However, it is more meaningful to look at the "sizes" of

Table 14.3 Analysis of closed-loop HDD performance using three approaches

Approach	Optimization Time (s)	$\mathscr{H}_2$ Guaranteed Cost	Optimal τ
`mincx`	147.0465	17.3455	3.3997
SeDuMi	88.1718	17.345	3.3994
DARE	0.8112	17.3451	3.3995

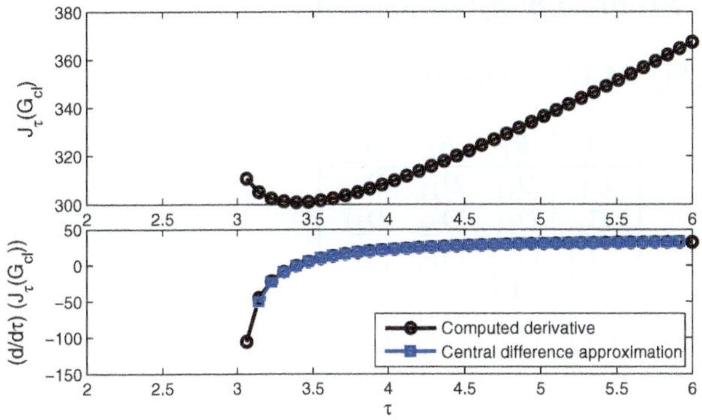

Fig. 14.11 $J_\tau(G_{cl})$, the computed value of $(\mathrm{d}/\mathrm{d}\tau)(J_\tau(G_{cl}))$, and the central difference approximation to $(\mathrm{d}/\mathrm{d}\tau)(J_\tau(G_{cl}))$

these three signals separately. To do so, we removed the outputs we are not interested in and computed the $\mathscr{H}_2$ guaranteed cost. For example, when we were interested in the $\mathscr{H}_2$ guaranteed cost associated with y_h, we removed the outputs u_v and u_p from G_{cl} and then computed the $\mathscr{H}_2$ guaranteed cost of the resulting system using the DARE approach. Doing this for y_h, u_v, and u_p yielded the $\mathscr{H}_2$ guaranteed costs given in Table 14.4. So, for the controller K^o, there very little control effort is used to achieve this level of robust position error performance.

Table 14.4 Closed-loop HDD performance

Signal	$\mathscr{H}_2$ Guaranteed Cost
y_h	17.3450 nm
u_v	0.0440 V
u_m	0.0052 V

14.4 Full Information H2 Guaranteed Cost Control

In this section, we show how to use the analysis results of the previous section to design controllers which optimize the $\mathscr{H}_2$ guaranteed cost. The particular control design problem we consider is the full information control problem in which the controller has access to the state of the system and the disturbances acting on the system. We first present an SDP for determining an optimal controller and then show that, as in the previous section, this convex optimization can be efficiently solved using nonlinear convex optimization involving DARE solutions. The results in this section are taken from (3). Again, for the sake of brevity and clarity of presentation, we do not present the proofs here; interested readers should read the paper cited above.

14.4.1 Semi-definite Programming Approach

We begin by letting G_{fi} in Fig. 14.12 have the known state-space realization

$$G_{fi} \sim \left[\begin{array}{c|ccc} A & B_1 & B_2 & B_3 \\ \hline C_1 & D_{11} & D_{12} & D_{13} \\ C_2 & D_{21} & D_{22} & D_{23} \\ \begin{bmatrix} I \\ 0 \\ 0 \end{bmatrix} & \begin{bmatrix} 0 \\ I \\ 0 \end{bmatrix} & \begin{bmatrix} 0 \\ 0 \\ I \end{bmatrix} & \begin{bmatrix} 0 \\ 0 \\ 0 \end{bmatrix} \end{array}\right]. \tag{14.29}$$

This corresponds to letting

$$y_{fi} = \begin{bmatrix} x_{fi} \\ d \\ w \end{bmatrix} \tag{14.30}$$

in Fig. 14.12 where x_{fi} is the state variable of G_{fi}. In this context, we are interested in finding a controller K_{fi} that achieves the best possible $\mathscr{H}_2$ guaranteed cost using this information. We will refer to this as the full information control problem.

It has been shown for the full information control problem that, given any state-space controller, it is always possible to construct a static gain controller which achieves the same $\mathscr{H}_2$ guaranteed cost. This allows us, without any loss of closed-loop performance, to consider only controllers of the form

$$u = K_x x + K_d d + K_w w \tag{14.31}$$

where K_x, K_d, and K_w are static gains. Equivalently, we make the restriction $K_{fi} = [K_x\ K_d\ K_w]$. Thus, we are interested in closed-loop systems of the form

$$\mathscr{F}_l(G_{fi}, K_{fi}) \sim \left[\begin{array}{c|cc} A + B_3K_x & B_1 + B_3K_d & B_2 + B_3K_w \\ \hline C_1 + D_{13}K_x & D_{11} + D_{13}K_d & D_{12} + D_{13}K_w \\ C_2 + D_{23}K_x & D_{21} + D_{23}K_d & D_{22} + D_{23}K_w \end{array}\right]. \tag{14.32}$$

In particular, we would like to find K_x, K_d, and K_w such that $\mathscr{F}_l(G_{fi}, K_{fi})$ achieves the best possible $\mathscr{H}_2$ guaranteed cost.

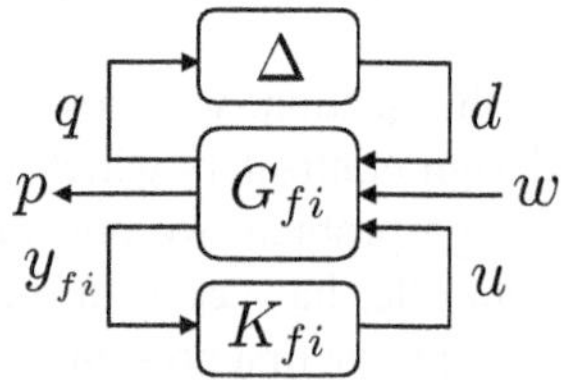

Fig. 14.12 LFT representation of $\mathscr{H}_2$ guaranteed cost control structure

Using the change of variables

$$\hat{P} := P^{-1} \tag{14.33a}$$
$$\hat{V} := VP^{-1} \tag{14.33b}$$
$$\varepsilon := \tau^{-1} \tag{14.33c}$$
$$\hat{K}_x := K_x P^{-1} \tag{14.33d}$$
$$\hat{K}_d := \tau^{-1} K_d \tag{14.33e}$$

with the $\mathscr{H}_2$ guaranteed cost characterization (14.15) applied to $\mathscr{F}_l(G_{fi}, K_{fi})$ yields, after multiplying $\bar{\mathscr{M}}$ on the left and right by $\mathrm{diag}[P^{-1}, \tau^{-1}I, I, P^{-1}, \tau^{-1}I, I]$, the optimization problem

$$\inf_{\varepsilon,\hat{P},W,\hat{V},\hat{K}_x,\hat{K}_d,K_w} \quad \mathrm{tr}\{W\} \quad \text{s.t.} \quad \mathscr{M}_{fi}(\varepsilon,\hat{P},W,\hat{V},\hat{K}_x,\hat{K}_d,K_w) \succ 0 \tag{14.34}$$

where

$$\mathscr{M}_{fi} := \begin{bmatrix} \hat{P} & \bullet & \bullet & \bullet & \bullet & \bullet \\ 0 & \varepsilon I & \bullet & \bullet & \bullet & \bullet \\ \hat{V} & 0 & W & \bullet & \bullet & \bullet \\ A\hat{P} + B_3\hat{K}_x & \varepsilon B_1 + B_3\hat{K}_d & B_2 + B_3 K_w & \hat{P} & \bullet & \bullet \\ C_1\hat{P} + D_{13}\hat{K}_x & \varepsilon D_{11} + D_{13}\hat{K}_d & D_{12} + D_{13}K_w & 0 & \varepsilon I & \bullet \\ C_2\hat{P} + D_{23}\hat{K}_x & \varepsilon D_{21} + D_{23}\hat{K}_d & D_{22} + D_{23}K_w & 0 & 0 & I \end{bmatrix}. \tag{14.35}$$

For any feasible $\varepsilon, \hat{P}, W, \hat{V}, \hat{K}_x, \hat{K}_d, K_w$, a controller which achieves the squared $\mathscr{H}_2$ guaranteed cost $\mathrm{tr}\{W\}$ (or better) is given by

$$K_{fi} = \begin{bmatrix} \hat{K}_x\hat{P}^{-1} & \varepsilon^{-1}\hat{K}_d & K_w \end{bmatrix}. \tag{14.36}$$

If the strict inequality in (14.34) is relaxed to a non-strict inequality, the optimization becomes a SDP. Thus, a reasonable way to solve the optimal full information control problem is to relax (14.34) to a SDP, solve the SDP using an appropriate solver, and then reconstruct the controller using (14.36).

14.4.2 Riccati Equation Approach

In this subsection, we examine the optimal full information control problem when the following so-called regularity conditions hold:

- $D_{13}^T D_{13} + D_{23}^T D_{23}$ is invertible
- (A, B_3) is d-stabilizable
- (A_{fi}, C_{fi}) is d-detectable for all $\varepsilon > 0$ where

$$A_{fi} := A - B_3(D_{13}^T D_{13} + \varepsilon D_{23}^T D_{23})^{-1}(D_{13}^T C_1 + \varepsilon D_{23}^T C_2) \tag{14.37a}$$
$$C_{fi} := \begin{bmatrix} C_1 \\ C_2 \end{bmatrix} - \begin{bmatrix} D_{13} \\ D_{23} \end{bmatrix} (D_{13}^T D_{13} + \varepsilon D_{23}^T D_{23})^{-1}(D_{13}^T C_1 + \varepsilon D_{23}^T C_2). \tag{14.37b}$$

As in the previous section, we begin by writing the relevant convex optimization problem as a nested optimization problem. In particular, we express (14.34) as

$$\inf_{\varepsilon>0} J_{fi,\varepsilon} \tag{14.38}$$

where

$$J_{fi,\varepsilon} := \inf_{\hat{P}\succ 0, W, \hat{V}, \hat{K}_x, \hat{K}_d, K_w} \operatorname{tr}\{W\} \quad \text{s.t.} \quad \mathscr{M}_{fi}(\varepsilon, \hat{P}, W, \hat{V}, \hat{K}_x, \hat{K}_d, K_w) \succ 0\,. \tag{14.39}$$

Whenever (14.39) is infeasible for a particular value of ε, we will say that $J_{fi,\varepsilon} = \infty$. Note that, since $\varepsilon = \tau^{-1}$, performing the optimization (14.39) is the same as optimizing $J_\tau(\mathscr{F}_l(G_{fi}, K_{fi}))$ via choice of K_{fi} for $\tau = \varepsilon^{-1}$.

Defining

$$\begin{bmatrix} Q | S \end{bmatrix} := C_1^T \begin{bmatrix} C_1 | D_{11}\ D_{13} \end{bmatrix} + \varepsilon C_2^T \begin{bmatrix} C_2 | D_{21}\ D_{23} \end{bmatrix} \tag{14.40a}$$

$$\begin{bmatrix} Q_W | S_W \end{bmatrix} := D_{12}^T \begin{bmatrix} D_{12} | D_{11}\ D_{13} \end{bmatrix} + \varepsilon D_{22}^T \begin{bmatrix} D_{22} | D_{21}\ D_{23} \end{bmatrix} \tag{14.40b}$$

$$R := \begin{bmatrix} R_{11} & \bullet \\ R_{21} & R_{22} \end{bmatrix} := \begin{bmatrix} D_{11}^T D_{11} - I & \bullet \\ D_{13}^T D_{11} & D_{13}^T D_{13} \end{bmatrix} + \varepsilon \begin{bmatrix} D_{21}^T D_{21} & \bullet \\ D_{23}^T D_{21} & D_{23}^T D_{23} \end{bmatrix} \tag{14.40c}$$

$$\phi := (A, [B_1\ B_3], Q, R, S) \tag{14.40d}$$

$$\psi := (B_2, [B_1\ B_3], Q_W, R, S_W) \tag{14.40e}$$

it was shown (3) that $J_{fi,\varepsilon} \neq \infty$ if and only if the DARE $\mathscr{R}_\phi(P) = P$ has a stabilizing solution $P_0 \succeq 0$ such that the factorization

$$\begin{bmatrix} B_1^T \\ B_3^T \end{bmatrix} P_0 \begin{bmatrix} B_1\ B_3 \end{bmatrix} + \begin{bmatrix} R_{11} & \bullet \\ R_{21} & R_{22} \end{bmatrix} = \begin{bmatrix} T_{21}^T T_{21} - T_{11}^T T_{11} & \bullet \\ T_{22}^T T_{21} & T_{22}^T T_{22} \end{bmatrix} \tag{14.41}$$

exists with T_{11} and T_{22} invertible. In this case, $J_{fi,\varepsilon} = \varepsilon^{-1} \operatorname{tr}\{\mathscr{R}_\psi(P_0)\}$ and the optimal values of the controller parameters are

$$\begin{bmatrix} K_x^o\ K_d^o\ K_w^o \end{bmatrix} := -(B_3^T P_0 B_3 + R_{22})^{-1} \begin{bmatrix} B_3^T P_0\ D_{13}^T\ \varepsilon D_{23}^T \end{bmatrix} \begin{bmatrix} A & B_1 & B_2 \\ C_1 & D_{11} & D_{12} \\ C_2 & D_{21} & D_{22} \end{bmatrix}. \tag{14.42}$$

If the factorization (14.41) exists, then it can be formed using the following steps:

1. Perform the Cholesky factorization $T_{22}^T T_{22} = B_3^T P_0 B_3 + R_{22}$.
2. Choose $T_{21} = T_{22}^{-T} (B_3^T P_0 B_1 + R_{21})$.
3. Perform the Cholesky factorization $T_{11}^T T_{11} = T_{21}^T T_{21} - B_1^T P_0 B_1 - R_{11}$.

One of the key steps in deriving this result from (14.38) was using matrix variable elimination to eliminate the controller parameters from the optimization problem. This would not have been possible if we had used the $\mathscr{H}_2$ norm characterization (14.7) instead of the characterization (14.8).

It was also shown (3) that if $J_{fi,\varepsilon_0} \neq \infty$, then $J_{fi,\varepsilon} \neq \infty$ for all $\varepsilon \in (0, \varepsilon_0)$. Note that this corresponds to $J_{fi,\varepsilon} \neq \infty$ for all $\varepsilon^{-1} > \varepsilon_0^{-1}$. Unlike the previous section, there is no Riccati equation result that allows us to check whether or not there exists a value of ε such that $J_{fi,\varepsilon} \neq \infty$. This is the case because the existence of such a value of ε is equivalent to the solvability of a full information $\mathscr{H}_\infty$ control problem which might not be solvable via a Riccati equation. Despite this, if (14.34) is feasible, we can always find values of $\varepsilon > 0$ for which the inner optimization problem in (14.38) is feasible simply by decreasing $\varepsilon > 0$ until $J_{fi,\varepsilon} \neq \infty$.

With this in mind, we would like to know how $J_{fi,\varepsilon}$ varies as ε varies. Using the same reasoning as in the previous section, we see that $J_{fi,\varepsilon}$ is an analytic function of ε. We will therefore find the global optimal value of $J_{fi,\varepsilon}$ by finding a value of ε such that $(\mathrm{d}/\mathrm{d}\varepsilon)(J_{fi,\varepsilon}) = 0$. Implicitly differentiating the DARE $\mathscr{R}_\phi(P_0) = P_0$ and differentiating the expression for $\mathscr{R}_\psi(P_0)$ with respect to ε yields

$$\begin{aligned}\frac{\mathrm{d}}{\mathrm{d}\varepsilon}(P_0) &= \left(A + \left[B_1\ B_3\right]\mathscr{K}_\phi(P_0)\right)^T \frac{\mathrm{d}}{\mathrm{d}\varepsilon}(P_0)\left(A + \left[B_1\ B_3\right]\mathscr{K}_\phi(P_0)\right)\\ &\quad + \left(C_2 + \left[D_{21}\ D_{23}\right]\mathscr{K}_\phi(P_0)\right)^T\left(C_2 + \left[D_{21}\ D_{23}\right]\mathscr{K}_\phi(P_0)\right) \qquad (14.43a)\\ \frac{\mathrm{d}}{\mathrm{d}\varepsilon}(\mathscr{R}_\psi(P_0)) &= \left(B_2 + \left[B_1\ B_3\right]\mathscr{K}_\psi(P_0)\right)^T \frac{\mathrm{d}}{\mathrm{d}\varepsilon}(P_0)\left(B_2 + \left[B_1\ B_3\right]\mathscr{K}_\psi(P_0)\right)\\ &\quad + \left(D_{22} + \left[D_{21}\ D_{23}\right]\mathscr{K}_\psi(P_0)\right)^T\left(D_{22} + \left[D_{21}\ D_{23}\right]\mathscr{K}_\psi(P_0)\right) \qquad (14.43b)\end{aligned}$$

The first of these equations is a discrete Lyapunov equation for $(\mathrm{d}/\mathrm{d}\varepsilon)(P_0)$. Since $A + [B_1\ B_3]\mathscr{K}_\phi(P_0)$ is stable (by the definition of a stabilizing solution of a DARE), we see that there exists upper triangular U such that $U^T U = (\mathrm{d}/\mathrm{d}\varepsilon)(\varepsilon P_0)$ and we can directly solve for U using the `dlyapchol` function in MATLAB. Using this, we express

$$\frac{\mathrm{d}}{\mathrm{d}\varepsilon}(J_{fi,\varepsilon}) = -\varepsilon^{-2}\mathrm{tr}\{\mathscr{R}_\psi(P_0)\} + \varepsilon^{-1}\mathrm{tr}\left\{\frac{\mathrm{d}}{\mathrm{d}\varepsilon}(\mathscr{R}_\psi(P_0))\right\} \qquad (14.44)$$

which implies that

$$\begin{aligned}\frac{\mathrm{d}}{\mathrm{d}\varepsilon}(J_{fi,\varepsilon}) &= \varepsilon^{-1}\Big(\left\|U\left(B_2 + \left[B_1\ B_3\right]\mathscr{K}_\phi(P_0)\right)\right\|_F^2\\ &\quad + \left\|D_{22} + \left[D_{21}\ D_{23}\right]\mathscr{K}_\phi(P_0)\right\|_F^2 - J_{fi,\varepsilon}\Big)\,. \qquad (14.45)\end{aligned}$$

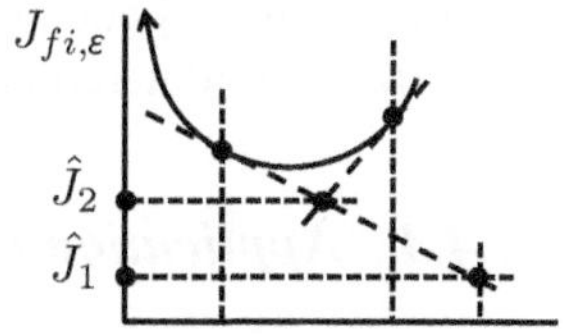

Fig. 14.13 Illustration of lower bound computation in optimal full information $\mathscr{H}_2$ guaranteed cost control

The value and derivative of $J_{fi,\varepsilon}$ is also useful for generating a lower bound on the optimal value of (14.38). Consider Fig. 14.13, which shows a representative graph of $J_{fi,\varepsilon}$ in which ε_0 is known to be a lower bound on the minimizing value of ε. By convexity, if ε_1 is known to be an upper bound on the minimizing value of ε, the value and derivative of $J_{fi,\varepsilon}$ at ε_0 gives us the lower bound $\hat{J}_1$. If instead, the value and derivative of $J_{fi,\varepsilon}$ at ε_0 and ε_2 is known, we have the lower bound $\hat{J}_2$. These lower bounds are respectively given by

$$\hat{J}_1 = J_{fi,\varepsilon_0} + \bar{m}_0(\varepsilon_1 - \varepsilon_2) \tag{14.46a}$$

$$\hat{J}_2 = \frac{\bar{m}_2[\bar{m}_0(\varepsilon_2 - \varepsilon_0) - (J_{fi,\varepsilon_2} - J_{fi,\varepsilon_0})]}{\bar{m}_2 - \bar{m}_0} + J_{fi,\varepsilon_2} \tag{14.46b}$$

where J_{fi,ε_i} is interpreted as $J_{fi,\varepsilon}$ evaluated at $\varepsilon = \varepsilon_i$ and $\bar{m}_i$ is $(\mathrm{d}/\mathrm{d}\varepsilon)(J_{fi,\varepsilon})$ evaluated at $\varepsilon = \varepsilon_i$. It should be noted that the second of these lower bounds is less conservative when it is applicable.

With these results in place, we can easily solve (14.38) using the following methodology:

1. **Find Initial Interval:** Choose $\alpha > 1$. Check if $J_{fi,\varepsilon} \neq \infty$ and $(\mathrm{d}/\mathrm{d}\varepsilon)(J_{fi,\varepsilon}) < 0$ when $\varepsilon = 1$. If so, start from $k = 1$ and increment k until either of these conditions fails to be met when $\varepsilon = \alpha^k$. Denoting the corresponding value of ε as ε_u, there exists an optimal value of ε in the interval $(\alpha^{-1}\varepsilon_u, \varepsilon_u)$.
   If instead either $J_{fi,\varepsilon} = \infty$ or $(\mathrm{d}/\mathrm{d}\varepsilon)(J_{fi,\varepsilon}) > 0$ when $\varepsilon = 1$, start from $k = 1$ and increment k until $J_{fi,\varepsilon} \neq \infty$ and $(\mathrm{d}/\mathrm{d}\varepsilon)(J_{fi,\varepsilon}) < 0$ when $\varepsilon = \alpha^{-k}$. Denoting the corresponding value of ε as ε_l, there exists an optimal value of ε in the interval $(\varepsilon_l, \alpha\varepsilon_l)$.
2. **Bisection:** Solve the equation $(\mathrm{d}/\mathrm{d}\varepsilon)(J_{fi,\varepsilon}) = 0$ over ε using bisection. Use (14.45) for evaluating $(\mathrm{d}/\mathrm{d}\varepsilon)(J_{fi,\varepsilon})$. Whenever, for a particular value of ε, the DARE $\mathscr{R}_\phi(P) = P$ does not have a stabilizing solution $P_0 \succeq 0$ such that the factorization (14.41) exists for invertible T_{11} and T_{22}, this value of ε is an upper bound on the optimal value of ε.
3. **Controller Construction:** For the value of ε which yielded the smallest cost, reconstruct the corresponding controller using (14.42).

In our implementation, we use $\alpha = 100$ and, in the bisection step, we use the geometric mean instead of the arithmetic mean. We use two stopping criteria in our implementation. If we define the relative error as $\nu_{fi} := 1 - \underline{J}_{fi}/J_{fi,\varepsilon}$ where $\underline{J}_{fi}$ is the lower bound computed by (14.46), we terminate the algorithm when either $\nu_{fi} < 10^{-10}$ or 30 iterations have been executed in steps 1–2.

14.4.3 Application to Hard Disk Drives

So far in this section, we have developed two methodologies for designing an optimal full information controller in terms of its $\mathscr{H}_2$ guaranteed cost—one based on the solution of a SDP and another based on nonlinear convex optimization involving Riccati equation solutions.

Although we would like to apply these methodologies to the model G_H introduced in Sect. 14.2, it is not in the form of a full information control problem. However, we can create a full information control problem by replacing the measurements generated by the model, y, by the vector $[x^T \; d^T \; w_a^T \; w_r^T \; w_n^T]^T$ where x is the state of G_H. The model was not balanced after redefining the measurement vector.

We used three approaches to find optimal full information $\mathscr{H}_2$ guaranteed cost controllers for this system: solving the SDP (14.34) using the `mincx` command, solving (14.34) using SeDuMi and YALMIP, and solving (14.38) using the methodology at the end of Sect. 14.4.2. We will refer to the latter of these approaches as the DARE approach. Using these three approaches on the model of G_H with the redefined measurement vector yielded the results listed in Table 14.5. Although the DARE approach and the `mincx` approach yielded similar accuracy, the SeDuMi approach seemed to yield better accuracy. However, analyzing the closed-loop systems using the DARE approach in Sect. 14.3.2 showed that the actual $\mathscr{H}_2$ guaranteed cost performance achieved by controller designed using the SeDuMi approach was only 5.9671, whereas the other two controllers achieved the costs stated in Table 14.5. Thus, for this system, SeDuMi actually had the worst numerical accuracy of the three approaches. In terms of efficiency, the DARE approach was more than 10 times faster than the other two approaches. The difference in computation time is less dramatic than the difference in Sect. 14.3.3 because the closed-loop systems here have 19 states whereas the system analyzed in Sect. 14.3.3 had 38 states.

We now look more closely at the DARE approach to this problem. In particular, we are interested in the values of $J_{fi,\varepsilon}$ and $(\mathrm{d}/\mathrm{d}\varepsilon)(J_{fi,\varepsilon})$ as functions of ε. Figure 14.14 shows a plot of these two quantities for 50 linearly spaced points in the interval $[0, 1.2]$ along with an estimate of $(\mathrm{d}/\mathrm{d}\varepsilon)(J_{fi,\varepsilon})$ obtained by applying the central difference approximation to $J_{fi,\varepsilon}$. As we would expect, the curves are smooth, $(\mathrm{d}/\mathrm{d}\varepsilon)(J_{fi,\varepsilon})$ is monotonic non-decreasing, and the computed values of $(\mathrm{d}/\mathrm{d}\varepsilon)(J_{fi,\varepsilon})$ agree with the central difference approximations. Also, for this particular example, $J_{fi,\varepsilon}$ is monotonic decreasing over the chosen values of ε.

The $\mathscr{H}_2$ guaranteed costs computed so far reflect a combination of the "sizes" of the signals y_h, u_v, and u_p. However, it is more meaningful to look at the "sizes" of these three signals separately. As in Sect. 14.3.3, we remove the outputs we are not interested in and compute the relevant $\mathscr{H}_2$ guaranteed costs associated with each output signal. Doing this for the closed-loop system designed using the DARE approach yields the $\mathscr{H}_2$ guaranteed cost given in Table 14.6. Thus, we see that the

Table 14.5 Analysis of closed-loop HDD performance using three approaches

Approach	Optimization Time (s)	$\mathscr{H}_2$ Guaranteed Cost	Optimal ε
DARE	0.37	5.9604	1
SeDuMi	4.4616	5.938	1.0005
`mincx`	27.5342	5.9604	1

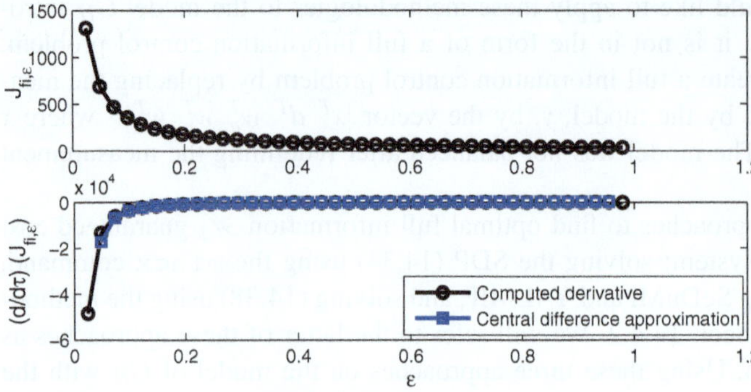

Fig. 14.14 $J_{fi,\varepsilon}$, the computed value of $(\mathrm{d}/\mathrm{d}\varepsilon)(J_{fi,\varepsilon})$, and the central difference approximation to $(\mathrm{d}/\mathrm{d}\varepsilon)(J_{fi,\varepsilon})$

Table 14.6 Analysis of closed-loop HDD performance for each closed-loop signal

Signal	$\mathscr{H}_2$ Guaranteed Cost
y_h	5.9551 nm
u_v	0.2439 V
u_p	0.0635 V

control effort is rather small for this controller. This suggests that we should deemphasize the control effort in the cost function in order to design controllers which achieve a higher level of performance.

14.5 Output Feedback H2 Guaranteed Cost Control

In this section, we consider the optimal output feedback $\mathscr{H}_2$ guaranteed cost control problem. We first present a non-convex optimization problem for determining an optimal controller and a solution heuristic which is based on the solution of SDPs. We then give an algorithm which exploits Riccati equation structure to give a more computationally efficient heuristic for finding an optimal controller. For the sake of brevity and clarity of presentation, we do not give full proofs here; these will be presented in future papers.

14.5.1 Sequential Semi-definite Programming Approach

We now consider an optimal $\mathscr{H}_2$ guaranteed cost control problem of the form shown in Fig. 14.15(a) on p. 262 in which G has the known state-space realization

$$G \sim \left[\begin{array}{c|ccc} A & B_1 & B_2 & B_3 \\ \hline C_1 & D_{11} & D_{12} & D_{13} \\ C_2 & D_{21} & D_{22} & D_{23} \\ C_3 & D_{31} & D_{32} & 0 \end{array}\right] . \tag{14.47}$$

Note that we are choosing many of the state-space entries to be the same as in the previous section. The only difference is the choice of information that is available to the controller. In this context, we are interested in finding a controller K which achieves the best possible $\mathscr{H}_2$ guaranteed cost using only the measurements y. We will refer to this as the output feedback control problem.

Using techniques similar to the one used in (3), it can be shown that, given any state-space controller, it is always possible to construct a state-space controller with the same number of states as G that achieves the same $\mathscr{H}_2$ guaranteed cost. This allows us, without any loss of closed-loop performance, to consider only state-space controllers of the form

$$K \sim \left[\begin{array}{c|c} A_c & B_c \\ \hline C_c & D_c \end{array}\right] \tag{14.48}$$

where A_c has the same dimensions as A.

14.5.1.1 Coordinate Descent Approach

We are interested in solving the optimization problem

$$\inf_{\tau>0,K} J_\tau(\mathscr{F}_l(G,K)) \tag{14.49}$$

where K has the realization (14.48). Applying the Lyapunov shaping paradigm (15) to the characterization (14.15) for the system $\mathscr{F}_l(G,K)$ yields the optimization problem

$$\inf_{\tau,X,Y,W,\hat{V}_1,\hat{V}_2,\hat{A},\hat{B},\hat{C},\hat{D}} \operatorname{tr}\{W\} \quad \text{s.t.} \quad \check{\mathscr{M}}(\tau,X,Y,W,\hat{V}_1,\hat{V}_2,\hat{A},\hat{B},\hat{C},\hat{D}) \succ 0 \tag{14.50}$$

where

$$\check{\mathscr{M}} := \begin{bmatrix} X & \bullet & \bullet & \bullet & \bullet & \bullet & \bullet & \bullet \\ I & Y & \bullet & \bullet & \bullet & \bullet & \bullet & \bullet \\ 0 & 0 & \tau I & \bullet & \bullet & \bullet & \bullet & \bullet \\ \hat{V}_1 & \hat{V}_2 & 0 & W & \bullet & \bullet & \bullet & \bullet \\ \check{M}_{11} & \check{M}_{12} & \check{M}_{13} & \check{M}_{14} & X & \bullet & \bullet & \bullet \\ \check{M}_{21} & \check{M}_{22} & \check{M}_{23} & \check{M}_{24} & I & Y & \bullet & \bullet \\ \tau\check{M}_{31} & \tau\check{M}_{32} & \tau\check{M}_{33} & \tau\check{M}_{34} & 0 & 0 & \tau I & \bullet \\ \check{M}_{41} & \check{M}_{42} & \check{M}_{43} & \check{M}_{44} & 0 & 0 & 0 & I \end{bmatrix} \tag{14.51a}$$

$$\begin{bmatrix} \check{M}_{11} & \cdots & \check{M}_{14} \\ \vdots & \ddots & \vdots \\ \check{M}_{41} & \cdots & \check{M}_{44} \end{bmatrix} := \begin{bmatrix} AX + B_3\hat{C} & A + B_3\hat{D}C_3 & B_1 + B_3\hat{D}D_{31} & B_2 + B_3\hat{D}D_{32} \\ \hat{A} & YA + \hat{B}C_3 & YB_1 + \hat{B}D_{31} & YB_2 + \hat{B}D_{32} \\ C_1X + D_{13}\hat{C} & C_1 + D_{13}\hat{D}C_3 & D_{11} + D_{13}\hat{D}D_{31} & D_{12} + D_{13}\hat{D}D_{32} \\ C_2X + D_{23}\hat{C} & C_2 + D_{23}\hat{D}C_3 & D_{21} + D_{23}\hat{D}D_{31} & D_{22} + D_{23}\hat{D}D_{32} \end{bmatrix}. \tag{14.51b}$$

For any feasible $\tau, X, Y, W, \hat{V}_1, \hat{V}_2, \hat{A}, \hat{B}, \hat{C}, \hat{D}$, a controller that achieves the squared $\mathscr{H}_2$ guaranteed cost $\mathrm{tr}\{W\}$ (or better) is given by

$$K \sim \left[\begin{array}{c|c} N^{-1}[\hat{A} - YAX - \hat{B}C_3X - YB_3(\hat{C} - \hat{D}C_3X)]M^{-T} & N^{-1}(\hat{B} - YB_3\hat{D}) \\ \hline (\hat{C} - \hat{D}C_3X)M^{-T} & \hat{D} \end{array}\right] \tag{14.52}$$

where M and N are chosen so that

$$MN^T = I - XY \,. \tag{14.53}$$

Although we use the QR decomposition to factorize $I - XY$ in our implementation, a pivoted LU decomposition would be equally suitable.

The optimization (14.50) is a non-convex optimization because the matrix inequality is nonlinear in the optimization parameters; the products τX, $\tau\hat{C}$ and $\tau\hat{D}$ appear in the matrix inequality. Thus, the matrix inequality is a bilinear matrix inequality (BMI) and the optimization (14.50) is a BMI optimization problem. If the value of τ is fixed or the values of X, $\hat{C}$, and $\hat{D}$ are fixed, then the BMI becomes an LMI. In either of these cases, if the strict inequalities in (14.50) are relaxed to non-strict inequalities, the optimization becomes a SDP. Thus, for a given initial guess for τ, a reasonable heuristic for solving (14.50) is to alternate between solving (14.50) for fixed τ and solving (14.50) for fixed X, $\hat{C}$, and $\hat{D}$.

There are two challenges in using this approach. The first difficultly we encounter is the difficulty of selecting the initial value of τ; since BMI optimization problems are non-convex, the selection of a "good" initial iterate is especially critical. The second difficulty that we encounter is in reconstructing the controller using (14.52). In particular, since the controller reconstruction depends on both M^{-1} and N^{-1}, we see that the controller reconstruction will be ill-conditioned if $I - XY$ is ill-conditioned with respect to inversion. We will show in the next two sections that it is possible to deal with both of these problem using semi-definite programming.

14.5.1.2 Initial Controller Design

We now examine the problem of finding an initial value of τ. To deal with this problem, we follow the approach used in (10) in which the solution of two SDPs yield initial values of all optimization parameters. Since we are only interested in the initial value of τ, we will not explicitly construct initial values for the remaining optimization parameters in this paper.

The first convex optimization is a state feedback control design. This is done by relaxing (14.34) to an SDP, and then solving it with the additional constraints that $\hat{K}_d = 0$ and $K_w = 0$. This yields a state feedback control law $u = K_x x$, where K_x is a static gain and x is the state variable of G.

In the second convex optimization, we first restrict the class of controllers to ones which have $C_c = K_x$ and $D_c = 0$. If the state variable of the controller is interpreted as an estimate of the state of G, this restriction can be interpreted as imposing a "separation structure" on the controller. We then form a realization of $\mathscr{F}_l(G,K)$ whose state is given by $[x^T \, (x - x_c)^T]^T$ where x_c is the state variable of K. Using the characterization (14.15) for this realization with the restriction $P = \mathrm{diag}[\tilde{X}, \tilde{Y}]$ and the change of variables

$$\begin{bmatrix} \tilde{A} & \tilde{B} \end{bmatrix} := \tilde{Y} \begin{bmatrix} A_c & B_c \end{bmatrix} \tag{14.54}$$

yields the optimization problem

$$\inf_{\tau, \tilde{X}, \tilde{Y}, W, V_1, V_2, \tilde{A}, \tilde{B}} \mathrm{tr}\{W\} \quad \text{s.t.} \quad \begin{bmatrix} \tilde{X} & \bullet & \bullet & \bullet & \bullet & \bullet & \bullet & \bullet \\ 0 & \tilde{Y} & \bullet & \bullet & \bullet & \bullet & \bullet & \bullet \\ 0 & 0 & \tau I & \bullet & \bullet & \bullet & \bullet & \bullet \\ V_1 & V_2 & 0 & W & \bullet & \bullet & \bullet & \bullet \\ \tilde{M}_{11} & \tilde{M}_{12} & \tilde{M}_{13} & \tilde{M}_{14} & \tilde{X} & \bullet & \bullet & \bullet \\ \tilde{M}_{21} & \tilde{M}_{22} & \tilde{M}_{23} & \tilde{M}_{24} & 0 & \tilde{Y} & \bullet & \bullet \\ \tilde{M}_{31} & \tilde{M}_{32} & \tilde{M}_{33} & \tilde{M}_{34} & 0 & 0 & \tau I & \bullet \\ \tilde{M}_{41} & \tilde{M}_{42} & \tilde{M}_{43} & \tilde{M}_{44} & 0 & 0 & 0 & I \end{bmatrix} \succ 0 \tag{14.55}$$

where

$$\begin{bmatrix} \tilde{M}_{11} & \cdots & \tilde{M}_{14} \\ \vdots & \ddots & \vdots \\ \tilde{M}_{41} & \cdots & \tilde{M}_{44} \end{bmatrix} := \begin{bmatrix} \tilde{X} A_{cl} & -\tilde{X} B_3 K_x & \tilde{X} B_1 & \tilde{X} B_2 \\ \tilde{Y} A_{cl} - \tilde{A} - \tilde{B} C_3 & \tilde{A} - \tilde{Y} B_3 K_x & \tilde{Y} B_1 - \tilde{B} D_{31} & \tilde{Y} B_2 - \tilde{B} D_{32} \\ \tau(C_1 + D_{13} K_x) & -\tau D_{13} K_x & \tau D_{11} & \tau D_{12} \\ C_2 + D_{23} K_x & -D_{23} K_x & D_{21} & D_{22} \end{bmatrix} \tag{14.56}$$

and $A_{cl} := A + B_3 K_x$. Although it is not used by our algorithm developed here, it is worth mentioning that, for any feasible $\tau, \tilde{X}, \tilde{Y}, W, V_1, V_2, \tilde{A}, \tilde{B}$, a controller which achieves the squared $\mathscr{H}_2$ guaranteed cost $\mathrm{tr}\{W\}$ or better is given by

$$K \sim \left[\begin{array}{c|c} \tilde{Y}^{-1} \tilde{A} & \tilde{Y}^{-1} \tilde{B} \\ \hline K_x & 0 \end{array} \right] . \tag{14.57}$$

If the strict inequalities in (14.55) are relaxed to non-strict inequalities, the optimization becomes a SDP. The value of τ which results from solving this optimization problem is a suitable initial value of τ for the BMI optimization problem (14.50).

14.5.1.3 Conditioning of the Controller Reconstruction Step

As mentioned in Sect. 14.5.1.1, we would like to make the matrix $I - XY$ well-conditioned with respect to inversion. Define

$$S := \begin{bmatrix} X & I \\ I & Y \end{bmatrix} . \tag{14.58}$$

Since $S \succ 0$ for any feasible iterate of (14.50), we see by Schur complements that $Y \succ 0$ and $X - Y^{-1} \succ 0$ for any feasible iterate of (14.50). In particular, Y and $X - Y^{-1}$ are invertible. Exploiting the invertibility of these matrices yields

$$S^{-1} = \begin{bmatrix} (X - Y^{-1})^{-1} & (I - YX)^{-1} \\ (I - XY)^{-1} & Y^{-1} + Y^{-1}(X - Y^{-1})^{-1}Y^{-1} \end{bmatrix}. \tag{14.59}$$

Since $(I - XY)^{-1}$ explicitly appears in the expression for S^{-1}, we see that if S is easy to invert then $I - XY$ must also be easy to invert. Since S is a positive definite matrix for any feasible iterate of (14.50), we would therefore like to make the ratio of its largest eigenvalue to its smallest eigenvalue (i.e. its condition number) as small as possible.

Since the condition number of $S \succ 0$ is less than κ if and only if there exists $t > 0$ such that $I \prec tS \prec \kappa I$, we see that a reasonable way to improve the conditioning of the controller reconstruction is to solve

$$\inf_{t,\kappa,tX,tY,tW,t\hat{V}_1,t\hat{V}_2,t\hat{A},t\hat{B},t\hat{C},t\hat{D}} \kappa \quad \text{s.t.} \quad t\check{\mathcal{M}} \succ 0, \; I \prec tS \prec \kappa I, \; \operatorname{tr}\{tW\} < t\gamma \tag{14.60}$$

where γ is some acceptable level of $\mathcal{H}_2$ guaranteed cost performance for the closed-loop system. Note that we have fixed the value of τ and the level of $\mathcal{H}_2$ guaranteed cost performance in this optimization. Also note that $t\check{\mathcal{M}} \succ 0$ implies that $t > 0$. Thus, this optimization problem minimizes the condition number of S subject to a constraint on the closed-loop $\mathcal{H}_2$ guaranteed cost and a fixed value of τ.

When the strict inequalities in (14.60) are relaxed to non-strict inequalities, it becomes an SDP in the variables t, κ, tX, tY, tW, $t\hat{V}_1$, $t\hat{V}_2$, $t\hat{A}$, $t\hat{B}$, $t\hat{C}$, and $t\hat{D}$. Thus, improving the conditioning of the controller reconstruction process for a fixed value of τ and a fixed closed-loop $\mathcal{H}_2$ guaranteed cost can be solved using this SDP.

14.5.1.4 Solution Methodology

With the results of the previous sections in place, a reasonable methodology for trying to solve (14.49) is:

1. **Find Initial Value of τ**
   a. **State Feedback Controller Design:** Solve (14.34) with the additional constraints that $\hat{K}_d = 0$ and $K_w = 0$ using an SDP solver. From the resulting set of optimization parameter values, reconstruct the state feedback gain $K_x := \hat{K}_x\hat{P}^{-1}$.
   b. **"Separation Principle" Controller Design:** Solve (14.55) using an SDP solver.

2. **Controller Design**
   a. **Controller Design (Fixed τ):** Fix τ to be the value obtained in the previous optimization. Solve (14.50) using an SDP solver.
   b. **Controller Design (Fixed $X, \hat{C}, \hat{D}$):** Fix X, $\hat{C}$, and $\hat{D}$ to be the values obtained in the previous optimization. Solve (14.50) using an SDP solver. If the stopping criteria have not been met (see below), return to step 2a.

3. **Conditioning:** Choose a value of $\beta > 0$. Fix τ to be the value which yielded the smallest cost γ_0 in the preceding optimizations[1] and fix γ to be $(1+\beta)\gamma_0$. Solve (14.60) using an SDP solver.
4. **Controller Reconstruction:** Reconstruct the controller using (14.52).

In our implementation of the algorithm, although we choose the default $\beta = 0.05$, we allow this to be specified by the user. This allows the user to trade off closed-loop performance versus conditioning in the controller reconstruction step. We use two stopping criteria in step 2. If 6 total optimizations have been performed in step 2 or if the cost has decreased less than 1% in the last two optimizations, we exit step 2 and move on to step 3.

14.5.2 Riccati Equation and Semi-definite Programming Approach

In this section, we consider the output feedback control problem when the regularity conditions of Sect. 14.4.2 hold. For now, we fix $\tau > 0$ and consider the problem of optimizing $J_\tau(\mathscr{F}_l(G,K))$. The approach we follow is similar to the approach taken in solving the discrete-time $\mathscr{H}_\infty$ control problem (9). However, since $\mathscr{H}_2$ guaranteed cost control does not have a frequency-domain interpretation, we directly manipulate Riccati equations to establish the relevant results.

If there exists an output feedback controller which achieves a finite value of $J_\tau(\mathscr{F}_l(G,K))$, then there exists a full information controller which achieves finite cost, i.e. $J_{fi,\varepsilon} \neq \infty$ for $\varepsilon = \tau^{-1}$. Thus, we begin by solving the full information problem for $\varepsilon = \tau^{-1}$. We will denote the stabilizing solution of the DARE $\mathscr{R}_\phi(P) = P$ (which is assumed to satisfy the relevant properties) as P_0. In addition to the notation from Sect. 14.4.2, we also define the notation

$$K_{dx} := T_{11}^{-1}T_{11}^{-T}\left[B_1^T P_0 A + D_{11}^T C_1 + \varepsilon D_{21}^T C_2 + (B_3^T P_0 B_1 + R_{21})^T K_x\right] \quad (14.61a)$$

$$K_{dw} := T_{11}^{-1}T_{11}^{-T}\left[B_1^T P_0 B_2 + D_{11}^T D_{12} + \varepsilon D_{21}^T D_{22} + (B_3^T P_0 B_1 + R_{21})^T K_w\right] \quad (14.61b)$$

$$\bar{K}_x := K_x + K_d K_{dx} \quad (14.61c)$$

$$\bar{K}_w := K_w + K_d K_{dw}\,. \quad (14.61d)$$

In this notation, it can be shown that

$$\mathscr{K}_\phi(P_0) = \begin{bmatrix} K_{dx} \\ \bar{K}_x \end{bmatrix}. \quad (14.62)$$

[1] Although the value of the optimization problem should decrease every time step 2a or 2b is executed, numerical inaccuracies might cause this not to be the case.

14.5.2.1 Reduction to Output Estimation Problem

We now consider Fig. 14.15(b) where the state-space realizations of G_1 and G_2 are respectively given by

$$G_1 \sim \left[\begin{array}{c|cccc} A+B_3K_x & B_1+B_3K_d & B_2+B_3K_w & B_3T_{22}^{-1} & 0 \\ \hline C_1+D_{13}K_x & D_{11}+D_{13}K_d & D_{12}+D_{13}K_w & D_{13}T_{22}^{-1} & 0 \\ C_2+D_{23}K_x & D_{21}+D_{23}K_d & D_{22}+D_{23}K_w & D_{23}T_{22}^{-1} & 0 \\ -T_{11}K_{dx} & T_{11} & -T_{11}K_{dw} & 0 & 0 \\ 0 & 0 & I & 0 & 0 \end{array}\right] \tag{14.63a}$$

$$G_2 \sim \left[\begin{array}{c|ccc} A+B_1K_{dx} & B_1T_{11}^{-1} & B_2+B_1K_{dw} & B_3 \\ \hline -T_{22}\bar{K}_x & -T_{22}K_dT_{11}^{-1} & -T_{22}\bar{K}_w & T_{22} \\ 0 & 0 & 0 & 0 \\ C_3+D_{31}K_{dx} & D_{31}T_{11}^{-1} & D_{32}+D_{31}K_{dw} & 0 \end{array}\right] . \tag{14.63b}$$

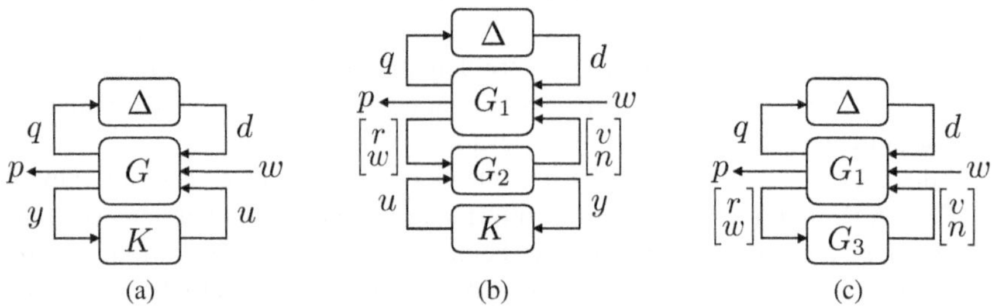

Fig. 14.15 Three equivalent block diagrams for output feedback $\mathscr{H}_2$ guaranteed cost control structure

Note that, with these definitions of G_1 and G_2, the signal n is zero. Although n does not play a role in the dynamics of the system, we will see later in this section that it serves a structural role. Using the fact that $A+[B_1\ B_3]\mathscr{K}_\phi(P_0)$ is Schur (by the definition of a stabilizing DARE solution), it can be shown that combining G_1 and G_2 into a single block yields the block diagram in Fig. 14.15(a). If we instead combine G_2 and K into a single block, G_3, it yields the block diagram in Fig. 14.15(c). Thus, the three block diagrams in Fig. 14.15 are equivalent.

Now suppose that G_3 has the realization

$$G_3 \sim \left[\begin{array}{c|cc} \tilde{A} & \tilde{B}_1 & \tilde{B}_2 \\ \hline \tilde{C} & \tilde{D}_1 & \tilde{D}_2 \\ 0 & 0 & 0 \end{array}\right] . \tag{14.64}$$

Note that although we are utilizing the fact that $n=0$, we are not explicitly exploiting any other structure of G_3. With this in place, we are interested in evaluating $J_\tau(\mathscr{F}_l(G_1,G_3))$. It can be shown that

$$J_\tau(\mathscr{F}_l(G_1,G_3)) = J_{fi,\varepsilon} + \varepsilon^{-1} J_1(G_3) \tag{14.65}$$

where $J_1(G_3)$ is interpreted as $J_\tau(G_3)$ at $\tau = 1$. Using the techniques of Sect. 14.3.2, $J_1(G_3) = \operatorname{tr}\{\mathscr{R}_{\tilde{\psi}}(\tilde{P}_0)\}$ where $\tilde{P}_0$ is the stabilizing solution of the DARE $\mathscr{R}_{\tilde{\phi}}(P) = P$ such that $\tilde{B}_1^T \tilde{P}_0 \tilde{B}_1 + \tilde{D}_1^T \tilde{D}_1 - I \prec 0$ where

$$\tilde{\phi} := (\tilde{A},\ \tilde{B}_1, \tilde{C}^T\tilde{C},\ \tilde{D}_1^T\tilde{D}_1 - I,\ \tilde{C}^T\tilde{D}_1) \tag{14.66a}$$

$$\tilde{\psi} := (\tilde{B}_2,\ \tilde{B}_1, \tilde{D}_2^T\tilde{D}_2,\ \tilde{D}_1^T\tilde{D}_1 - I,\ \tilde{D}_2^T\tilde{D}_1)\,. \tag{14.66b}$$

It should be noted that (14.65) could only be written in such a compact form due to the placeholder signal n; without that placeholder, G_3 would only have one output and the expression $J_\tau(G_3)$ would not be well-defined.

The proof of (14.65) proceeds along the following lines. First, find a realization of $\mathscr{F}_l(G_1,G_3)$ with $[x_1^T\ x_3^T]^T$ as the state where x_1 and x_3 are respectively the states of G_1 and G_3 and then apply the techniques of Sect. 14.3.2. This yields that $J_\tau(\mathscr{F}_l(G_1,G_3)) = \operatorname{tr}\{\mathscr{R}_{\check{\psi}}(\check{P}_0)\}$ where $\check{P}_0$ is the stabilizing solution of the DARE $\mathscr{R}_{\check{\phi}}(P) = P$ such that the relevant matrix is negative definite. (The exact expressions of the relevant quantities are omitted for brevity.) Using the matrix pencil method for analyzing DAREs (11), it can be shown that $\check{P}_0$ has the form $\operatorname{diag}[P_0, \varepsilon^{-1}P_{22}]$ for some matrix P_{22}. Using this expression, the DARE $\mathscr{R}_{\check{\phi}}(P) = P$ can be written

$$\begin{bmatrix} P_0 & 0 \\ 0 & \varepsilon^{-1}P_{22} \end{bmatrix} = \begin{bmatrix} P_0 & 0 \\ 0 & \varepsilon^{-1}\mathscr{R}_{\tilde{\phi}}(P_{22}) \end{bmatrix}. \tag{14.67}$$

Thus, with a little more algebra, it can be shown that the existence of the stabilizing DARE solution such that the relevant matrix is negative definite is equivalent to the existence of $\tilde{P}_0$ with the relevant properties. Moreover, $\check{P}_0 = \operatorname{diag}[P_0, \varepsilon^{-1}\tilde{P}_0]$. Plugging this into the expression for $\mathscr{R}_{\check{\psi}}(\check{P}_0)$ yields after some algebra that $\mathscr{R}_{\check{\psi}}(\check{P}_0) = \varepsilon^{-1}\mathscr{R}_{\psi}(P_0) + \varepsilon^{-1}\mathscr{R}_{\tilde{\psi}}(\tilde{P}_0)$. This immediately gives (14.65).

Using (14.65), we see that the optimization problem we are interested in solving is

$$\begin{aligned} \inf_K J_\tau(\mathscr{F}_l(G,K)) &= \inf_K J_\tau(\mathscr{F}_l(G_1,\mathscr{F}_l(G_2,K))) = \inf_K \{J_{fi,\varepsilon} + \varepsilon^{-1} J_1(\mathscr{F}_l(G_2,K))\} \\ &= J_{fi,\varepsilon} + \varepsilon^{-1} \inf_K J_1(\mathscr{F}_l(G_2,K))\,. \end{aligned} \tag{14.68}$$

The remaining optimal control problem

$$\inf_K J_1(\mathscr{F}_l(G_2,K)) \tag{14.69}$$

is analogous to the output estimation problem in the $\mathscr{H}_\infty$ literature.

14.5.2.2 Reduction to Full Control Problem

For $\mathscr{H}_\infty$ control, the output estimation problem is solved by applying duality—transposing the closed-loop system transfer function matrix—to transform the problem into a disturbance feedforward problem, reducing this to a full information control problem, then applying the optimal full information controller. However, in our approach, this approach does not work because there is no known duality result. Thus, we now diverge slightly from the approach taken in (9).

At this point, we restrict the class of controllers to ones which can be expressed as $K = \mathscr{F}_l(\hat{K}, \tilde{K})$ where

$$\hat{K} \sim \left[\begin{array}{c|ccc} A + [B_1\ B_3]\mathscr{K}_\phi(P_0) & 0 & -I & B_3 \\ \hline \bar{K}_x & 0 & 0 & I \\ -(C_3 + D_{31}K_{dx}) & I & 0 & 0 \end{array}\right]. \tag{14.70}$$

For this control structure, the block diagrams in Figs. 14.16(a) and 14.16(b) are equivalent. Combining G_2 and $\hat{K}$ in Fig. 14.16(b) into a single block, G_4, yields Fig. 14.16(c). Thus, the three block diagrams in Fig. 14.16 are equivalent for this control structure.

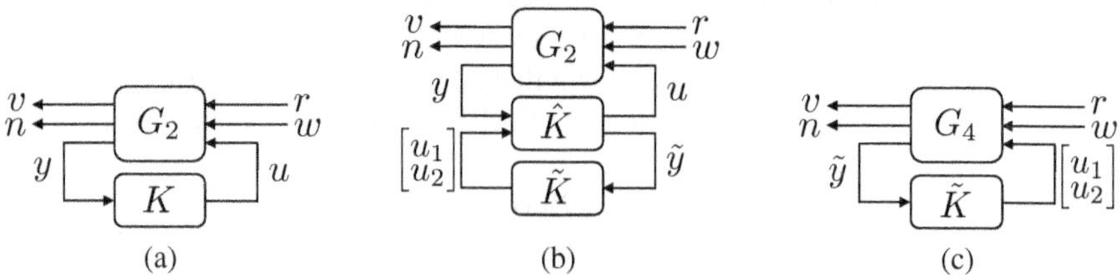

Fig. 14.16 LFT representation of $\mathscr{H}_2$ guaranteed cost control structure

The obvious question is whether or not this control structure worsens the level of achievable performance of the closed-loop system. To answer this, we first note that, since $A + [B_1\ B_3]\mathscr{K}_\phi(P_0)$ is Schur, G_4 has the realization

$$G_4 \sim \left[\begin{array}{c|ccc} A + B_1K_{dx} & B_1T_{11}^{-1} & B_2 + B_1K_{dw} & I\ 0 \\ \hline -T_{22}\bar{K}_x & -T_{22}K_dT_{11}^{-1} & -T_{22}\bar{K}_w & 0\ T_{22} \\ 0 & 0 & 0 & 0\ 0 \\ C_3 + D_{31}K_{dx} & D_{31}T_{11}^{-1} & D_{32} + D_{31}K_{dw} & 0\ 0 \end{array}\right]. \tag{14.71}$$

Thus, we see that if we make the restriction $u_1 = B_3u_2$ in Fig. 14.16(c), we exactly recover the control problem shown in Fig. 14.16(a). This means that choosing this special control structure does not affect the achievable performance of the closed-loop system. The remaining optimal control problem

$$\inf_{\tilde{K}} J_1(\mathscr{F}_l(G_4, \tilde{K})) \tag{14.72}$$

is analogous to the full control problem in the $\mathscr{H}_\infty$ literature.

14.5.2.3 Solving the Full Control Problem

Using techniques similar to the one used in (3), it can be shown that, given any state-space controller for the block diagram in Fig. 14.16(c), it is always possible to construct a static gain controller for the block diagram in Fig. 14.16(c) that achieves the same performance in terms of $J_1(\mathscr{F}_l(G_4,\tilde{K}))$. We therefore express

$$\tilde{K} = \begin{bmatrix} L_x \\ L_v \end{bmatrix} \tag{14.73}$$

where L_x and L_v are static gains.

It is not currently known whether or not the full control problem can be solved using Riccati equations. Therefore, to solve this problem, we will resort to the SDP approach. We first apply the change of variables $\hat{L}_x := PL_x$. Applying the characterization (14.15) with $\tau = 1$ to $\mathscr{F}_l(G_4,\tilde{K})$ yields, after multiplying $\tilde{\mathscr{M}}$ on the right by $\Phi := \mathrm{diag}[I,T_{11},I,I,T_{22},I]$ and on the left by Φ^T, the optimization problem

$$\inf_{P,W,V,\hat{L}_x,L_v} \mathrm{tr}\{W\} \quad \text{s.t.} \quad \begin{bmatrix} \tilde{\mathscr{M}} & 0 \\ 0 & I \end{bmatrix} \succ 0 \tag{14.74}$$

where

$$\tilde{\mathscr{M}} := \begin{bmatrix} P & \bullet & \bullet & \bullet & \bullet \\ 0 & T_{21}^T T_{21} - B_1^T P_0 B_1 - R_{11} & \bullet & \bullet & \bullet \\ V & 0 & W & \bullet & \bullet \\ P\check{A} + \hat{L}_x\check{C}_3 & PB_1 + \hat{L}_x D_{31} & P\check{B} + \hat{L}_x\check{D}_3 & P & \bullet \\ \check{M}_1 & \check{M}_2 & \check{M}_3 & 0 & B_3^T P_0 B_3 + R_{22} \end{bmatrix} \tag{14.75}$$

and

$$\begin{bmatrix} \check{A} & \check{B} \\ \check{C}_1 & \check{D}_1 \\ \check{C}_2 & \check{D}_2 \\ \check{C}_3 & \check{D}_3 \end{bmatrix} := \begin{bmatrix} A + B_1 K_{dx} & B_2 + B_1 K_{dw} \\ C_1 + D_{11} K_{dx} & D_{12} + D_{11} K_{dw} \\ C_2 + D_{21} K_{dx} & D_{22} + D_{21} K_{dw} \\ C_3 + D_{31} K_{dx} & D_{32} + D_{31} K_{dw} \end{bmatrix} \tag{14.76a}$$

$$\check{M}_1 := (B_3^T P_0 B_3 + R_{22}) L_v \check{C}_3 + B_3^T P_0 \check{A} + D_{13}^T \check{C}_1 + \varepsilon D_{23}^T \check{C}_2 \tag{14.76b}$$

$$\check{M}_2 := (B_3^T P_0 B_3 + R_{22}) L_v D_{31} + B_3^T P_0 B_1 + D_{13}^T D_{11} + \varepsilon D_{23}^T D_{21} \tag{14.76c}$$

$$\check{M}_3 := (B_3^T P_0 B_3 + R_{22}) L_v \check{D}_3 + B_3^T P_0 \check{B} + D_{13}^T \check{D}_1 + \varepsilon D_{23}^T \check{D}_2 \,. \tag{14.76d}$$

(Recall that, in these definitions, $\varepsilon = \tau^{-1}$.) Since $\mathrm{diag}[\tilde{\mathscr{M}},I] \succ 0$ if and only if $\tilde{\mathscr{M}} \succ 0$, we see that (14.74) is equivalent to

$$\inf_{P,W,V,\hat{L}_x,L_v} \mathrm{tr}\{W\} \quad \text{s.t.} \quad \tilde{\mathscr{M}} \succ 0 \,. \tag{14.77}$$

For any feasible $P,W,V,\hat{L}_x,L_v$, a controller which achieves $J_1(\mathscr{F}_l(G_4,\tilde{K})) \leq \mathrm{tr}\{W\}$ is given by

$$\tilde{K} = \begin{bmatrix} L_x \\ L_v \end{bmatrix} = \begin{bmatrix} P^{-1}\hat{L}_x \\ L_v \end{bmatrix}. \tag{14.78}$$

Putting this result together with (14.65) and the structure $K = \mathscr{F}_l(\hat{K},\tilde{K})$ yields the following result: for any feasible $P,W,V,\hat{L}_x,L_v$ of (14.77), an output feedback controller which achieves $J_\tau(\mathscr{F}_l(G,K)) \leq J_{fi,\varepsilon} + \mathrm{tr}\{W\}$ is given by

$$K \sim \left[\begin{array}{c|c} A + B_1K_{dx} + B_3\bar{K}_x + (P^{-1}\hat{L}_x - B_3L_v)(C_3 + D_{31}K_{dx}) & P^{-1}\hat{L}_x - B_3L_v \\ \hline L_v(C_3 + D_{31}K_{dx}) - \bar{K}_x & L_v \end{array}\right] \tag{14.79}$$

If the strict inequality in (14.77) is relaxed to a non-strict inequality, the optimization becomes a SDP. Thus, a reasonable way to solve the output feedback control problem (for fixed τ) is to relax (14.77) to a SDP, solve the SDP using an appropriate solver, then reconstruct the output feedback controller using (14.79).

14.5.2.4 Solution Methodology

With the results of the previous sections in place, a reasonable methodology for trying to solve (14.49) is:

1. **Find Initial Value of τ**

   a. **Full Information Controller Design:** Using the methodology of Sect. 14.4.2, design an optimal full information controller.
   b. **Find Feasible Value of τ:** Choose $\alpha > 0$. For the final values determined during the last full information controller design, solve (14.77) using an SDP solver. If the optimization was feasible, reconstruct the corresponding output feedback controller K using (14.79). If the optimization was not feasible, set $\tau \leftarrow \alpha\tau$, design a full information controller for $\varepsilon = \tau^{-1}$ using the methods of Sect. 14.4.2, and redo this step.
   c. **Closed-Loop System Analysis (Fixed K):** Form the closed-loop system $\mathscr{F}_l(G,K)$ and analyze its $\mathscr{H}_2$ guaranteed cost performance using the methodology of Sect. 14.3.2.

2. **Controller Design**

   a. **Output Feedback Controller Design (Fixed τ):** For the value of $\tau > 0$ found in the previous closed-loop system analysis step, solve (14.77) using an SDP solver and reconstruct the corresponding controller K using (14.79).
   b. **Closed-Loop System Analysis (Fixed K):** Form the closed-loop system $\mathscr{F}_l(G,K)$ and analyze its $\mathscr{H}_2$ guaranteed cost performance using the methodology of Sect. 14.3.2. Return to step 2a.

In our implementation, we use $\alpha = 100$. We use two stopping criteria in this algorithm. If the number of output feedback controller optimizations (i.e. the number of times steps 1b and 2a have been executed) exceeds 30 or if $J_{of}^{[i-1]}/J_{of}^{[i]} - 1 < 10^{-4}$ where $J_{of}^{[i]}$

is the cost reported the ith time step 2b executes, we terminate the algorithm. We also terminate the algorithm if the SDP solver claims infeasibility in step 2a.

We now take a minute to explain step 1b in the above methodology. By the results in Sect. 14.3.2, $J_\tau(\mathscr{F}_l(G,K)) \neq \infty$ implies that $J_{\bar{\tau}}(\mathscr{F}_l(G,K)) \neq \infty$ whenever $\bar{\tau} > \tau$. Therefore to find a value of τ for which there exists K satisfying $J_\tau(\mathscr{F}_l(G,K)) \neq \infty$, we should make τ increasingly large.

14.5.3 Application to Hard Disk Drives

So far in this section, we have developed two methodologies for designing output feedback controllers—one based on solving a sequence of SDPs and another which exploits the solution of Riccati equations to yield simplified SDPs. We will respectively call these approaches the SSDP approach and the DARE/SDP approach. In this section, we will design controllers for the HDD model presented in Sect. 14.2. As we saw in Sect. 14.4.3, the controllers designed for this HDD model tend to use very little control effort. In an effort to boost the PES performance of the closed-loop system (at the expense of the control effort), we will deemphasize the control effort in the cost function by applying our solution heuristics to the plant

$$\hat{G}_H := \text{diag}[1,1,0.01,0.01,1]G_H\,. \tag{14.80}$$

We first applied the SSDP approach in Sect. 14.5.1.4 to the design of an output feedback controller for $\hat{G}_H$. Doing so yielded the results shown in Table 14.7, which breaks down the $\mathscr{H}_2$ guaranteed cost and the cumulative optimization time at each optimization step. The first thing to note is that the SSDP approach took just over 1 hour to design a controller for this system. The next thing to note is that, at steps 1b and 3, there are degradations in performance. However, these are both expected. At step 1a, since the designed controller has direct access to the state of $\hat{G}_H$, it is likely that the cost reported after this step is smaller than is achievable by an output feedback controller. In step 3, we allowed the cost to increase by up to 2% in order to optimize the conditioning of the controller reconstruction process; the

Table 14.7 Closed-loop performance and cumulative optimization after each optimization step in SSDP approach

Step	$\mathscr{H}_2$ Guaranteed Cost	Cumulative Optimization Time (s)
1a	6.255	10.811
1b	9.730	239.89
2a	8.095	561.74
2b	8.095	1188.12
2a	8.095	1552.77
2b	8.095	2091.08
3	8.257	3610.92

$\mathscr{H}_2$ guaranteed cost was a design parameter rather than the objective function to be optimized.

Before performing step 3, we tried to reconstruct a controller which achieved the $\mathscr{H}_2$ guaranteed cost 8.095. However, the condition number of the matrix $I - XY$ (i.e. the ratio of its largest and smallest singular values) was 3.802×10^{22}. This resulted in large numerical inaccuracies when reconstructing the controller, which in turn resulted in an unstable closed-loop system. After performing step 3, the condition number of the matrix $I - XY$ was improved to 7.933×10^8. With the resulting values of the optimization parameters, a controller was reconstructed which achieves the $\mathscr{H}_2$ guaranteed cost 7.958 (as computed by the methodology in Sect. 14.3.2). Interestingly, this is 4% better than the cost reported by the solver in step 3. Figure 14.17 shows the Bode magnitude plot of the nominal closed-loop sensitivity function from r to y_h along with pointwise upper and lower bounds on its Bode magnitude plot over modeled uncertainty. The nominal Bode magnitude plot has the peak value 1.25 dB, and the upper bound on the Bode magnitude plot has the peak value 5.57 dB. Thus, even in the worst case, the Bode magnitude plot of the sensitivity function from r to y_h has a low peak.

After designing a controller using the SSDP approach, we designed a controller using the DARE/SDP approach. The algorithm took 39.03 seconds to run and reported a closed-loop $\mathscr{H}_2$ guaranteed cost of 7.747. By construction, this value of the closed-loop $\mathscr{H}_2$ guaranteed cost is exactly equal to the $\mathscr{H}_2$ guaranteed cost computed by applying the methods of Sect. 14.3.2 to analyze the closed-loop performance. Thus, this controller performs 2% better than the controller designed using the SSDP approach. Also note that the DARE/SDP approach was more than 90 times faster than the SSDP approach.

Figure 14.18 shows the Bode magnitude plot of the nominal closed-loop sensitivity function from r to y_h along with pointwise upper and lower bounds on its Bode magnitude plot over modeled uncertainty. The nominal Bode magnitude plot has the peak value 1.01 dB, and the upper bound on the Bode magnitude plot has the peak value 4.16 dB. Thus, like the closed-loop system designed using the SSDP approach, the Bode magnitude plot of the sensitivity function from r to y_h has a low "hump" in both the nominal and the worst case.

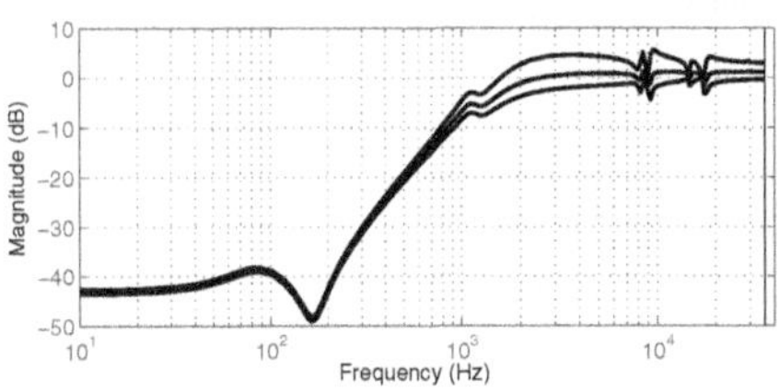

(a)

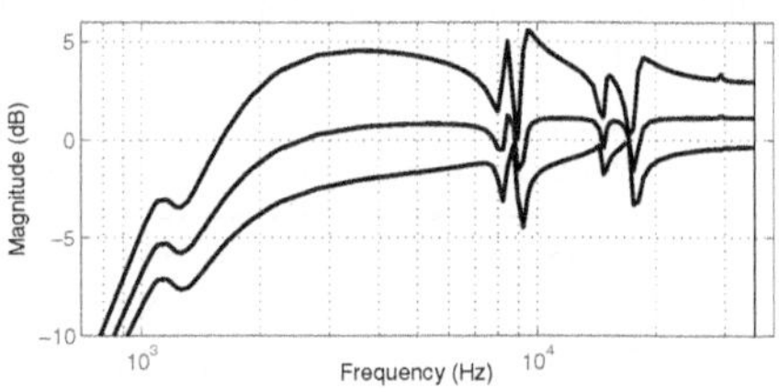

(b)

Fig. 14.17 Bode magnitude plot of the nominal closed-loop sensitivity function from r to y_h for the controller designed using the SSDP approach along with its pointwise upper and lower bounds over all modeled uncertainty

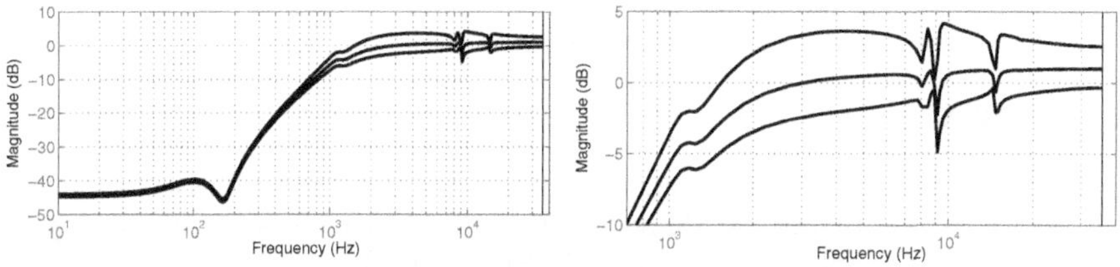

(a) (b)

Fig. 14.18 Bode magnitude plot of the nominal closed-loop sensitivity function from r to y_h for the controller designed using the DARE/SDP approach along with its pointwise upper and lower bounds over all modeled uncertainty

It should be noted that the $\mathscr{H}_2$ guaranteed cost performance of a system is an upper bound on the worst-case $\mathscr{H}_2$ performance of the system over all unmodeled uncertainty—not necessarily the actual worst-case performance. It is thus useful to perform a Monte Carlo analysis of the closed-loop system. Using the function `usample` in the Robust Control Toolbox, we fist chose 3000 random samples of the closed-loop system with Δ restricted to be a stable causal 3rd-order system satisfying $\|\Delta\|_\infty \leq 1$. For each of the 3000 systems, we then found the standard deviation of each of the outputs by computing the relevant $\mathscr{H}_2$ norm. The worst-case standard deviation of each signal is summarized in Table 14.8. We see that these results are significantly smaller than predicted by $\mathscr{H}_2$ guaranteed cost analysis of the closed-loop system.

Table 14.8 Worst-case standard deviation of closed-loop signals over 3000 random closed-loop samples with controller designed using the DARE/SDP approach

Signal	Standard Deviation
y_h	2.466 nm
u_v	0.316 V
u_p	0.279 V

14.6 Conclusion

In this chapter, we have developed two heuristics for solving the output feedback $\mathscr{H}_2$ guaranteed cost control problem. In the first method, a series of SDPs are solved to design a controller. The second method exploits the solutions of Riccati equations to simplify the output feedback control problem (for a fixed value of τ) to a full control problem, which can be solved using an SDP. These methods were applied to a track-following HDD control problem in which the HDD had a PZT-actuated suspension. It was shown that although both methods yielded controllers with a reasonable level of robust performance, the algorithm that exploited Riccati equation structure was more than 90 times faster and yielded a slightly better value of the cost function.

References

[1] Boyd, S., El Ghaoui, L., Feron, E., Balakrishnan, V.: Linear Matrix Inequalities in System and Control Theory. SIAM Studies in Applied Mathematics, vol. 15. Society for Industrial and Applied Mathematics, Philadelphia (1994)

[2] Conway, R., Horowitz, R.: Analysis of discrete-time H2 guaranteed cost performance. In: Proc. ASME Dyn. Syst. Control Conf., DSCC 2009. PART A, pp. 427–435 (2010)

[3] Conway, R., Horowitz, R.: Optimal full information H2 guaranteed cost control of discrete-time systems. In: Proc. ASME Dyn. Syst. Control Conf., DSCC 2010, vol. 2, pp. 523–532 (2010); Paper DSCC2010-4220

[4] Conway, R., Felix, S., Horowitz, R.: Model reduction and parametric uncertainty identification for robust H_2 control synthesis for dual-stage hard disk drives. IEEE Trans. Magn. 43(9), 3763–3768 (2007)

[5] Delchamps, D.F.: Analytic stabilization and the algebraic Riccati equation. In: Proc. IEEE Conf. Decis. Control, vol. 22, pp. 1396–1401 (1983)

[6] Evans, R.B., Griesbach, J.S., Messner, W.C.: Piezoelectric microactuator for dual stage control. IEEE Trans. Magn. 35(2), 977–982 (1999)

[7] Hirano, T., Fan, L.S., Semba, T., Lee, W.Y., Hong, J., Pattanaik, S., Webb, P., Juan, W.H., Chan, S.: High-bandwidth HDD tracking servo by a moving-slider microactuator. IEEE Trans. Magn. 35(5), 3670–3672 (1999)

[8] Huang, X., Nagamune, R., Horowitz, R.: A comparison of multirate robust track-following control synthesis techniques for dual-stage and multi-sensing servo systems in hard disk drives. IEEE Trans. Magn. 42(7), 1896–1904 (2006)

[9] Iglesias, P.A., Glover, K.: State-space approach to discrete-time $\mathscr{H}_\infty$ control. Int. J. Control 54(5), 1031–1073 (1991)

[10] Kanev, S., Scherer, C., Verhaegen, M., De Schutter, B.: Robust output-feedback controller design via local BMI optimization. Automatica 40(7), 1115–1127 (2004)

[11] Lancaster, P., Rodman, L.: Algebraic Riccati Equations. Oxford Science Publications, Clarendon Press, Oxford (1995)

[12] Löfberg, J.: YALMIP: A toolbox for modeling and optimization in MATLAB. In: Proc. IEEE Int. Symp. Comput. Aided Control Syst. Des., pp. 284–289 (2004), `http://control.ee.ethz.ch/~joloef/yalmip.php`

[13] Nie, J., Horowitz, R.: Design and implementation of dual-stage track-following control for hard disk drives. In: Proc. ASME Dyn. Syst. Control Conf., DSCC 2009. PART B, pp. 1477–1484 (2010)

[14] Petersen, I.R., McFarlane, D.C., Rotea, M.A.: Optimal guaranteed cost control of discrete-time uncertain linear systems. Int. J. Robust Nonlinear Control 8(8), 649–657 (1998)

[15] Scherer, C., Gahinet, P., Chilali, M.: Multiobjective output-feedback control via LMI optimization. IEEE Trans. Autom. Control 42(7), 896–911 (1997)

[16] Sturm, J.F.: Using SeDuMi 1.02, a MATLAB toolbox for optimization over symmetric cones. Optim. Methods Software 11, 625–653 (1999)

[17] Toshiyoshi, H., Mita, M., Fujita, H.: A MEMS piggyback actuator for hard-disk drives. J. Microelectromech. Syst. 11(6), 648–654 (2002)

[18] Yu, L., Gao, F.: Output feedback guaranteed cost control for uncertain discrete-time systems using linear matrix inequalities. J. Optim. Theory Appl. 113(3), 621–634 (2002)

Chapter 15
Lateral Tape Motion and Control Systems Design in Tape Storage

Angeliki Pantazi

Abstract. In tape storage systems, lateral tape motion (LTM) of the flexible moving medium is the main limiting factor for increasing the track density on a magnetic tape. A track-follow control system is used to reduce the misalignment between the medium and the recording head by compensating for the lateral tape vibrations and thereby determine the achievable positioning accuracies. In this chapter, improvements in the track-following capabilities of tape storage systems are discussed. Specifically, the characteristics of LTM, the main disturbance in the tape track-follow system, are described. Also evaluated are control system concepts that improve the track-follow performance either by reducing the LTM or by enhancing the track-follow control system.

15.1 Introduction

High areal recording density and the subsequent storage capacity increase are the future requirements for data storage systems ranging from probe-based storage devices (1; 2) to hard disk drives (HDDs) (3) and tape storage (4). The positioning accuracy during track following is one of the key requirements for achieving high linear and track densities. For example, probe-based storage devices have the most stringent requirements in positioning resolution, reaching values of a fraction of a nanometer (5; 6). Nanometer-scale precision is also required for the servo system of HDDs (7; 8), as well as for the servo system of future tape drives (4; 9; 10). Furthermore, a comparison of the areal density of today's HDDs and tape drives shows that although there is a rather small difference in the linear density, the tape drives have a much smaller track density because of the moving medium. Therefore, in tape drive systems, storage capacity increase can be achieved by an aggressive scaling of the track density.

Angeliki Pantazi
IBM Research - Zurich,
8803 Rüschlikon, Switzerland
e-mail: agp@zurich.ibm.com

E. Eleftheriou & S.O.R. Moheimani (Eds.): Cntrl. Tech. for Emerging Micro/Nanoscale Sys. LNCIS 413, pp. 271–287.
springerlink.com

Achieving higher tape track densities requires significant improvements in the position error signal (PES) used to characterize the track-follow performance. The main contributor to the PES is lateral tape motion (LTM), which mainly depends on tape guiding and tape path characteristics. This chapter focuses on the characterization of the LTM in tape systems with flanged and flangeless rollers. The frequency characteristics of the LTM and their effect on the servo performance are presented. One of the main characteristics of LTM is the periodic components related to the roller frequencies. Another contributor to LTM are the stack shifts, which are low-frequency disturbances. Nonetheless, they affect the track follow performance, especially in tape paths with flangeless rollers.

Control system design concepts can be employed either to reduce the LTM or to enhance the capabilities of the track-follow system to follow the LTM. Specifically, this chapter presents the development of a tape transport enhancement that reduces the LTM, eliminating the need for flanges on the rollers. The concept relies on controlling the tilt of the rolling elements in the tape path based on the measured LTM. In addition, track-follow control architectures that provide enhanced performance in the presence of periodic disturbances and low frequency stack shift disturbances are presented. Experimental and simulation results evaluate the efficacy of the proposed control concepts.

15.2 System Description

In Linear Tape Open (LTO) drives, the data track layout is based on the LTO format specifications defined by the LTO consortium (11; 12). The LTO format is based on a multi-channel linear serpentine recording. Specifically, an LTO Ultrium tape contains five servo bands located at a specified lateral position that are recorded on the tape during the cartridge manufacturing. Within each servo band, the servo pattern consisting of servo stripes with two azimuthal slopes is written, as shown in Fig. 15.1. Servo stripes are grouped together to create the servo bursts. The information from the servo pattern is used by the track-follow system so that the head element becomes aligned in the lateral direction with respect to the tape. For higher

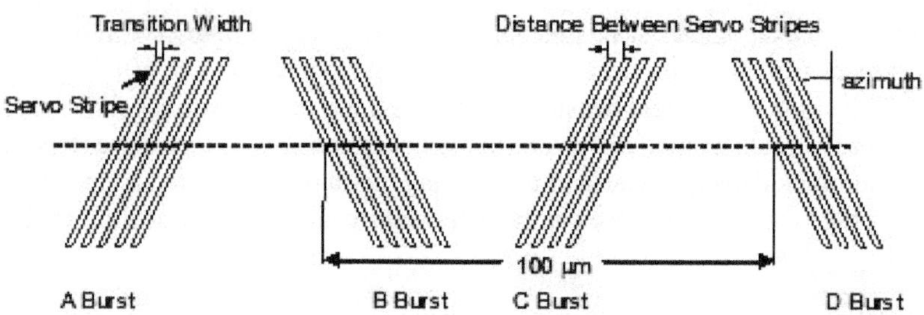

Fig. 15.1 Timing-based servo pattern. From (18)

servo accuracy, the information from two servo bands can be used simultaneously. In addition to the lateral position information used in the track-follow system, longitudinal position information (LPOS) is encoded into the servo frame using pulse-position modulation to indicate the absolute position along the length of the tape. The data bands are placed between pairs of servo bands and are written individually. A group of data tracks, called a wrap, is recorded simultaneously while the tape moves in the forward or backward direction. The forward direction of motion is defined as the motion from the beginning of tape (BOT) to the end of tape (EOT), whereas the backward direction is the opposite.

During track-follow, the position error signal, which is created based on the servo pattern, is used to position the head actuator in the lateral direction. The timing-based servo (TBS) technology developed in the mid-1990s for linear tape drives is employed for PES generation (13). In the TBS servo channel, the lateral position is derived from the relative timing of pulses generated by the head while reading the pattern. To improve the resolution of the position estimate, the synchronous servo channel architecture has been proposed (14). The synchronous servo channel achieves optimal filtering of the servo signal by a digital matched-filter interpolator/correlator. In addition to the lateral position estimate, the servo channel provides information about the tape velocity used in the tape transport system.

In tape storage systems, the tape is transported through a tape path from one reel to the other. Rollers guide the tape from the cartridge reel over the head element to the take-up reel. The head is mounted on an actuator, and dedicated servo transducers on the head read the pre-formatted servo track information. The track follow and the tape transport are the main control systems in tape storage. The main task of the track-follow controller is to accurately move the head actuator in such a way that it follows the centerline of the data tracks during read/write operations. The tape transport system is responsible for maintaining a constant tape velocity and a constant tape tension, while the tape is being transported from one reel to the other. The supported tape speeds are in the range of 1 m/s to 10 m/s.

Tape recording systems must deal with effects originating from the moving flexible media, guiding system imperfections and other tape path characteristics that create LTM (15; 16; 17). This LTM leads to misalignment of the head read/write transducers with respect to the track locations. This is one of the key challenges that the closed-loop track-follow control system has to resolve to push the limits of tape storage capacity and performance. In conventional tape paths, flanged rollers transport the tape between the reels. Although the use of flanges constrains the tape and limits LTM, it introduces debris accumulation on the flanges that affects the lifetime of the tape as well as creates undesirable dynamic effects (10). These effects can be alleviated by removing the flanges from the rollers; however this approach introduces new challenges which need to be addressed. Removing the flanges also removes the constraint on the motion of the tape, which results in an increase in the amplitude of the low-frequency LTM.

15.3 Lateral Tape Motion

Several techniques can be utilized to measure the lateral tape motion, defined as the displacement of the tape in a direction perpendicular to the tape transport (18). The lateral tape displacement can be measured using optical edge sensors that detect the position of the bottom edge of the tape as it is transported from one reel to the other. This technique is useful, as it provides measurements of the LTM at various locations along the tape path and because it is not affected by errors during servo writing that may appear in LTM measurements that use the PES signal. One drawback of this technique is that tape edge defects contribute to the LTM measured, although in general these effects are less pronounced than the LTM. For the experiments, the optical sensors were custom build and based on an Optek OPB819Z slotted optical switch. They provide a linear range of approximately 1.5 mm and a 1 σ noise of about 50 nm in a 10 kHz bandwidth.

In the standard track-follow operation, the PES generated from the servo pattern is used to derive the lateral displacement of the tape with respect to the head position. By keeping the actuator in a fixed position, therefore operating in an open-loop configuration, the PES measured by the servo channel is a direct measurement of the LTM. This operation is shown schematically in Fig. 15.2, where d is the LTM disturbance and y_G is the position of the actuator measured with respect to a fixed reference. Although this measurement is a true description of the tape disturbances introduced in the track-follow control system and not affected by tape-edge effects, the LTM measurement range is limited to the servo pattern height, i.e. $\sim 100\,\mu$m.

In addition, by operating the actuator in a closed-loop configuration, another estimate of the LTM can be provided, without the measurement range limitation present in the previous technique described above. The LTM can be estimated using the closed-loop PES information or the control signal information provided to the actuator. Given that the closed-loop system has a standard one-degree-of-freedom feedback configuration, the disturbance estimate can either be given by $d = (1+GK)$PES, or by $d = uG+$PES, where G and K are transfer functions describing the actuator and the track-follow controller, respectively. This technique overcomes the measurement range limitation but any model mismatches with respect to the actual system will introduce errors in the LTM estimation.

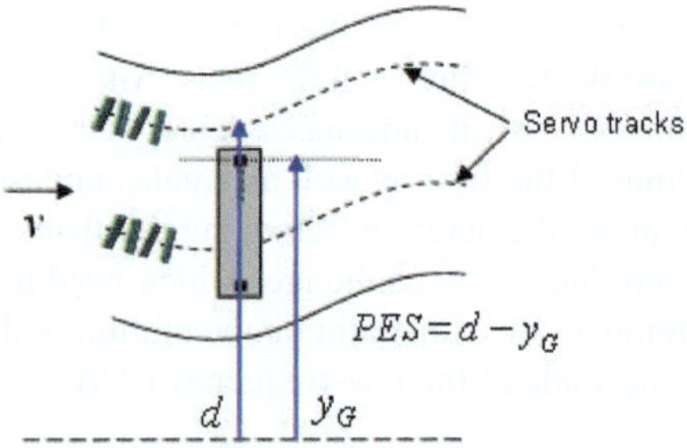

Fig. 15.2 LTM measurement methodology. From (18).

15.3.1 Stationary Periodic Components

The rollers used to transport the tape from one reel to the other exhibit run-outs due to imperfections in their geometry or other defects. These effects appear on the LTM as stationary periodic disturbances. The run-out effects on the LTM were investigated, using a tape path with flanged rollers, with open-loop and closed-loop PES data. Figure 15.3 shows the power spectral density of the captured waveforms. Many stationary periodic components appear in the PES spectrum with frequencies that emerge at the rolling frequency and its harmonics. By increasing the tape speed, the periodic disturbances are shifting to higher frequencies. Therefore, given that the power of the control system is limited at higher frequencies, it is expected that at the higher speeds the periodic disturbances will affect the closed-loop PES more strongly. From the experimental data, the standard deviation, 1 σ, of the closed-loop PES is 172 nm for 5.36 m/s and 151 nm for 2.57 m/s. This further validates the fact that a degradation of the track-follow performance occurs with increasing tape speed that is caused mostly by the periodic disturbances.

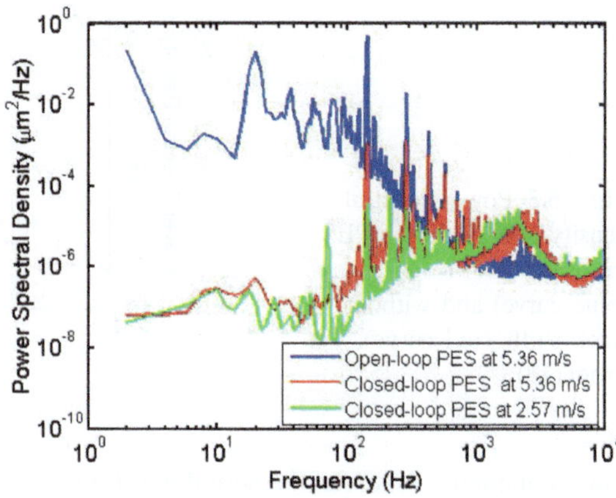

Fig. 15.3 Power spectral density of open-loop and closed-loop PES data. From (18).

15.3.2 Stack Shifts

The contact of the roller flange with the tape edge introduces high-frequency dynamic effects on the LTM. These effects can be reduced by using a tape path with flangeless rollers; however this results in an increase of the low-frequency LTM. Specifically, effects called stack shifts are becoming more pronounced. A lateral displacement of a portion of the tape inside the tape pack can create a stack shift during normal operation, which appears as a low-frequency lateral displacement with high amplitude. In general, when using the same cartridge, the stack shifts are repeated at the same longitudinal position. Figure 15.4 depicts a stack shift occurrence measured by means of an optical edge sensor and estimated from closed-loop PES. The frequency content of the LTM during stack shift occurrences is at the

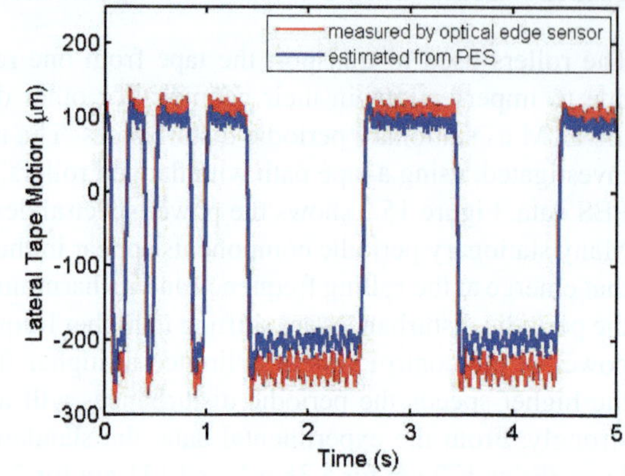

Fig. 15.4 Lateral tape displacement in the presence of stack shifts. From (18).

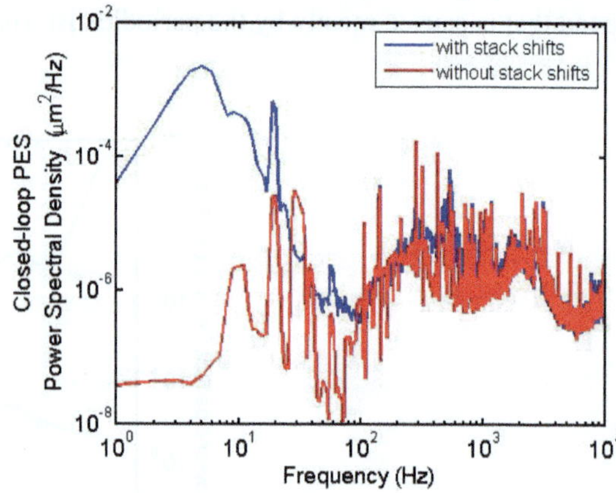

Fig. 15.5 Power spectral density of closed-loop PES in cases of LTM with many (blue curve) and without stack shifts (red curve). From (18).

low frequencies, whereas the actual duration depends on the tape speed and the reel radius. Figure 15.5 compares the closed-loop PES spectrum for two experimental cases: one with many stack shifts and the other without stack shifts. From the experimental data, the 1 σ PES is 165 nm with many stack shifts and 112 nm without stack shifts. Therefore, even though the stack shifts are low-frequency disturbances, they cause a degradation of the track-follow performance because of their high amplitude.

15.4 Active Tape Guiding

A tape transport enhancement can be utilized to mitigate the effects of the flangeless rollers in the tape path. The concept relies on controlling the tilt of one or more rolling elements in the tape path based on the measured LTM (19). Using this active tape-guiding concept eliminates the need for flanges to constrain the tape.

Figure 15.6 is a schematic view of the experimental tape transport that includes the active tape-guiding system. The tape guiding through the tape path and across the head element is performed by four rollers. Rollers R2 and R3 are flangeless, stationary rollers. The active guiding is based on rollers R1 and R4, which are tilting rollers. To measure LTM, the active guiding system uses optical edge sensors placed between rollers R1 and R2 and between rollers R3 and R4.

Figure 15.7 shows a photograph of the tilting roller element. As shown in the photograph, "safety" flanges have been installed on the roller to limit the tape motion in case of large tape excursions. During the normal track-follow operation, there is no contact between the tape and these flanges. The actuator housed within the tilting roller sleeve uses a voice coil motor (VCM) actuation mechanism. A reluctance motor design was used that consists of an 'H'-shaped soft magnetic yoke element and two pairs of permanent magnets mounted on a fixture. A coil is wrapped around the central bar of the yoke. Passing a current through the coil generates a magnetic flux in the yoke, which in turn generates a force on the magnets. The force on each pair of magnets is approximately equal in magnitude, but directed in opposite directions because of the orientation of the magnets. This pair of forces generates a torque, causing the magnet holder to tilt relative to the yoke. This design results in a high power density, generating about 25 mN of force with an input power of 50 mW. Even without a current applied to the coil, there is an attractive force between the magnets and the yoke, which centers the magnets in the gap and gives rise to an effective stiffness of the actuator of approximately 0.6 N/degree.

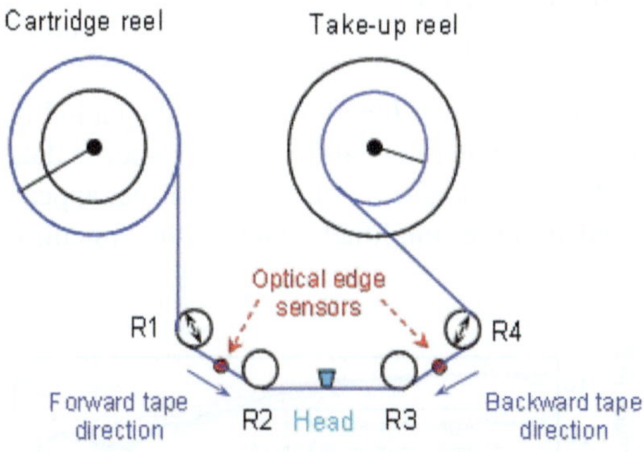

Fig. 15.6 Schematic of tape path with active tape-guiding system. From (19).

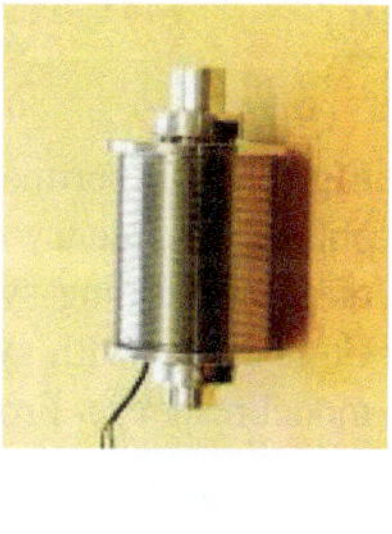

Fig. 15.7 Photograph of a tilting roller element. From (19).

To characterize the active guiding system, the frequency response was measured using the optical edge sensors. A chirp excitation has been applied to the VCM tilting actuator, and the LTM has been captured at two different speeds, 3.9 m/s and 2.3 m/s. The tape tension was at a nominal value of 1 N. Figure 15.8 shows the experimentally obtained frequency responses. To model the system behavior, the existing mathematical models that describe the lateral dynamics of moving webs were used (20; 21). Specifically, for the simplest case, where the mass and stiffness of the tape are considered negligible, the kinematics relations are used to describe the tape motion. The transfer functions describing the lateral tape displacement as a function of the angle of the steering element, as well as the response of the tape at the point where the sensor is located are given by (15.1) and (15.2), respectively. Specifically, the transfer function $P_{y\theta}$ relates the lateral tape position with the tilting angle, whereas the transfer function P_{sy} relates the lateral tape position at the optical edge sensor location with the tape position at a given roller. In these equations, L_1 is the length of the entering span, L_2 is the length of the exiting span, $T_1 = L_1/v$ and $T_2 = L_2/v$ are the corresponding time constants, v is the velocity of the tape, and x is the distance from the roller to the optical edge sensor. Assuming that the transfer function $P_{\theta i}$ of the titling roller actuator is a constant, the system transfer function relating the input current to the actuator with the lateral tape position at the sensor location can be described by $G = P_{\theta i} P_{y\theta} P_{sy}$, where

$$P_{y\theta} = \frac{L_1}{T_1 s + 1} \quad \text{and} \tag{15.1}$$

$$P_{sy} = \frac{T_2(1 - \frac{x}{L_2}) + 1}{T_2 s + 1}. \tag{15.2}$$

This model approximation described by the transfer function G (also shown in Fig. 15.8) proves to be an excellent match to the experimental frequency response data. Note that at frequencies beyond 100 Hz, the resonance of the actuator that appears in the frequency response is not included in the simplified model of the system. In

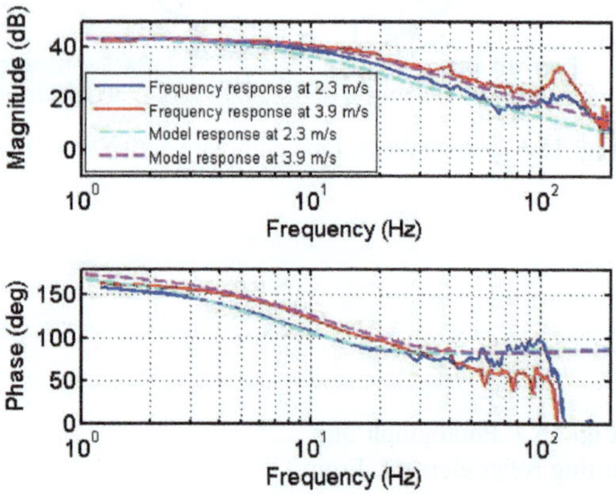

Fig. 15.8 Experimentally obtained frequency response of active guiding system. Comparison with system model response. From (19).

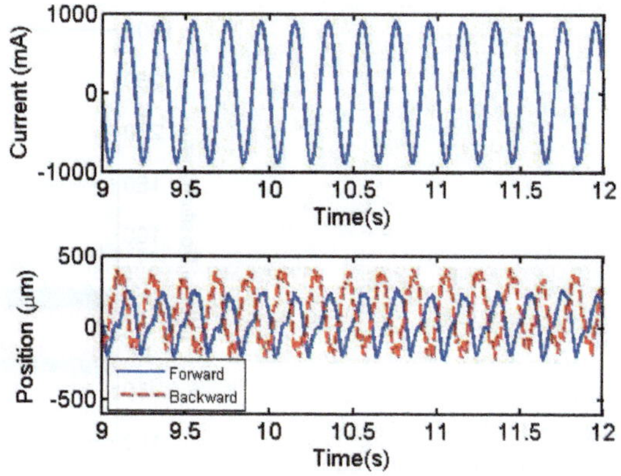

Fig. 15.9 Effect of tape motion direction on system response. From (19).

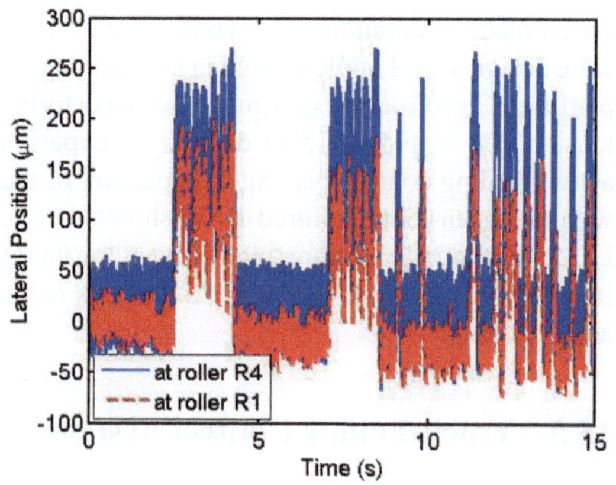

Fig. 15.10 LTM measurements without active tape guiding. From (19).

addition, the direction of tape motion affects the response of the system. Figure 15.9 shows the LTM measured at the sensor location as a response to a sinusoidal input to the actuator. Reversing the direction of the tape motion results in an 180° phase shift that needs to be taken into account during the closed-loop operation of the system.

The active tape-guiding control system has been evaluated in the tape path with flangeless rollers, in which the sudden lateral displacements due to stack shifts are more pronounced. Figure 15.10 shows LTM measurements from the optical edge sensors located next to rollers R1 and R4. The measurements are taken with no current applied to the tilting rollers. As shown in Fig. 15.10, in this open-loop operation the LTM contains high-amplitude components due to the presence of stack shifts. The closed-loop design concept of the active tape guiding is first to detect the LTM and then to actively tilt the roller to shift the tape laterally, thus reducing the LTM effects. As mentioned in Section 15.3.2, stack shifts are low-frequency disturbances, therefore a low-bandwidth controller should be able to correct the LTM. The transfer function G described above is used as the nominal model of the system.

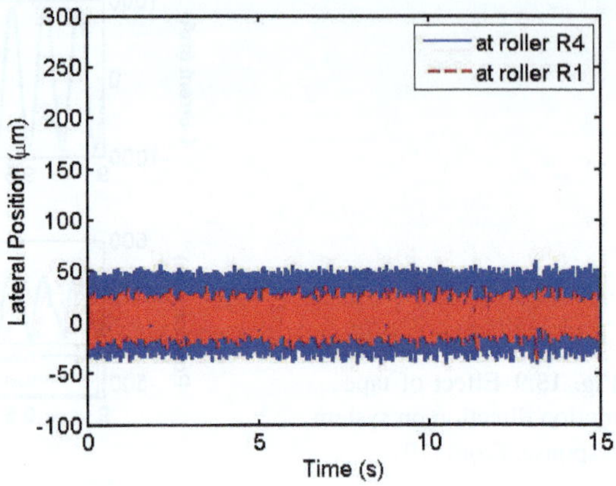

Fig. 15.11 LTM measurements with active tape guiding. From (19).

Based on this system model, two independent proportional-integral (PI) controllers were designed and implemented in a digital-signal-processor (DSP)-based hardware platform. Closed-loop experiments were performed with a tape speed of 2.3 m/s and tension of 1 N. Figure 15.11 depicts the experimental LTM data captured with the active guiding control system. Comparison of the experimental closed-loop results with the open-loop captured LTM shows that active tape guiding can reduce the LTM substantially, eliminating the need for flanges on the rollers. The performance improvement in terms of the 1 σ LTM is a reduction of over 70% with the use of active guiding.

15.5 Track-Follow Control System

Reducing the LTM by optimizing the tape path characteristics is desirable, because it reduces the disturbances that enter the track-follow control system. It is subsequently the task of the track-follow controller to follow the LTM and position the head read/write elements at the centerline of the tracks. The bandwidth of the actuator and the frequency characteristics of the LTM disturbance determine the achievable closed-loop bandwidth and thereby the track-follow performance. Several control methods as well as high-bandwidth actuator designs have been proposed to increase the performance of the track-follow control system (22; 23; 24; 25).

Figure 15.12 shows a block diagram of the one-degree-of-freedom track-follow controller implementation. In this figure, G relates the coil current (u in mA) to the output lateral displacement (y_G in μm). To identify the transfer function G, the frequency response of the VCM actuator was measured by applying a chirp excitation to the actuator and then measuring the position of the head using a laser Doppler vibrometer and by estimating the lateral position using the servo channel measurement while the tape moves. The experimentally obtained frequency responses are shown in Fig. 15.13. The plant dynamics are dominated by the fundamental resonance mode, which can be accurately captured by a second-order model. The delay

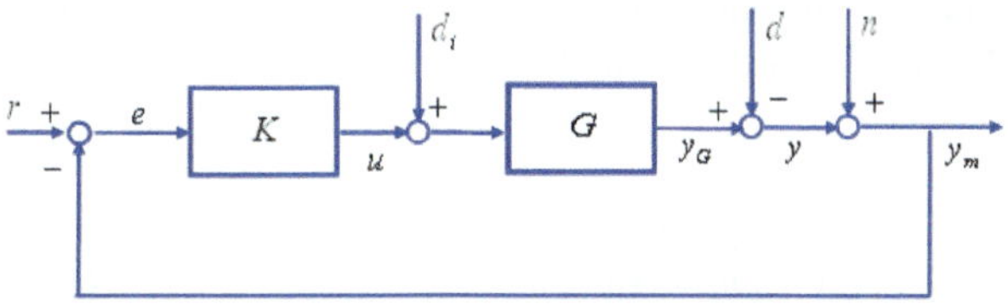

Fig. 15.12 Schematic diagram of the track-follow control system

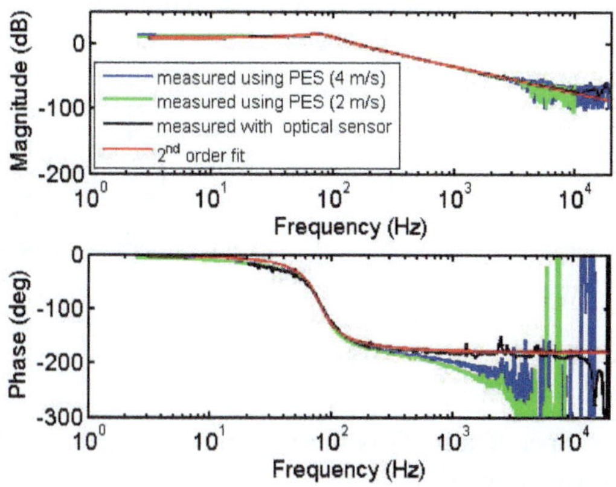

Fig. 15.13 Voice-coil actuator frequency response

in the phase response data, measured using the servo pattern, corresponds mainly to the position measurement delay, which depends on the servo pattern format and the tape speed.

15.5.1 Track-Follow Controller for Periodic Disturbances

In linear control systems, the system's tracking capability and its sensitivity to measurement noise cannot be easily decoupled. Following Bode's integral theorem, increasing the bandwidth of the system increases its sensitivity to measurement noise. This constraint may result in an amplification of the measurement noise in some frequency regions. Knowledge of the frequency characteristics of the disturbances, can be beneficial in increasing the closed-loop system's capability. The LTM characterization presented earlier showed that the stationary periodic components of the LTM are among the main contributors to the PES. Next, an architecture with enhanced track-follow capabilities at the frequencies dictated by the periodic components of the LTM is presented (18).

Track-follow architectures based on Linear Quadratic Gaussian (LQG) controller have been successfully used in many track-following applications (see for example (5)). The LQG controller consists of an optimal state feedback regulator and a

Kalman filter for state estimation. The equations for the plant states and the output measurement used for the Kalman filter design in the general case are given by

$$\begin{aligned} \dot{x} &= Ax + Bu + Ew \\ y &= Cx + Du + Hw + n, \end{aligned} \tag{15.3}$$

where w and n are white noise sources representing the process noise and the measurement noise, respectively. The system model can be extended to include the LTM disturbance together with the actuator response as shown in Fig. 15.14.

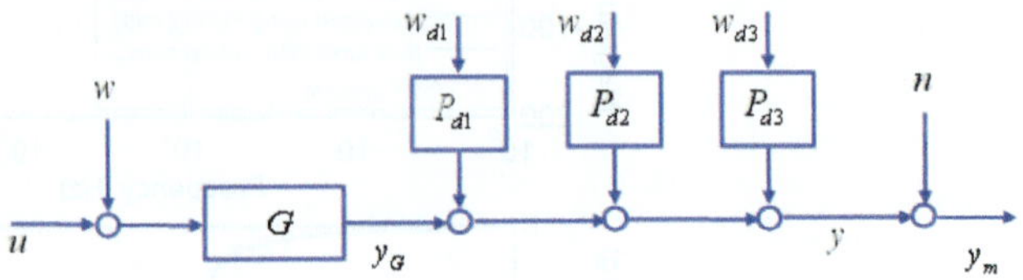

Fig. 15.14 System configuration

Specifically, G is the model of the actuator that in state space form can be described by

$$\begin{aligned} \dot{x}_G &= A_G x_G + \begin{bmatrix} B_G & B_G \end{bmatrix} \begin{bmatrix} u \\ w \end{bmatrix} \\ y_G &= C_G x_G + \begin{bmatrix} D_G & D_G \end{bmatrix} \begin{bmatrix} u \\ w \end{bmatrix}. \end{aligned} \tag{15.4}$$

The frequency characteristics of the LTM disturbance are included in the extended model of the system with the transfer functions P_{d1}, P_{d2} and P_{d3}. These transfer functions model the periodic LTM disturbances and are described by second-order bandpass filters driven by the white noise sources w_{d1}, w_{d2} and w_{d3}. The state-space form of the extended system model is given by

$$\begin{aligned} \begin{bmatrix} \dot{x}_G \\ \dot{x}_{d1} \\ \dot{x}_{d2} \\ \dot{x}_{d3} \end{bmatrix} &= \begin{bmatrix} A_G & 0 & 0 & 0 \\ 0 & A_{d1} & 0 & 0 \\ 0 & 0 & A_{d2} & 0 \\ 0 & 0 & 0 & A_{d3} \end{bmatrix} \begin{bmatrix} x_G \\ x_{d1} \\ x_{d2} \\ x_{d3} \end{bmatrix} + \begin{bmatrix} B_G \\ 0 \\ 0 \\ 0 \end{bmatrix} u \\ &+ \begin{bmatrix} B_G & 0 & 0 & 0 \\ 0 & B_{d1} & 0 & 0 \\ 0 & 0 & B_{d2} & 0 \\ 0 & 0 & 0 & B_{d3} \end{bmatrix} \begin{bmatrix} w \\ w_{d1} \\ w_{d2} \\ w_{d3} \end{bmatrix} \end{aligned} \tag{15.5}$$

$$y = \begin{bmatrix} C_G \, C_{d1} \, C_{d2} \, C_{d3} \end{bmatrix} \begin{bmatrix} x_G \\ x_{d1} \\ x_{d2} \\ x_{d3} \end{bmatrix} + D_G u + \begin{bmatrix} D_G \, D_{d1} \, D_{d2} \, D_{d3} \end{bmatrix} \begin{bmatrix} w \\ w_{d1} \\ w_{d2} \\ w_{d3} \end{bmatrix}. \quad (15.6)$$

This extended system model is enhanced with the states of the disturbance models and the disturbance inputs describing the periodic components are included in the LQG controller design along with the process noise. The proposed control architecture is captured by the block diagram shown in Fig. 15.15. In addition, an integrator is used as a tracking controller. Figure 15.16 compares the resulting closed-loop transfer function from the tape disturbance input to the output, with that obtained using a standard controller. It shows that the proposed LQG controller can provide improved disturbance rejection capabilities at the frequencies of the periodic LTM components. Note that because the periodic components depend on the tape speed, the closed-loop transfer function will be different depending on the tape speed.

The control architecture was implemented in an experimental set-up consisting of a tape path with flanged rollers. Several experiments were performed at different longitudinal positions and a tape speed of 5.36 m/s. The experimental results are summarized in Fig. 15.17, where the 1 σ PES for a standard controller is compared

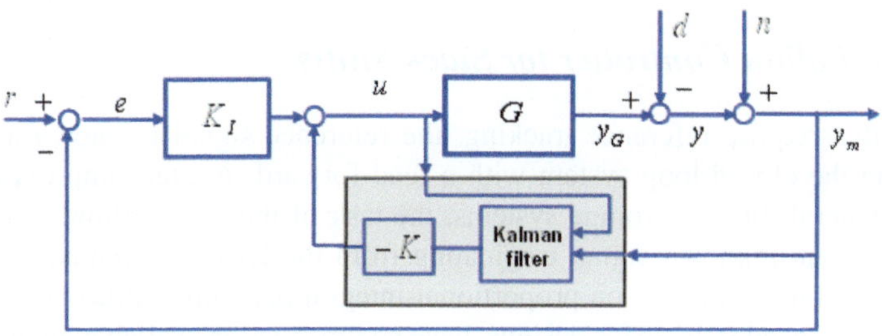

Fig. 15.15 Block diagram of the track-follow control system

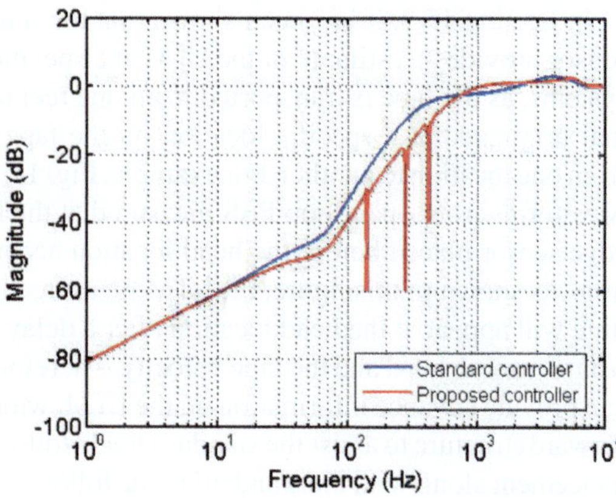

Fig. 15.16 Magnitude response of the closed-loop transfer functions. From (18).

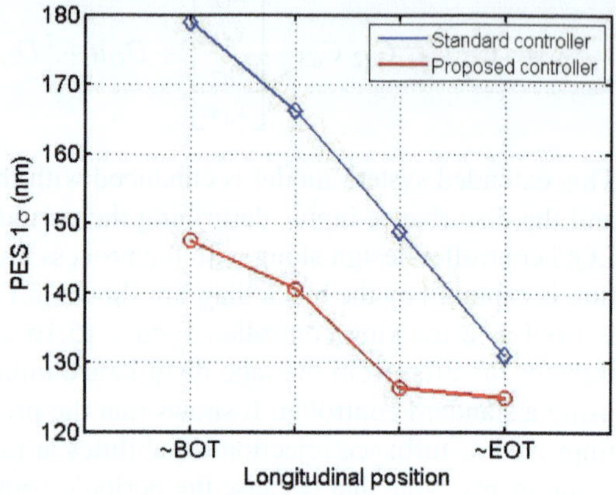

Fig. 15.17 Experimental results at different longitudinal positions ranging from approx. the beginning of tape (BOT) up to approx. the end of tape (EOT). From (18).

with that of the enhanced controller design for cancellation of stationary periodic disturbances. With the proposed controller, improved performance is demonstrated, also highlighting the fact that periodic disturbances are the main contributors to the closed-loop PES in tape paths with flanged rollers.

15.5.2 Track-Follow Controller for Stack Shifts

In applications that require reference tracking, the reference signal is known a-priori. Enhancing the closed-loop system with a feed-forward structure improves the tracking in general. In tape storage systems, the task of the track-follow controller is to follow an unknown signal originating from the LTM. Currently, the feedback control techniques based on proportional-integral-derivative (PID), lead-lag or even more advanced state-space designs, have no information available on the LTM during track-follow. In this section, an estimate of the LTM is generated, and a feed-forward control scheme is presented that can provide significant improvement, especially for low-frequency disturbances, such as stack shifts (18).

In Section 15.3, it has been shown that the measurements using an optical edge sensor provide an estimate of the LTM at a specific location in the tape path. In tape systems, as the tape is transported from one reel to the other, the LTM disturbances that originated at a specific location on the tape path will propagate downstream from one location to another. For example, Fig. 15.18 shows experimentally captured waveforms from open-loop PES measured at the head location and from an optical edge sensor placed before the head location according to the direction of tape motion. As shown in this figure, the LTM measured by the sensor at a specific point in time will appear at the head location after a delay time that depends on the distance from sensor to head and the tape velocity. Therefore, the optical sensor measurement can provide advance information on the LTM, which therefore can be used in a feed-forward structure to assist the standard track-follow controller. This feed-forward enhancement along with the standard track-follow scheme are illustrated in Fig. 15.19.

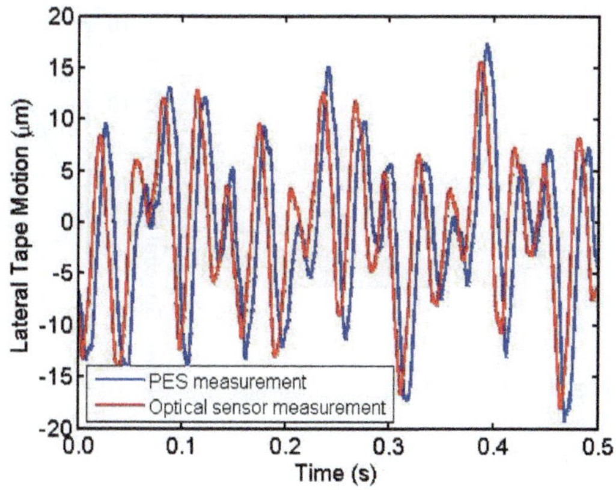

Fig. 15.18 Downstream LTM propagation

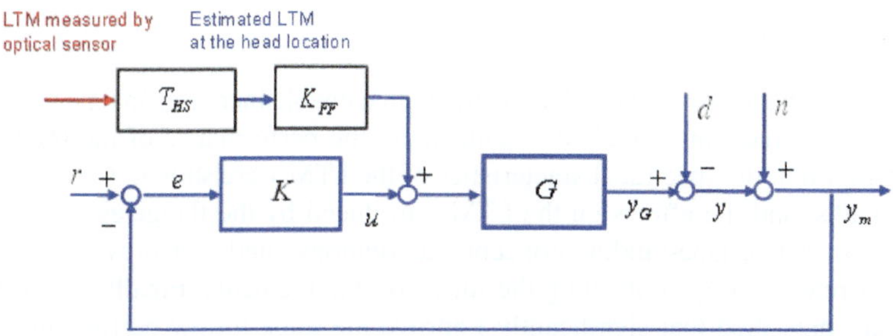

Fig. 15.19 Feed-forward control architecture

The system, T_{HS}, relates the estimated LTM at the head location to the measurement from the optical sensor. A simple first-order model or a higher-order structure that captures the tape transport dynamics more accurately, can be employed. Given that the stack-shifts are low-frequency disturbances, the component K_{FF} is selected in this case to be equal to the inverse of the dc gain of G.

A realistic simulation using experimentally captured waveforms has been performed to evaluate the performance of the control scheme described. Specifically, an optical-edge-sensor measurement and the LTM estimated from the closed-loop PES were captured concurrently and were used as feed-forward input and disturbance input, respectively. The transfer functions of the models identified for the actuator and the track-follow controller were used as G and K, respectively. Figure 15.20 shows the simulation results of the closed-loop system with and without the feed-forward control enhancement. The proposed method provides significant improvement in the cancellation of the low-frequency disturbances, especially of those due to stack shifts in tape paths with flangeless rollers. The performance improvement in terms of the 1 σ PES is a reduction from 165 nm, without the feed-forward scheme, to 128 nm using the feed-forward scheme.

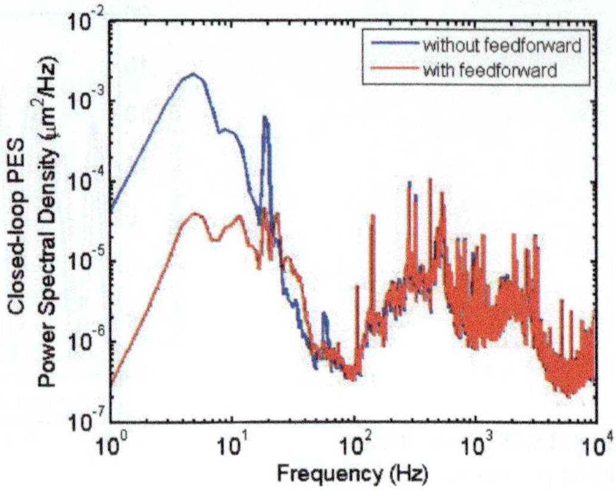

Fig. 15.20 Power spectral density of closed-loop PES with and without feed-forward control. From (18).

15.6 Conclusion

Control-system enhancements were demonstrated that enable increases in tape track densities by either reducing the LTM or enhancing the performance of the track-follow system. First, the spectral characteristics of the LTM were shown in conventional tape paths, and the effects on the LTM introduced by the flangeless rollers highlighted. An active tape-guiding concept was demonstrated that provides significant LTM reduction by controlling the tilt of rolling elements. Finally, control schemes were presented that significantly improve the track-follow performance. Specifically, an enhanced state-space controller has achieved improved performance in the presence of the stationary periodic LTM components. In addition, a feedforward control scheme using an optical sensor measurement was developed that efficiently alleviates the stack shift effects generated in flangeless roller tape paths. The control methodologies presented show that there exists potential for further scaling the track density and storage capacity of tape systems.

Acknowledgements. The author would like to thank her colleagues N. Bui, G. Cherubini, E. Eleftheriou, W. Haeberle, J. Jelitto, M. Lantz, W. Imaino, H. Rothuizen and K. Tsuruta for their contributions in this work.

References

[1] Eleftheriou, E., et al.: IEEE Trans. Magn. 39(2), 938–945 (2003)
[2] Pantazi, A., et al.: IBM J. Res. Develop. 52(4/5), 493–511 (2008)
[3] Bandic, Z.Z., Victoria, R.H.: Proc. IEEE 96(11), 1749–1753 (2008)
[4] Cherubini, G., et al.: IEEE Trans. Magn. 47(1), 137–147 (2011)
[5] Pantazi, A., et al.: IEEE Trans. Control Syst. Technol. 15(5), 824–841 (2007)
[6] Sebastian, A., Pantazi, A., Pozidis, H., Eleftheriou, E.: IEEE Control Systems Mag. 28(4), 26–35 (2008)

[7] Jianxu, M., Ang Jr., M.H.: Microrobotics and Microassembly II. In: Proc. SPIE, vol. 4194, pp. 94–102 (2000)
[8] Horowitz, R., et al.: In: Bhushan, B. (ed.) Springer Handbook of Nanotechnology, ch. 31, pp. 921–950. Springer, Berlin (2004)
[9] Berman, D., et al.: IEEE Trans. Magn. 43(8), 3502–3508 (2007)
[10] Argumedo, A.J., et al.: IBM J. Res. Develop. 52(4/5), 513–527 (2008)
[11] Standard ECMA-319. Data interchange on 12.7 mm 384-track magnetic tape cartridges - Ultrium-1 format (2001)
[12] Jaquette, G.A.: IBM J. Res. Develop. 47(4), 429–444 (2003)
[13] Barrett, R.C., Klaassen, E.H., Albrecht, T.R., Jaquette, G.A., Eaton, J.H.: IEEE Trans. Magn. 34(4), Part 1, 1872–1877 (1998)
[14] Cherubini, G., Eleftheriou, E., Jelitto, J., Hutchins, R.: In: Proc. ASME Information Storage and Processing Systems Conf., Santa Clara, CA, pp. 160–162 (2007)
[15] Taylor, R.J., Talke, F.E.: Tribology International 38(6-7), 599–605 (2005)
[16] Hansen, W.S., Bhushan, B.: J. Magn. Mater. 293, 826–848 (2005)
[17] Raeymaekers, B., Talke, F.E.: J. Trib. 131(1), 011903 (2009)
[18] Pantazi, A., Jelitto, J., Bui, N., Eleftheriou, E.: In: Proc. 5th IFAC Symposium on Mechatronics Systems, Cambridge, MA, pp. 532–537 (2010)
[19] Pantazi, A., et al.: In: Proc. ASME Information Storage and Processing Systems Conf., Santa Clara, CA, pp. 304–306 (2010)
[20] Shelton, J.J., Reid, K.N.: Trans. ASME, J. Dyn. Syst. Meas. Control, Ser. G (3), 187–192 (1971)
[21] Shelton, J.J., Reid, K.N.: Trans. ASME, J. Dyn. Syst. Meas. Control, Ser. G (3), 180–186 (1971)
[22] Panda, S.P., Engelmann, A.P.: Microsystem Technol. 10, 11–16 (2003)
[23] Kinney, C.E., de Callafon, R.A.: In: Proc. JSME-IIP/ASME-ISPS Conf., Ibaraki, Japan (2009)
[24] Boettcher, U., Raeymaekers, B., de Callafon, R.A., Talke, F.E.: IEEE Trans. Magn. 45(7), 3017–3024 (2009)
[25] Kartik, V., Pantazi, A., Lantz, M.: In: Proc. ASME Information Storage and Processing Systems Conf., Santa Clara, CA, pp. 265–267 (2010)

Author Index

Lecture Notes in Control and Information Sciences

Edited by M. Thoma, F. Allgöwer, M. Morari

Further volumes of this series can be found on our homepage:
springer.com

Vol. 413: Eleftheriou, E.; Moheimani, S.O.R.:
Control Technologies for Emerging Micro and Nanoscale Systems
289 p. 2011 [978-3-642-22172-9]

Vol. 412: Fridman, L.; Moreno, J.; Iriarte, R.:
Sliding Modes after the first Decade of the 21st Century
XXX p. 2011 [978-3-642-22163-7]

Vol. 411: Kaczorek, T.;
Selected Problems of Fractional Systems Theory
344 p. 2011 [978-3-642-20501-9]

Vol. 410: Bourlès, H., Marinescu, B.:
Linear Time-Varying Systems
637 p. 2011 [978-3-642-19726-0]

Vol. 409: Xia, Y., Fu, M., Liu, G.-P.:
Analysis and Synthesis of Networked Control Systems
198 p. 2011 [978-3-642-17924-2]

Vol. 408: Richter, J.H.;
Reconfigurable Control of Nonlinear Dynamical Systems
291 p. 2011 [978-3-642-17627-2]

Vol. 407: Lévine, J., Müllhaupt, P.:
Advances in the Theory of Control, Signals and Systems with Physical Modeling
380 p. 2010 [978-3-642-16134-6]

Vol. 406: Bemporad, A., Heemels, M., Johansson, M.:
Networked Control Systems
appro. 371 p. 2010 [978-0-85729-032-8]

Vol. 405: Stefanovic, M., Safonov, M.G.:
Safe Adaptive Control
appro. 153 p. 2010 [978-1-84996-452-4]

Vol. 404: Giri, F.; Bai, E.-W. (Eds.):
Block-oriented Nonlinear System Identification
425 p. 2010 [978-1-84996-512-5]

Vol. 403: Tóth, R.;
Modeling and Identification of Linear Parameter-Varying Systems
319 p. 2010 [978-3-642-13811-9]

Vol. 402: del Re, L.; Allgöwer, F.; Glielmo, L.; Guardiola, C.; Kolmanovsky, I. (Eds.):
Automotive Model Predictive Control
284 p. 2010 [978-1-84996-070-0]

Vol. 401: Chesi, G.; Hashimoto, K. (Eds.):
Visual Servoing via Advanced Numerical Methods
393 p. 2010 [978-1-84996-088-5]

Vol. 400: Tomás-Rodríguez, M.; Banks, S.P.:
Linear, Time-varying Approximations to Nonlinear Dynamical Systems
298 p. 2010 [978-1-84996-100-4]

Vol. 399: Edwards, C.; Lombaerts, T.; Smaili, H. (Eds.):
Fault Tolerant Flight Control
appro. 350 p. 2010 [978-3-642-11689-6]

Vol. 398: Hara, S.; Ohta, Y.; Willems, J.C.; Hisaya, F. (Eds.):
Perspectives in Mathematical System Theory, Control, and Signal Processing
appro. 370 p. 2010 [978-3-540-93917-7]

Vol. 397: Yang, H.; Jiang, B.; Cocquempot, V.:
Fault Tolerant Control Design for Hybrid Systems
191 p. 2010 [978-3-642-10680-4]

Vol. 396: Kozlowski, K. (Ed.):
Robot Motion and Control 2009
475 p. 2009 [978-1-84882-984-8]

Vol. 395: Talebi, H.A.; Abdollahi, F.; Patel, R.V.; Khorasani, K.:
Neural Network-Based State Estimation of Nonlinear Systems
appro. 175 p. 2010 [978-1-4419-1437-8]

Vol. 394: Pipeleers, G.; Demeulenaere, B.; Swevers, J.:
Optimal Linear Controller Design for Periodic Inputs
177 p. 2009 [978-1-84882-974-9]

Vol. 393: Ghosh, B.K.; Martin, C.F.; Zhou, Y.:
Emergent Problems in Nonlinear Systems and Control
285 p. 2009 [978-3-642-03626-2]

Vol. 392: Bandyopadhyay, B.; Deepak, F.; Kim, K.-S.:
Sliding Mode Control Using Novel Sliding Surfaces
137 p. 2009 [978-3-642-03447-3]

Vol. 391: Khaki-Sedigh, A.; Moaveni, B.:
Control Configuration Selection for Multivariable Plants
232 p. 2009 [978-3-642-03192-2]

Vol. 390: Chesi, G.; Garulli, A.; Tesi, A.; Vicino, A.:
Homogeneous Polynomial Forms for Robustness Analysis of Uncertain Systems
197 p. 2009 [978-1-84882-780-6]

Vol. 389: Bru, R.; Romero-Vivó, S. (Eds.):
Positive Systems
398 p. 2009 [978-3-642-02893-9]

Vol. 388: Jacques Loiseau, J.; Michiels, W.; Niculescu, S-I.; Sipahi, R. (Eds.):
Topics in Time Delay Systems
418 p. 2009 [978-3-642-02896-0]

Vol. 387: Xia, Y.; Fu, M.; Shi, P.:
Analysis and Synthesis of Dynamical Systems with Time-Delays
283 p. 2009 [978-3-642-02695-9]

Vol. 386: Huang, D.; Nguang, S.K.:
Robust Control for Uncertain Networked Control Systems with Random Delays
159 p. 2009 [978-1-84882-677-9]

Vol. 385: Jungers, R.:
The Joint Spectral Radius
144 p. 2009 [978-3-540-95979-3]

Vol. 384: Magni, L.; Raimondo, D.M.; Allgöwer, F. (Eds.):
Nonlinear Model Predictive Control
572 p. 2009 [978-3-642-01093-4]

Vol. 383: Sobhani-Tehrani E.; Khorasani K.;
Fault Diagnosis of Nonlinear Systems Using a Hybrid Approach
360 p. 2009 [978-0-387-92906-4]

Vol. 382: Bartoszewicz A.; Nowacka-Leverton A.;
Time-Varying Sliding Modes for Second and Third Order Systems
192 p. 2009 [978-3-540-92216-2]

Vol. 381: Hirsch M.J.; Commander C.W.; Pardalos P.M.; Murphey R. (Eds.)
Optimization and Cooperative Control Strategies: Proceedings of the 8th International Conference on Cooperative Control and Optimization
459 p. 2009 [978-3-540-88062-2]

Vol. 380: Basin M.
New Trends in Optimal Filtering and Control for Polynomial and Time-Delay Systems
206 p. 2008 [978-3-540-70802-5]

Vol. 379: Mellodge P.; Kachroo P.;
Model Abstraction in Dynamical Systems: Application to Mobile Robot Control
116 p. 2008 [978-3-540-70792-9]

Vol. 378: Femat R.; Solis-Perales G.;
Robust Synchronization of Chaotic Systems Via Feedback
199 p. 2008 [978-3-540-69306-2]

Vol. 377: Patan K.
Artificial Neural Networks for the Modelling and Fault Diagnosis of Technical Processes
206 p. 2008 [978-3-540-79871-2]

Vol. 376: Hasegawa Y.
Approximate and Noisy Realization of Discrete-Time Dynamical Systems
245 p. 2008 [978-3-540-79433-2]

Vol. 375: Bartolini G.; Fridman L.; Pisano A.; Usai E. (Eds.)
Modern Sliding Mode Control Theory
465 p. 2008 [978-3-540-79015-0]

Vol. 374: Huang B.; Kadali R.
Dynamic Modeling, Predictive Control and Performance Monitoring
240 p. 2008 [978-1-84800-232-6]

Vol. 373: Wang Q.-G.; Ye Z.; Cai W.-J.; Hang C.-C.
PID Control for Multivariable Processes
264 p. 2008 [978-3-540-78481-4]

Vol. 372: Zhou J.; Wen C.
Adaptive Backstepping Control of Uncertain Systems
241 p. 2008 [978-3-540-77806-6]

Vol. 371: Blondel V.D.; Boyd S.P.; Kimura H. (Eds.)
Recent Advances in Learning and Control
279 p. 2008 [978-1-84800-154-1]

GPSR Compliance

The European Union's (EU) General Product Safety Regulation (GPSR) is a set of rules that requires consumer products to be safe and our obligations to ensure this.

If you have any concerns about our products, you can contact us on ProductSafety@springernature.com

In case Publisher is established outside the EU, the EU authorized representative is:

Springer Nature Customer Service Center GmbH
Europaplatz 3
69115 Heidelberg, Germany

Batch number: 10621434

Printed by Printforce, the Netherlands